Geotechnical Modelling

Geotechnical Modelling

David Muir Wood

Spon Press
Taylor & Francis Group

LONDON AND NEW YORK

First published 2004
by Spon Press
2 Park Square, Milton Park, Abingdon, Oxfordshire OX14 4RN

Simultaneously published in the USA and Canada
by Spon Press
711 Third Avenue, New York NY 10017

*Spon Press is an imprint of the Taylor & Francis Group,
an informa business*

Publisher's Note
This book has been produced from camera-ready copy supplied
by the authors

The publisher makes no representation, express or implied, with regard to
the accuracy of the information contained in this book and cannot accept any
legal responsibility or liability for any errors or omissions that may be made.

British Library Cataloguing in Publication Data
A catalogue record for this book is available
from the British Library

Library of Congress Cataloging in Publication Data
Muir Wood, David.
  Geotechnical modelling / David Muir Wood.
    p.   cm.
  Includes bibliographical references and index.
  ISBN 0-415-34304-6 (hbk.: alk. paper) — ISBN 0-419-23730-5 (pbk.: alk. paper)
  1. Soil mechanics—Mathematical models.   2. Engineering
  geology—Mathematical models.   I. Title.
  TA710 .W596 2004
  624.1'5136—dc22

                                                            2003024081
ISBN 13: 978-0-415-34304-6 (hbk)
ISBN 13: 978-0-419-23730-3 (pbk)

Contents

Preface

Modelling forms an implicit part of all engineering design but many engineers are not aware either of the fact that they are making assumptions as part of the modelling or of the nature and consequences of those assumptions. Many engineers make use of numerical modelling but may not have stopped to think about the approximations and assumptions that are implicit in that modelling—still less about the nature of the constitutive models that may have been invoked. Many engineers are probably not aware of the possibilities and implications of physical modelling either at single gravity or on a centrifuge at multiple gravities.

I have worked for many years at the interface between research and industry in developing numerical models of soil behaviour and in attempting to explain to practising engineers the possibilities of soil modelling. In particular, in 1996 and 1997 I held a Royal Society Industry Fellowship to be seconded to work within Babtie Group with both geotechnical and structural engineers and as a result became more aware of the realities of the conditions within typical consulting engineering companies. This book was conceived during that secondment. The scope of the book attempts to cover the range of guidance that I believe that engineers who are undertaking geotechnical modelling need. I hope that they will find the approach accessible.

Much of the material in this book has been developed during courses given to final year MEng and postgraduate students at Bristol University and elsewhere. The reader is assumed to have a familiarity with basic soil mechanics and with traditional methods of geotechnical design. Some modest mathematical ability is expected: this is not intended to deter, but rather to indicate the nature of the theoretical understanding that is necessary if geotechnical modelling is to be safely undertaken.

My previous book *Soil behaviour and critical state soil mechanics* (Muir Wood, 1990) used a particular constitutive model for soil behaviour, Cam clay, as a vehicle for describing the mechanical behaviour of soils and of some simple geotechnical structures. While Cam clay is presented briefly in section §3.4.2, this present book deliberately tries not to repeat too much of the material in that earlier book: there is more description of simple alternative constitutive models and of the modelling of a range of geotechnical systems. The two books should be seen as complementary.

I am grateful to the Royal Society for the Industry Fellowship and to Babtie Group for welcoming me into their midst: they may not have anticipated that this would have been the outcome. This book project has inevitably lingered and I am grateful to Bristol University for giving me a University Research Fellowship during academic year 2002-2003 in order to give me slightly more time to work on the manuscript. The final surge to completion was greatly helped by a Visiting Professorship funded by the Foundation for the Promotion of Industrial Science which Kazuo Konagai arranged for me at the Institute of Industrial Science of the University of Tokyo. Osamu Kusakabe gave me particular assistance in locating references at Tokyo Institute of Technology. Jacques Garnier, Charles Ng and Sarah Springman were also generous with information and images.

Erdin Ibraim and Adrian Russell provided some helpful suggestions for improvement. However, the rapid march to complete the manuscript—*schnell zum Schluß*—will surely have left errors for which I apologise and accept full responsibility. I can only hope that the irritation attendant on their discovery will be more than compensated by the educational benefit associated with the working through to the correct results.

I am grateful to editorial staff at E & FN Spon for their patience.

I thank Helen for tolerating my obsessive work on a second book.

David Muir Wood
Abbots Leigh
April 2004

1

Introduction to modelling

1.1 Introduction

In the same way that we may be surprised to find that prose is what we have been speaking all our lives, so scientists and engineers are often unaware that almost everything that they do is concerned with modelling. This book is concerned with the application of principles of modelling to soil mechanics and geotechnical engineering.

A model is an appropriate simplification of reality. The skill in modelling is to spot the appropriate level of simplification—to recognise those features which are important and those which are unimportant. Very often engineers are unaware of the simplifications that they have made and problems may arise precisely because the assumptions that have been made are inappropriate in a particular application.

Engineering is fundamentally concerned with modelling. Engineering is concerned with finding solutions to real problems—we cannot simply look around until we find problems that we think we can solve. We need to be able to see through to the essence of the problem and identify the key features which need to be modelled—which is to say those features of which we need to take account and include in the design. One aspect of engineering judgement is the identification of those features which we believe it safe to ignore.

In this chapter the theme of modelling is introduced by reference to modelling activities that are familiar from early and standard courses in soil mechanics and geotechnical engineering within degree programmes in civil engineering and which form the basis for the development of geotechnical design. The scope of subsequent chapters of the book is then defined.

1.2 Empirical models

Although the preference in this book is for models which have a sound analytical or theoretical basis there is a long history of empirical modelling in geotechnical engineering. The dictionary tells us that empiricism rests solely on experience

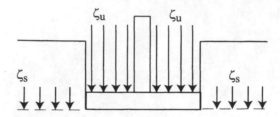

Figure 1.1: Bearing capacity of shallow foundation on clay

and rejects all prior knowledge (and defines an empiric as a 'quack'). Precisely because soils are tricky materials to deal with, a lot of geotechnical engineering has had to be based on experience—because the more rigorous modelling tools have tended to lag behind the demands of the industry. Many of the techniques have been semi-empirical rather than purely empirical. A few examples are given here. It may be objected that these are empirical procedures rather than empirical models: the distinction is somewhat semantic. The key is that these procedures have been found to provide satisfactory answers even though the logical thread cannot always be continuously traced. (The prescription of many medicines is based on knowledge *that* they work without necessarily being able to state exactly *why* they work.)

1.2.1   Vane strength correction

Much of geotechnical design has hitherto relied upon ultimate limit state calculations which are driven by estimates of soil strength hoping, thus, to guard against geotechnical collapse. (Classically, a factored design based on an ultimate limit state calculation, with the factor chosen from experience, might be used to guarantee satisfactory serviceability without performing a separate serviceability calculation.) Thus the ultimate bearing capacity ζ_u of a footing on a clay of undrained strength c_u might classically be written as

$$\zeta_u = N_c c_u + \zeta_s \qquad (1.1)$$

where N_c is a so called bearing capacity factor and ζ_s is the surcharge on the surface of the clay at the level of the foundation—which might simply be due to the weight of overburden at this level (Fig 1.1). Then, if we can find values of undrained strength, we can estimate capacities of footings; similar calculations can be performed for other classes of geotechnical structure, such as embankments and excavations.

   Given a strength model it needs to be populated with values of soil strength determined from laboratory or *in-situ* testing. Most of the widely used strength models lack the subtleties of, for example, rate effects and anisotropy with which the ground itself is certainly familiar. Particular tests measure soil strengths in particular ways: if short term undrained strength of clay is of concern then the *in-situ* vane is commonly used to estimate the undrained strength. The

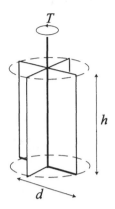

Figure 1.2: Shear vane

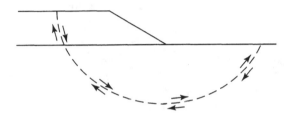

Figure 1.3: Shear strength mobilised along slip surface in clay

vane (Fig 1.2) measures a mixed strength, combining shearing on horizontal and vertical surfaces in the soil. A strength model is required to extract an estimate of undrained soil strength from the actual measurement of the torque required to rotate the vane and hence to generate a failure mechanism through the clay. A simple assumption of uniform soil strength on all surfaces of the failing block of clay of height h and diameter d indicates that the torque T is given by

$$T = \frac{1}{6}\pi c_u d^3 (1 + 3\frac{h}{d})$$

(1.2)

Any actual failure mechanism of the geotechnical structure will require the clay to shear along surfaces having completely different alignments (Fig 1.3).

The vane measures the strength in a matter of seconds—in practice a geotechnical structure may take weeks or months to complete and yet the permeability of the ground may be sufficiently low for the behaviour still to be described as undrained.

Comparisons of estimates of failure conditions (or margin of safety against collapse) of embankments and excavations in soft clay (Bjerrum, 1972) indicate that a 'correction' factor, μ, dependent on soil plasticity (section §1.8) must be applied to the strength emerging from the vane test (Fig 1.4). That is:

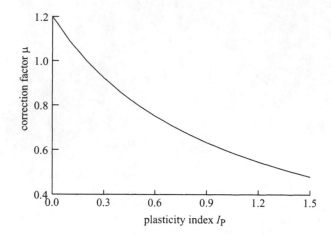

Figure 1.4: Correction factor for vane strength (after Bjerrum, 1972)

$$\frac{c_{u(design)}}{c_{u(fieldvane)}} = \mu \tag{1.3}$$

Typically (at any rate for $I_P > 0.2$) the field vane overestimates the strength which is actually mobilised at failure.

These correction factors have been determined empirically and can be applied with some confidence to future ultimate limit state designs of embankments and excavations which share the same generic character of the bank of observations from which they were deduced. However, they do not provide a secure route for extrapolation from vane strengths (or, more precisely, from the torques required to rotate field vanes) to design calculations in other circumstances. This is an empirical correction to strength measured with a particular device to provide input to traditional design calculations: the application is thus very specific.

1.2.2   Consolidation settlement

Consolidation settlement beneath foundations in clays occurs as a result of dissipation of pore pressures generated by the original loading. A logical argument suggests that, if we can estimate these pore pressures and estimate a soil stiffness which controls the vertical strains that develop as the pore pressures dissipate, then we can combine these estimates to calculate the expected settlement. A tractable procedure which cuts a few corners at each stage was proposed by Skempton and Bjerrum (1957) and has been widely used. We will discuss it for the simple case of an axisymmetric loaded area but it is evident that an essentially similar procedure could be used for other shapes of foundation.

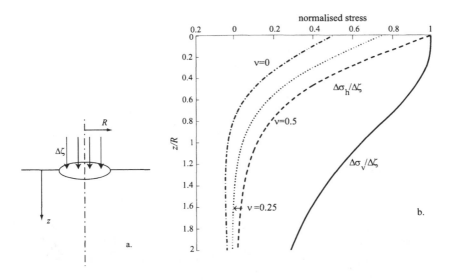

Figure 1.5: (a) Uniformly loaded circular area on the surface of an elastic half-space; (b) stresses on centreline beneath circular load

The magnitude of the pore pressures will depend on the constitutive behaviour of the soil. Typical characteristics of soil behaviour will be described in Chapter 2 and various constitutive modelling possibilities will be presented in Chapter 3. Let us suppose that we can estimate the changes in total mean stress Δp and deviator stress Δq from an elastic analysis, applying the equations of Boussinesq[1]. For loads on the surface of an isotropic elastic half space (to which we approximate our soil layer) the stress changes are independent of elastic stiffness but are dependent on the Poisson's ratio of the soil which for undrained loading (of an isotropic elastic material) we can propose to be $1/2$. Although there are already several assumptions here it turns out that for vertical loading of the halfspace the changes in vertical stress are rather insensitive to the details of the constitutive description of the soil, being largely controlled by a dispersed equilibrium. However, the changes in horizontal stress are extremely sensitive to the details of the soil model. Stress distributions for the centreline of a uniformly loaded circular area on an elastic half space are shown in Fig 1.5: the effect of Poisson's ratio on horizontal stress is very apparent.

From these changes in total stress we can then use our experience with similar soils, or our observations of behaviour of the actual soil in undrained triaxial tests, to estimate the pore pressure using a pore pressure parameter a (§2.6.2):

$$\Delta u = \Delta p + a\Delta q \qquad (1.4)$$

For an isotropic elastic material $a = 0$ and the change in pore pressure is equal to the change in total mean stress.

---

[1]The stress variables p and q are defined in §2.4.

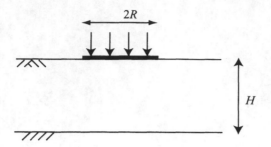

Figure 1.6: Circular foundation on surface of elastic layer

Next it is proposed that the deformations that result from pore pressure dissipation will be largely one dimensional so that the stiffness to be used, E_{oed}, can be obtained from oedometer tests and this stiffness can be used directly to convert pore pressure changes into settlement ρ:

$$\rho = \int_0^H \frac{\Delta u}{E_{oed}}\, \mathrm{d}z \qquad (1.5)$$

integrating the pore pressure changes with depth z over a layer of thickness H (Fig 1.6). Substituting for the pore pressure (1.4) this becomes:

$$\rho = \int_0^H \frac{(\Delta p + a\Delta q)}{E_{oed}}\, \mathrm{d}z \qquad (1.6)$$

If it were assumed that the settlements resulted purely from the change in vertical stress caused by the loading, then we could calculate an oedometric settlement ρ_{oed}:

$$\rho_{oed} = \int_0^H \frac{(\Delta p + \frac{2}{3}\Delta q)}{E_{oed}}\, \mathrm{d}z \qquad (1.7)$$

and the ratio of 'actual' settlement ρ to the oedometric settlement ρ_{oed} is, neglecting variation of the stiffness E_{oed}, a function of pore pressure parameter a and the geometry of the problem which controls the ratio of elastic total stress changes Δp and Δq:

$$\frac{\rho}{\rho_{oed}} = \frac{\int_0^H \left(1 + a\frac{\Delta q}{\Delta p}\right)\mathrm{d}z}{\int_0^H \left(1 + \frac{2}{3}\frac{\Delta q}{\Delta p}\right)\mathrm{d}z} \qquad (1.8)$$

or

$$\frac{\rho}{\rho_{oed}} = \frac{1 + a\mathcal{I}}{1 + \frac{2}{3}\mathcal{I}} \qquad (1.9)$$

where

$$\mathcal{I} = \frac{1}{H} \int_0^H \frac{\Delta q}{\Delta p}\, \mathrm{d}z \qquad (1.10)$$

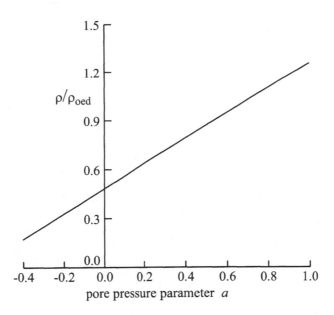

Figure 1.7: Ratio of consolidation settlement and one-dimensional settlement

Evidently the two settlements are equal if the pore pressure parameter $a = 2/3$. The variation of the settlement ratio with a is shown in Fig 1.7 for a circular footing on an elastic layer of thickness H where, for simplicity and illustration, the effect of the finite thickness of the layer on the stress distribution within the layer has been neglected. Skempton and Bjerrum give a more detailed analysis of the settlement taking account of the elastic stress distribution within a finite layer: the ratio remains linearly dependent on a.

This procedure for calculation of consolidation settlement combines elements of a number of quite distinct soil models: an elastic model to calculate the total stress changes (which, for a clay layer of uniform stiffness and of either infinite depth or underlain by a rigid layer, are not actually dependent on that stiffness); an empirical model to link total stress changes with changes in pore pressure; and a one-dimensional model for conversion of pore pressure change to settlement. Each of these models introduces its own simplifications which could in principle be relaxed: isotropic elasticity, variable stiffness layered soil, constant pore pressure parameter, constant oedometric stiffness. There *is* an underlying logic and the success that the method has enjoyed is itself indicative of the insensitivity of some aspects of soil response to the details and accuracy of the calculation procedure: this is a conclusion that will be discovered again in later chapters.

Figure 1.8: Cone penetrometer

1.2.3   Cone penetration test and settlement of footings on sand

Clearly the ease with which a conical object can be made to penetrate the ground (Fig 1.8) will depend on the strength of the ground which will in turn depend on the stresses in the ground and the density of packing of the soils. It is logical then that correlations should be possible between penetration resistance and basic material characteristics: these are empirical correlations but the results can then be used as input to more general design procedures. Alternative empirical rules jump directly from the penetration resistances to geotechnical designs: here the degree of possible extrapolation must be borne in mind. Correlations have been produced both for the Standard Penetration Test in which a rather blunt object is hammered into the ground and for the Cone Penetration Test in which a cylindrical object with a standard 60° conical tip is pushed steadily into the ground at a standard rate. The latter seems to have more scope for rational interpretation and one example of empirical modelling using the cone penetration test will be briefly described.

Let us restrict ourselves to the estimation of the settlements at the centre of a circular footing on sand. With the z axis vertical (Fig 1.5a) we observe that settlement $\Delta\rho$ is an integration of vertical strain increments $\Delta\epsilon_z$

$$\Delta\rho = \int_0^\infty \Delta\epsilon_z \, \mathrm{d}z \tag{1.11}$$

If the soil behaves isotropically and elastically then, from Hooke's law, at any depth

$$\Delta\epsilon_z = \frac{1}{E}(\Delta\sigma_z - 2\nu\Delta\sigma_r) \tag{1.12}$$

where $\Delta\sigma_z$ and $\Delta\sigma_r$ are the increments in vertical and radial (horizontal) stress produced by the footing (Fig 1.5b) and E and ν are Young's modulus and Poisson's ratio for the soil. This can be written

$$\Delta\epsilon_z = \frac{1}{E}[(1 - 2\nu)\Delta\sigma_z + 2\nu(\Delta\sigma_z - \Delta\sigma_r)] \tag{1.13}$$

in order to emphasise that settlement results both from increase in vertical stress ($\Delta\sigma_z$) and from increase in deviator stress ($\Delta q = \Delta\sigma_z - \Delta\sigma_r$). The

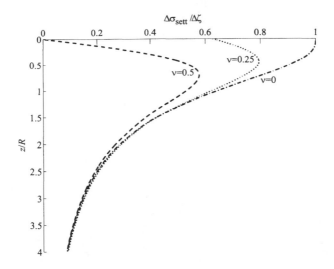

Figure 1.9: Distribution of stress $\Delta\sigma_{sett}$ driving settlement beneath circular footing on elastic half-space

relative contributions of these terms to the settlement will depend on the value of Poisson's ratio. The variation with depth of the composite stress $\Delta\sigma_{sett}$

$$\Delta\sigma_{sett} = [(1 - 2\nu)\Delta\sigma_z + 2\nu(\Delta\sigma_z - \Delta\sigma_r)] \tag{1.14}$$

that actually drives the settlement is shown in Fig 1.9 normalised with the applied pressure $\Delta\zeta$ and as a function of normalised depth z/R for different values of Poisson's ratio for a uniformly loaded circular area of radius R.

Even though the soil is unlikely to be elastic, (1.13) and Fig 1.9 show clearly the contribution that *shearing* is likely to make to settlement: it is not enough to consider only the change in vertical stress. (Of course, this same message formed part of the Skempton/Bjerrum procedure for estimating consolidation settlement that was discussed earlier.) Fig 1.9 also shows that the maximum value of $\Delta\sigma_{sett}$ does not occur at the surface. Indeed for high values of Poisson's ratio $\nu \to 0.5$, the peak shear stress and hence peak value of $\Delta\sigma_{sett}$ occur at a depth of about $0.7R$.

So far this is a coherent modelling strategy based on an elastic analysis for a uniform isotropic soil. Schmertmann (1970) takes inspiration from this to devise a procedure which can be applied more generally. If we write

$$\Delta\sigma_{sett} = I_z\Delta\zeta \tag{1.15}$$

where I_z is a dimensionless influence factor that varies with depth, then

$$\Delta\rho = \Delta\zeta \int_0^\infty \frac{I_z}{E}\,\mathrm{d}z \tag{1.16}$$

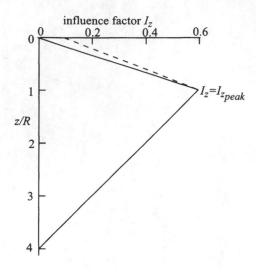

Figure 1.10: Influence factor for calculation of settlement (adapted from Meigh, 1987)

A simplified profile is assumed for I_z (Fig 1.10), partly inspired by analysis (Fig 1.9) and partly inspired by experiment. This assumes that the influence of the footing peaks ($I_{z_{peak}} = 0.6$) at a depth equal to the radius of the footing ($z/R = 1$) and becomes negligible for depths greater than twice the diameter $z > 4R$. This is a logical engineering approach which acknowledges both that, quite apart from the fact that the stress changes induced by the loaded footing fall with depth (Fig 1.9), soils tend to become stiffer with depth and, also, that we now understand that the stiffness of soils increases as the magnitude of the applied strain increment reduces (see §2.5). The contribution to the footing settlement of the actual strains in the deeper soils is expected to be insignificant.

The value of Young's modulus is assumed to be correlated with cone penetration tip resistance q_c

$$E = 2.5q_c \qquad (1.17)$$

Then the settlement becomes

$$\Delta\rho = \frac{\Delta\zeta}{2.5} \int_0^{4a} \frac{I_z}{q_c}\, dz \qquad (1.18)$$

which is more usually evaluated as a sum over a finite number of layers (Fig 1.11). For each layer the quotient is calculated of the average value of I_z (Fig 1.10) and the average value of q_c for that layer (Fig 1.11). The sum of these quotients, weighted by the thickness of each layer replaces the integral in (1.18):

$$\Delta\rho \approx \frac{\Delta\zeta}{2.5} \sum_{i=1}^{n} \left(\frac{\bar{I}_{z_i}}{\bar{q}_{c_i}} \Delta z_i \right) \qquad (1.19)$$

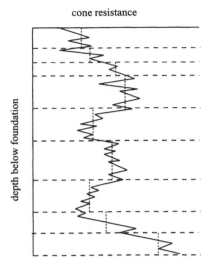

Figure 1.11: Cone penetration profile and division of soil into layers

Schmertmann, Hartman and Brown (1978) later introduced a slightly more elaborate empirical expression for the peak value of the influence factor

$$I_{z_{peak}} = 0.5 + 0.1\sqrt{\frac{\Delta\zeta}{\sigma'_{v_{z=R}}}} \qquad (1.20)$$

and assumed that $I_z = 0.1$ at $z = 0$ (Fig 1.10). There are also empirical correction factors for depth of embedment of the footing and for creep/time effects but the principle remains the same: a profile of stiffness variation deduced by an empirical correlation from *in-situ* testing is combined with a simplified assumed profile of stress change, inspired by theoretical analysis.

1.2.4 Pressuremeter

The pressuremeter is a device which can be used to determine the properties of the ground *in situ* (Mair and Wood, 1987). There are several different pressuremeter devices but in essence (Fig 1.12a): a cylindrical cavity lined with a rubber membrane is created in the ground; the cavity is expanded and the observed relationship between the cavity pressure and the cavity expansion provides some sort of stress:strain response from which certain properties of the ground can be deduced (Fig 1.12b, c). The nature of this response depends to a large extent on the degree of disturbance of the ground during the creation of the cylindrical cavity: earliest pressuremeters used a preformed hole which inevitably disturbed the surrounding soil because it was more or less completely unloaded; more recent pressuremeters use a self-boring technique—a sort of

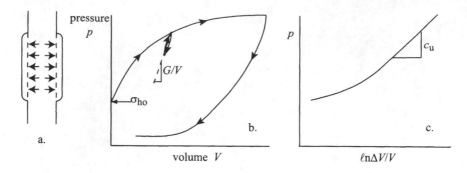

Figure 1.12: (a) Pressuremeter; (b) cavity expansion response of pressuremeter presented in terms of cavity volume; (c) cavity expansion response of pressuremeter presented in terms of logarithmic volume change

vertical tunnel boring machine—which attempts to reduce the disturbance to an absolute minimum.

There are two ways in which the results of pressuremeter tests are used. The expansion of a long cylindrical cavity in an infinite medium is a well defined boundary value problem which is capable of exact theoretical analysis[2]. Armed with such an analytical model, the cavity pressure:cavity expansion results can be interpreted to give values of soil stiffness, soil strength, *in-situ* stresses and perhaps even more detail of the mechanical response of the soil. These quantities can be seen as fundamental properties of the soil which can then be applied to the analysis of quite different geotechnical problems. Thus when a pressuremeter is rapidly expanded in clay the response curve (Fig 1.12b, c) can in principle be interpreted to give the *in-situ* horizontal total stress, σ_{ho}, the undrained strength, c_u, and the shear stiffness of the clay, G. The *in-situ* stress is deduced from the cavity pressure at which expansion of the pressuremeter begins. The interpretation of the results of a pressuremeter test in terms of *the* undrained strength and *the* shear stiffness implies a certain assumed elastic-perfectly plastic model for the shear response of the clay (Fig 1.13): discussion in subsequent chapters will show that the picture is not quite as simple as that. This model assumes that the soil is elastic as the shear stress increases until a limiting value c_u is attained. In such a material the response of the pressuremeter can be written

$$p = p_L + c_u \ln \frac{\Delta V}{V} \tag{1.21}$$

where the limit pressure p_L is the pressure developed at infinite cavity expansion, when the change in cavity volume from the start of the expansion, ΔV, is equal to the current volume V. This limit pressure is given by

$$p_L = \sigma_{ho} + c_u \left(1 + \ln \frac{G}{c_u} \right) \tag{1.22}$$

---

[2]The analysis is essentially similar to that of a collapsing circular tunnel (§8.8) but with the signs reversed.

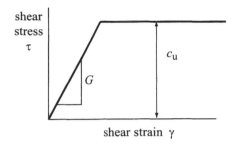

Figure 1.13: Elastic-perfectly plastic constitutive response

Thus (1.21) shows that the undrained strength c_u can be deduced from the ultimate slope of the pressuremeter expansion plotted in terms of p and $\ln\left(\frac{\Delta V}{V}\right)$ (Fig 1.12c) and the shear stiffness G can be subsequently deduced from p_L and (1.22). In practice it is preferred to determine the shear stiffness by performing unload-reload cycles during the cavity expansion (Fig 1.12b): the slope of such a cycle in a plot of cavity pressure p and cavity volume V is G/V.

So far the example of the pressuremeter merely indicates the use of an underlying model to provide a rational route to the interpretation of the test. The model may not be an accurate description of the way in which clay behaves but it is at least theoretically consistent and, if this is the world view that we have and intend to apply to other situations, then the logic is complete.

Knowing that the model is in fact too simplistic for all sorts of reasons (the constitutive behaviour of the soil is wrong; the boundary conditions imposed on the theoretical analysis are wrong—there is an assumption of plane strain and potential drainage of excess pore pressures is ignored) an alternative possibility is to regard the pressuremeter as some sort of index test (albeit a rather expensive one)—and here the interpretation of the pressuremeter becomes equivalent to the interpretation of cone penetration or even standard penetration test results. Empirical rules are used to convert index values from the test directly into geotechnical design: emerging with axial capacities of piles, response of piles under lateral loading, estimation of foundation capacity and settlement, for example. Thus Baguelin *et al.* (1979) work with the cavity pressures p_5 and p_{20} (corresponding to proportional cavity volume changes $\Delta V/V_o = 0.05$ and 0.2 respectively, where V_o is the original volume of the cavity $(= V - \Delta V)$) and define a parameter β

$$\beta = \frac{p_{20} - p_5}{p_{20} - \sigma_{ho}} \tag{1.23}$$

The value of this parameter β can be used to predict the soil type (in the same way as can the measurements made with a piezocone: tip resistance, shaft friction, pore pressure) but it can also be used as input to empirical design procedures.

Again, there is a tenuous logic to this empirical use of the pressuremeter. Such procedures are appropriate for interpolation within the range of experience:

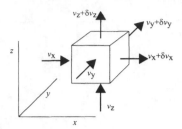

Figure 1.14: Seepage through three dimensional soil element

and, for known soils from which the empirical rules were originally generated, such procedures may well be secure. However, there is less security attached to extrapolation to new geological environments and new soil types, where the application of a more complete underlying theoretical model appears to give greater prospects of success.

1.3   Theoretical models

There are two reasons for the continued successful application of empirical models. On the one hand, geotechnical design cannot just come to a halt while more rigorous models are developed: experience provides a reassuring mode of proceeding. On the other hand, even when accepted theoretical models exist it may not be easy to apply them for the actual boundary conditions of a particular problem. Theoretical models can be seen as elegant solutions looking for problems to which they can be applied: an initial step is often to assess how the observed soil behaviour can best be fitted into the framework that the theoretical model imposes. Once a theoretical model has been formulated there are two possibilities for its application: either the boundary conditions of the problem can be massaged in such a way that an exact analytical result can be obtained; or a numerical solution is required. We will look at some of the issues associated with numerical analysis in Chapter 4. Here we will look at one of the many theoretical models which have been widely applied to geotechnical engineering—others will be presented in Chapter 7.

1.3.1   Steady seepage

The *steady* flow of an incompressible fluid through a porous medium is governed by a familiar partial differential equation. Conservation of mass (volume) requires (Fig 1.14) that the flows into and out of an element of the material must balance (assuming that the element does not contain either a source or a sink). If we assume that flow is driven by a potential gradient then Darcy's law applies:

$$v_x = \frac{k_x}{\gamma_w} \frac{\partial u_w}{\partial x} \qquad (1.24)$$

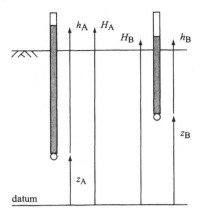

Figure 1.15: Total head H, elevation head z and pressure head h

where v_x is the velocity of flow in the x direction across the complete cross section of the material, k_x is a permeability for flow in the x direction, γ_w is the unit weight of the flowing fluid (assumed to be water) and u_w is the *total* water pressure. For a material with porosity n (ratio of void volume to total volume), the speed at which the water flows through the pores is actually v_x/n (assuming that the ratio of areas of void and of solid material in the direction of flow is also n).

The total water pressure $u_w = \gamma_w H$ (referred to some arbitrary datum, Fig 1.15) is made up of a pressure resulting from the elevation of the element $\gamma_w z$ (referred to the same datum, Fig 1.15) together with the actual pore water pressure $u = \gamma_w h$ (which is independent of the choice of datum). Clearly in a swimming pool the pressure varies with depth but the total pressure is everywhere the same and no flow occurs. The same conclusion is drawn if soil is shovelled into the swimming pool producing a soil with water in its pores. Seepage can only occur in the presence of gradients of *total* pressure.

The mass conservation equation for a three dimensional element (Fig 1.14), assuming constant soil permeability, is

$$k_x \frac{\partial^2 u_w}{\partial x^2} + k_y \frac{\partial^2 u_w}{\partial y^2} + k_z \frac{\partial^2 u_w}{\partial z^2} = 0 \tag{1.25}$$

The form of (1.25) allows for anisotropy of permeability ($k_x \neq k_y \neq k_z$) but can be simplified by working in a transformed coordinate space. If we write

$$x' = \left(\sqrt{\frac{k_z}{k_x}} \right) x \quad \text{and} \quad y' = \left(\sqrt{\frac{k_z}{k_y}} \right) y \quad \text{and} \quad z' = z \tag{1.26}$$

then the equation becomes

$$\frac{\partial^2 u_w}{\partial x'^2} + \frac{\partial^2 u_w}{\partial y'^2} + \frac{\partial^2 u_w}{\partial z'^2} = 0 \tag{1.27}$$

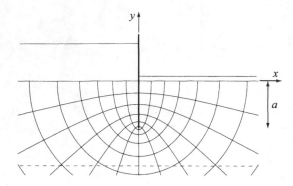

Figure 1.16: Seepage under a sheet pile wall in permeable half-space

This is Laplace's equation for steady seepage flow which occurs in many physical problems—stress analysis, heat flow, electricity flow—and for which standard methods of solution are available. Many of the geotechnical seepage problems that are considered are two dimensional: plane flow or radial flow. For these, exact analytical solutions can be obtained for certain simple boundary conditions and these may be near enough to the real boundary conditions for the results to be acceptable.

An example of flow through a permeable half-space under a sheet pile wall is shown in Fig 1.16 in the form of flow lines and equipotential curves along which the total pore pressure is constant (see, for example, Raudkivi and Callander, 1976). For this idealised problem, with the wall placed at $x = 0$ extending from $y = 0$ to $y = -a$ the flow lines are confocal ellipses:

$$\left(\frac{x}{a\sinh\alpha}\right)^2 + \left(\frac{y}{a\cosh\alpha}\right)^2 = 1 \qquad (1.28)$$

and the equipotentials are confocal hyperbolae:

$$-\left(\frac{x}{a\sin\beta}\right)^2 + \left(\frac{y}{a\cos\beta}\right)^2 = 1 \qquad (1.29)$$

where α and β take appropriate values: β varies from $+\pi/2$ to $-\pi/2$ from the upstream to the downstream side of the sheet pile wall and $\alpha = 0$ for the degenerate elliptical flowline that hugs the sheet pile wall.

The presence of a horizontal impermeable boundary at some depth cuts across the theoretical flow net. An alternative solution procedure is then to sketch a flow net which satisfies the actual boundary conditions more closely but starts from the theoretical net. It is usually found that the total flow rates are not greatly affected by the accuracy with which the net is generated (provided it is at least generally plausible): total flow rate is an integrated quantity. Estimation of pore pressure gradients, which may be important for stability of the geotechnical structure, is more sensitive to the detail of the solution: this requires differentiation of the results.

Because Laplace's equation governs many different physical problems, solutions for one case can also be obtained by studying physical analogues which are governed by the same equation. Thus measurements of electric potential at points on a sheet of conducting paper cut to correspond to the boundary conditions of the seepage problem will be directly translatable into values of total pore pressure driving seepage. Equally, full numerical solution of the equation will provide another route.

Let us remind ourselves of the assumptions that have underpinned the demonstration that Laplace's equation is appropriate for this geotechnical problem. The pore fluid has been assumed incompressible; the soil has been assumed homogeneous, although possibly anisotropic; the flow has been assumed to be governed by Darcy's law. So far as the derivation of the equations governing the flow is concerned, these assumptions can be readily relaxed, but this will lead to a more complex form of the equation which might include spatial variation of permeability, dependence of fluid density (or unit weight) on pore pressure, and an alternative flow law. Under such relaxed conditions numerical solution of the governing equations is likely to be the only option available.

1.4   Numerical modelling

There are several conclusions that can be drawn from discussion of theoretical models that are in common regular use in geotechnical engineering.

Understanding the controlling physical constraints on each problem is crucial. Within an understanding of the physics there is usually a need to idealise the material characterisation and the representation of the boundary conditions of the problem in order that a solution may be obtained. Exact, closed-form solutions are in general only obtainable for a rather limited set of conditions. There will always be a strong temptation to convince oneself that a problem can be fitted into one of these limited sets because of the ease with which a solution may thus be obtained. It is always necessary to consider whether the massaging of the problem to fit these constraints removes any key characteristics of the problem being considered. Where the departure from the ideal situation is clearly too great there is the possibility of using numerical techniques to obtain a solution, retaining the elegance of an underlying simple and widely accepted theoretical description of the physics of the problem on a local scale but using the numerical approximation to allow realistic boundary conditions to be accommodated.

Some of the implications of numerical solution of such problems are discussed in Chapter 4. Numerical solution usually implies the replacement of a continuous description of a problem by one in which the solution is only obtained at a finite number of points in space and time. The quality of the numerical modelling result can only be as good as the quality of the numerical approximation. Where key quantities are changing very rapidly with position or with time then it is necessary either to increase the density of the discretisation used

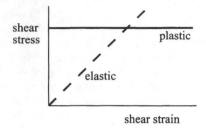

Figure 1.17: Perfectly linear elastic and rigid-perfectly plastic constitutive response

in the numerical modelling in order to be able to follow the changes or else to incorporate within the numerical description some mathematical interpolation which is able to follow the real variation between discrete modelling points. Of course the speed and cost of numerical modelling increases as the density of the modelling points increases. In general it should first be verified that a procedure that is developed for numerical solution of problems is indeed able to give correct results when applied to a situation for which an exact answer is known. It can then be applied with greater confidence to the problem of concern.

1.5 Constitutive modelling

Numerical modelling is not only needed in order to manage irregular or non-ideal boundary conditions. Much more serious often are the idealisations of material behaviour that are necessary in order that simple theoretical models can be developed. Elasticity is convenient because of the wide range of analytical results to which access can be gained for elastic materials. Chapter 2 contains discussion of typical aspects of the mechanical behaviour of soils and it will become clear that linear isotropic elasticity can only provide a very inadequate representation of the observed response (§2.5).

The nonlinearity that is observed in soil behaviour is usually an indication of plasticity: permanent, irrecoverable changes in the fabric of the soil. A simple illustration of the effects of soil plasticity on the character of the response of a geotechnical structure is provided by the schematic illustration of the pattern of deformation beneath a footing on a linear elastic soil and on a rigid-perfectly plastic soil. The stress-strain responses of these ideal soils are shown in Fig 1.17 and the deformation patterns in Fig 1.18. The elastic material clings together: a movement in one location is felt at great distance. The footing produces gradients of deformation, and hence strains, to great depth. The plastic material is happy to separate into separate blocks of soil as it gradually forms a failure mechanism (Fig 1.3). The displacements are entirely contained within this failure mechanism; gradients of displacement only occur at the boundaries between the sliding blocks, and here they are infinite; elements of soil at depths below

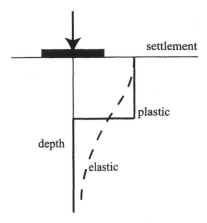

Figure 1.18: Schematic illustration of settlement profiles with depth beneath footing

the mechanism are completely unaware of the presence of the footing (the soil has been assumed rigid before failure).

A second example is provided by the vertical displacements around a group of piles in the same two contrasting materials (Fig 1.19). In the elastic material each pile tends to drag down the surrounding soil in which the adjacent piles are founded. In the plastic material, all displacement is concentrated at the interfaces between the piles and the soil: beyond this interface there is no effect. These are extreme types of soil model, but they serve to demonstrate the effect that nonlinearity of constitutive response can have and, in particular, to indicate that nonlinearity can have a major effect on the interaction between neighbouring geotechnical structures.

Chapter 3 describes some of the alternative possibilities for constitutive modelling of soils and attempts to open to the reader the vocabulary of constitutive modelling and remove some of the mysteries of this modelling. A constitutive model is still governed by equations which ultimately describe the link between changes in strain and changes in stress for any element of soil. Each constitutive model is itself certainly a simplification of soil behaviour but a simplification inspired by experimental observation. The art of constitutive modelling is to identify the features of soil behaviour that are vital in a particular application: the penalty for increased complexity in constitutive modelling is the increased number of material properties that must be defined from a greater number of laboratory or *in-situ* tests. (An isotropic linear elastic model is completely defined by just two material properties: Young's modulus and Poisson's ratio.)

Adequate complexity of constitutive modelling should be the goal in order that analysis of boundary value problems should be efficient. For most constitutive models it is impossible to obtain closed form estimates of the link between stress and strain for anything but the simplest of histories and for single uniform elements of soil. For realistic histories and boundary conditions numerical

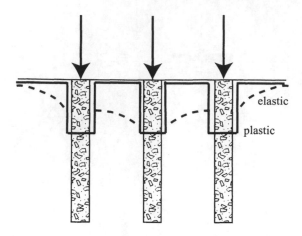

Figure 1.19: Schematic illustration of surface settlement profiles within pile group

analysis is again required in order to manage the basic integration of the constitutive relations.

1.6   Physical models

Physical modelling plays a fundamental role in the development of geotechnical understanding. In fact, taking the broadest possible interpretation of physical modelling, we can declare that every experiment is a physical model intended, if it is a good model, to advance our confidence in some supporting theoretical model which the experiment was designed to probe. We can see the physical modelling as forming the *observation* part of a 'reflective practice' cycle (Fig 1.20); theoretical modelling forms part of the *prediction*.

Physical modelling is performed in order to validate theoretical or empirical hypotheses. Geotechnical construction is thus also physical modelling: geotechnical design makes hypotheses about expected behaviour which may be tested in greater or lesser detail depending on the extent to which the response of the geotechnical system is observed. At the very least there will be a binary observation: has the geotechnical structure failed? A failure will be a pretty clear indication of inadequacy in the supporting models. If the designer is less confident in the supporting design models then more extensive *observation*—for example of displacements or pore pressures—may provide more secure information about the way in which the geotechnical materials are in fact behaving. *Reflection* on these observations then provides the route for improved future design or modelling.

Laboratory testing of small elements of soil (for example in triaxial apparatus, shear box, etc) and *in-situ* field testing (geophysical testing, penetration testing, pressuremeter testing, etc) presuppose some model for the way in which

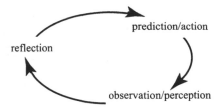

Figure 1.20: Reflective practice cycle

the soil is going to respond. In many cases the underlying models are hidden in experience: we have particular stiffness and strength models which we have used for similar materials in the past so we can choose particular test types—rates of loading, expected stress levels, ranges of transducers—almost without thinking about them. Routine testing is usually merely trying to scale an existing model to fit a given material or set of data—it is less usual that the testing sets out either to demonstrate that the model is inappropriate or to discover information which might be used to improve the model.

A well designed physical model provides an important opportunity in the modelling cycle. It is always tempting to assume that a theoretical model (particularly if, mathematically, it is a very elegant model) somehow encapsulates truth. We can never prove a theoretical model to be true; all we can say about a successful model, or a conjecture on which that model is based, is that it has not yet been falsified or refuted. In practice, all geotechnical models are probably very easily refuted and our interest as engineers is in identifying the range within which the refutation of individual models is weakest since it is this which defines the range of relevance of those models.

A well designed physical model—retaining the broad interpretation—can deliberately set out to probe rival conjectures. Poorly designed physical modelling is mere data gathering. If the models to be tested are not understood or recognised then it is unlikely that the correct data will be assembled: the physical modelling is then stuck in the *prediction/observation* part of the loop which is not closed by the need for precedent and subsequent *reflection*.

1.6.1 Physical models: full-scale

The term 'physical modelling' is usually associated with the performance of physical testing of complete geotechnical systems. Where there is a distrust of theory and analysis, because the assumptions are seen to be too sweeping or the relevant aspects of material response too complex or the realities of reliable numerical solution too far-fetched, physical modelling can seem an appropriate route. Physical modelling can use real geotechnical materials, so the need for theoretical modelling of their behaviour disappears. Physical modelling of geotechnical systems can (and indeed should) provide data for validation of analytical modelling approaches and can thus provide a basis for extrapolation from the physical model to the geotechnical prototype—although, as noted, an instrumented and monitored geotechnical prototype can itself be a physical model serving this validation purpose.

Logically, if we are performing physical modelling because we are unsure about the ways in which we might reproduce the detail of a geotechnical system, then our optimum strategy might be to perform the physical modelling at full scale. We will not concern ourselves particularly with full-scale models in this book but it may be helpful to see them as indeed examples of models. Given the uncertain and variable nature of the ground there is obvious value to be obtained from conducting trials at full scale which will load real soils under real loading conditions. Full scale testing is usually performed to evaluate geotechnical processes which it is believed may be so dependent on the detail of actual soil fabric and structure that it is imperative to use real soils as prepared by nature.

Trial embankments provide an obvious example: usually the need is to evaluate processes of ground improvement. For example, the use of different types and spacings of drains to speed the process of consolidation of soft ground may be studied. It is known that drain installation produces fabric change of the soils local to the drains; it is also known that the *in-situ* fabric of the ground has a strong influence on the flow characteristics which will also have a major influence on rates of flow and hence of consolidation.

Other processes of ground improvement might be considered as means of increasing embankment or structural stability without actually necessarily increasing the rate of consolidation: examples include ground reinforcement using grids and fabrics; cement treatment of the ground or sections of it; installation of columns of compacted granular material to provide local strengthening of the ground. Model testing at small scale may be possible in all cases but the details of the process may be best evaluated at full scale.

Our understanding of the behaviour of piled foundations is improving but there is still a general feeling that the supporting theoretical models of pile-ground interaction are not completely reliable. Again, the uncertainties may well attach to the process of pile installation—whether by driving/jacking or by boring and concreting—and to the detail of the interaction of this installation process with the ground. Test piles are consequently regularly required—these can use actual intended installation procedures and actual ground conditions and, of course, full scale component dimensions (Fig 1.21). The unreliability of theoretical models of pile response is such that Eurocode 7 (EC7, 1995) makes it clear that all pile design calculations must be related, directly or indirectly, to the results of static pile load tests which must be shown to be compatible with general experience. There is thus a Eurocode requirement to complete the *prediction, observation, reflection* loop (Fig 1.20): the design model should be modified in the light of the experience of the full scale physical modelling.

The principal advantage of full scale modelling is that we are working with real ground conditions, real soils, real loads, real stress levels, real stress histories: these are all things that need to be considered in any geotechnical modelling. Over some of these we have direct control: we can be sure of the dimensions of structures we create, heights of embankments, diameters of driven piles. However, we have no control over the ground conditions and the extreme

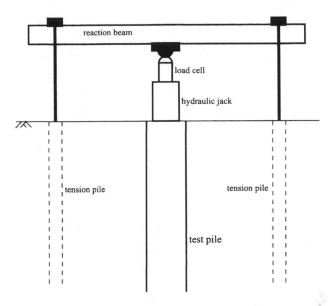

Figure 1.21: Full scale pile loading test

realism may in the end be a disadvantage. The physical modelling is seen as part of a coordinated cycle of theoretical and physical modelling. If we are not sure exactly what the ground conditions are then we cannot be sure how we should tune our theoretical model, and we cannot be sure whether the source of discrepancy with the theory lies in the theory itself or in some unknown detail of the ground conditions (the anisotropy of the mechanical behaviour or the drainage properties, for example).

There are obvious disadvantages of full scale modelling. Often smaller scale modelling leads to much more rapid results purely because of the smaller size. Construction of an embankment over soft soils, for example for a road or airport, may take years to complete. Full scale testing to study the rates at which the embankment can safely be built can occur no more rapidly (though one of the purposes of the full scale testing may be to explore ways in which the construction process can be safely accelerated); cost will increase with the scale of modelling. For both these reasons small scale modelling may be preferred because it permits more tests to be performed and more variables to be explored. Real conditions may be a problem as much as a benefit because if the physical modelling is to be used as part of a process of validation of theoretical modelling then it must be a fair competition—the physical and theoretical approaches must be considering the same problem.

1.6.2   Physical models: small-scale

If the physical modelling is to be performed at any scale other than full scale
then the key question is concerned with establishing the validity of the models
and ensuring that we have a secure route to extrapolation from the behaviour
we observe at model scale to the behaviour that we could expect at prototype
scale: eventually all geotechnical modelling is seeking to improve geotechnical
practice. The existence of supporting theoretical models is thus even more
important for interpretation of small scale physical models than for full scale
models. The understanding of relevant scaling laws and the dimensional analysis
which controls them is essential.

Seepage model

Let us consider the physical modelling of a seepage problem, such as the flow
under a sheet pile wall (Fig 1.16), as an example. In this case the underlying
theoretical model is quite well established: steady flow through a porous soil is
governed by Darcy's law. We know that flow rates will be directly dependent
on soil permeability so we need to know the representative field permeability,
k_f, and a corresponding model permeability, k_m. We note immediately that the
permeability may be anisotropic and we guess that we need to ensure that the
nature of this anisotropy is the same in the field and in the model. We note also
that the permeability is likely to depend on the porosity or density of packing
of the soil: the actual density of packing does not have to be the same in the
field and the model but we need to be confident that the spatial variation of
permeability is the same in both.

Applying Darcy's law at the system level rather than the element level tells
us that we can expect the flow velocities, v, to be proportional to the overall
hydraulic gradient and permeability

$$v \propto \frac{k}{\gamma_w} \frac{\Delta p}{L_1} \tag{1.30}$$

where Δp is the pressure drop across the sheet pile wall, and L_1 is a typical
dimension controlling the distance over which this pressure drop occurs (Fig
1.22). The volume flow rate per unit length of a long wall, Q, which is of
primary concern if the ground on one side of the wall is to be kept dry, is then
given by

$$Q = \lambda \frac{k}{\gamma_w} \frac{L_2}{L_1} \Delta p \tag{1.31}$$

where L_2 is a second typical dimension controlling the distance through which
the flow is occurring (Fig 1.22). The multiplier λ is likely to be a function of the
geometry of the problem. The theoretical model governing the quantity of flow
under the wall is thus extremely simple: the flow is proportional to a material
quantity, the permeability k, and to an input quantity, the pressure drop Δp;
the flow is controlled by a system quantity, a dimensionless geometrical property
of the problem $\lambda L_2/L_1$, which will be related to the depth of penetration of the
piles and might well be a variable whose influence would need to be studied

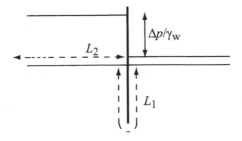

Figure 1.22: Parameters controlling seepage under a sheet pile wall

using the physical modelling. In fact, this simple theoretical analysis shows that the geometry is really the only variable to consider. If we write

$$\frac{Q\gamma_w}{k\Delta p} = f\left(\frac{L_2}{L_1}\right) \tag{1.32}$$

then we see that we need only make one measurement of flow for each geometry (L_2/L_1) and that it is entirely superfluous to vary Δp or k as well. Field flow rates can then be estimated knowing the field permeability and the field pressure differential combined with the geometrical factor deduced from the physical modelling.

This simple example is intended to illustrate the benefit of understanding the theoretical model which underlies the physical problem. It secures the extrapolation and can focus the modelling energies on the key controlling parameters of the problem. In this case too there is a well accepted route through sketching of flow nets which can be used both to predict the physical capacities required for the physical modelling (for example, pump capacities) and to ensure that the measured results are in accord with expectation.

Small scale models

The great advantage of small scale laboratory modelling is that we have full control over all the details of the model. We can choose the soils that we test and ensure that we have supporting data to characterise their mechanical behaviour. We can choose the boundary and loading conditions of the model so that we know exactly how the loads are being applied, and to what extent drainage is permitted or controlled at the boundaries. The nature of the problem to be modelled theoretically in parallel with the physical modelling is thus well defined. Small quantities of soil are required; drainage paths are short so test durations may also be short; and the possibility exists of performing many tests repeating observations and studying the effect of varying key parameters. The costs of individual tests will be correspondingly lower than full scale tests.

The size of the models is both an advantage and a disadvantage. If a particular prototype is to be modelled physically then a length scale must be chosen. A typical length scale might be 1:100 so that a 10 m high prototype structure becomes a 100 mm high model. Features of the fabric of the ground—for example, seasonal layering of silts and clays—having a prototype spacing of the order of a few millimetres would have to be modelled with spacings of a few tens of microns—or an alternative modelling decision would have to be made. A prototype granular material might have a typical particle size of the order of a few millimetres, so that the ratio of structure dimension to particle size is of the order of 10^3. Use of this same material in the physical model, which would be very desirable if continuity of mechanical behaviour were to be assured, would then lead to the ratio of structure dimension to particle size falling to only of the order of 10. This ratio might be too small to guarantee correct response in the physical model. There are ways in which such difficulties can be solved (see §5.3.6)—the important point is that they cannot merely be ignored.

1.7   Geological model

This is not a book about geology: the emphasis is on mechanical aspects of geotechnical modelling. The questions to be answered relate to the engineering characteristics and properties of the geotechnical materials. However, these materials have been placed either by geological and geomorphological history or by man. Knowledge of this history can help to focus our ideas about the likely mechanical characteristics and a geological or stratigraphic model is usually recommended as a precursor to and underpinning feature of the geotechnical model.

A reasonably well developed geological model can lead to economy and efficiency in subsequent site investigation to determine quantitative properties of the ground. Parallels can be drawn with past experience and with adjacent sites with similar geologies. The expected properties, the nature and mineralogy of soil particles, the appropriate constitutive models (which may in some ways predefine the *in-situ* or laboratory testing), and the likely pitfalls can be predicted. For example, fractured rock associated with faulting or irregular buried erosion features in weaker rocks may be anticipated. Although geophysical techniques can be used to obtain an overview of the structure of the ground, detailed knowledge usually comes from discrete boreholes. A geological model is necessary to be able to propose continuity (or lack of continuity) of stratigraphy between boreholes (Fig 1.23, Fig 1.24). The ground is usually not homogeneous. Vertically the inhomogeneities may primarily result from depositional layering: different rock layers at one scale—with spacings perhaps of the order of metres; varves resulting from seasonal variations in sediment transport and water velocities at another—with spacings perhaps of the order of millimetres. Horizontally there may also be variations. The geological model can help to understand the reason for and the nature of the spatial variations.

At the simplest level, the boundary between the soil-like materials which are expected to deform and control the behaviour of the geotechnical system, and the rock-like materials which are expected to be more or less rigid (and possibly impermeable) in comparison, is important in defining the extent of the ground that needs to be modelled either physically or numerically. Of course a rock layer is not always the boundary beyond which nothing of interest or concern

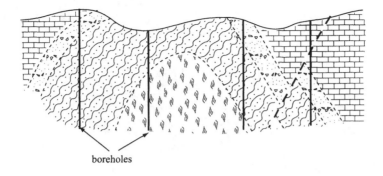

boreholes

Figure 1.23: Geological model deduced from borehole exploration (solid lines)

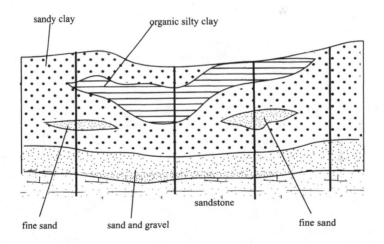

sandy clay

organic silty clay

sandstone

fine sand

sand and gravel

fine sand

Figure 1.24: Model of stratigraphy of surface deposits deduced from borehole exploration (solid lines)

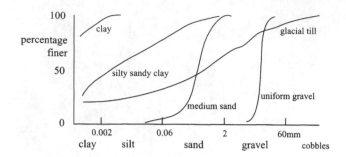

Figure 1.25: Particle size distributions

Figure 1.26: Particle sizes determined by sieving

will occur but very often it will indeed serve to limit the size of the physical or numerical model (see §4.10.2).

1.8   Classification model

Before any tests to determine the mechanical properties of the soils are performed, and as an almost subliminal guide to the sorts of models that might be used to inspire the design of a programme of testing, a classification model is usually invoked with which the soils that are encountered are popped into categories. Samples have been recovered; they are possibly disturbed but one can from simple visual inspection categorise the soil as broadly gravelly or sandy or silty or clayey. Particle size distributions can be obtained (Fig 1.25) which confirm this initial visual classification. The determination of these distributions itself invokes a simplified model of the soil in which the particles of which it is formed are replaced by equivalent spheres. For a soil with particles large enough to sieve, the size of the equivalent spheres is defined by the size of the mesh spacing through which the particles—of whatever actual shape—will fit (Fig 1.26). For a soil with finer particles, Stokes' law, which describes the terminal velocity of spheres falling through a viscous fluid (Fig 1.27), is used to define the size of the spheres to which the actual soil particles in their rate of descent through the fluid are equivalent. In addition, sometimes as a luxury or afterthought, some assessment may be made of the typical particle shapes (Fig 1.28) (and possibly also particle mineralogy).

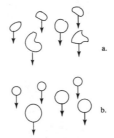

Figure 1.27: Particle sizes determined by sedimentation: (a) actual soil particles; (b) equivalent spherical particles

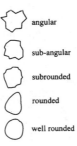

angular

sub-angular

subrounded

rounded

well rounded

Figure 1.28: Particle shapes

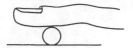

Figure 1.29: Plastic limit test

If the soil is basically sandy then standard index tests can be used to determine the typical range of densities over which the dry soil can exist (Kolbuszewski, 1948). A standard vibratory procedure is used to discover the maximum density, or minimum void ratio e_{min}, of the soil. A series of repeated inversions of a large tube containing a sample of the sandy soil is used to estimate the minimum density, or maximum void ratio, e_{max}. If the soil is then prepared or found to exist at any other void ratio, e, then a relative density, D_r, can be defined

$$D_r = \frac{e_{max} - e}{e_{max} - e_{min}} \tag{1.33}$$

It is to be expected that this range of void ratios will itself depend on the range of particle sizes and the typical particle shapes. Many empirical links have been proposed between relative density (often combined with some statement about stress level) and other soil properties—such as stiffness and strength. Some sophisticated soil models reckon to obtain all their constitutive parameters by correlation with this standard range of void ratios (Herle and Gudehus, 1999).

It is found that the maximum and minimum void ratios do not actually define the extremes of packing: they merely provide a useful index for the soil. There are some repeated shearings which can lead to even greater densities, lower void ratios, than the standard procedure. The standard procedures are evidently conducted at very low stress level: in the large tube used for estimating the maximum void ratio the vertical stress in the sand is unlikely to be greater than a few kilopascals.

If the soil is basically clayey and sticks together as a sample then the so-called Atterberg limits again provide an indication of the range of packings at which the soil can ideally exist (Atterberg, 1911). Atterberg's limits seem almost more relevant to the selection of clays for use in making pots:

> Then said another with a long-drawn Sigh
> 'My Clay with long oblivion is gone dry:
> But, fill me with the old familiar Juice,
> Methinks I might recover By-and-bye'

(stanza LXV: *Rubáiyát of Omar Khayyám*)

As water is added to a clay there is a range of water contents for which the clay can be readily moulded without cracking. If the water content is too high the clay becomes a slurry and flows like a liquid. If the water content is too low the clay tends to crumble when it is moulded. Thus broadly were defined the liquid limit and plastic limit for a clay soil.

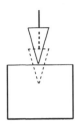

Figure 1.30: Liquid limit test

The plastic limit, w_P, is still determined by a procedure similar to that proposed by Atterberg: a thread of soil of diameter 3 mm is rolled out (Fig 1.29) until it starts to crumble. The rolling hand gradually draws off water, thus steadily moving the water content towards the plastic limit.

Atterberg's test to determine the liquid limit was indeed a test to spot the occurrence of flow in the soil. However, this test has been abandoned in many countries (because it was found to be somewhat operator sensitive) and replaced by a test which is actually a strength test (Wood, 1985)(§5.2.2). A cone of standard geometry and standard mass is allowed to fall into the soil under its own weight from contact with the flat surface of the soil (Fig 1.30). If the cone angle is 30°, the cone mass 80 g and the depth of penetration 20 mm then, according to the British Standard (BS1377, 1990), the soil must be exactly at its liquid limit, w_L. (In Scandinavian countries the cone angle is 60°, the mass 60 g and the depth of penetration 10 mm (Karlsson, 1977).)

This cone test is exactly equivalent to the indentation hardness tests used, in a nondestructive way, to estimate the yield strength of metals. A test which was originally required to set a standard limit to the volumetric packing of the clay has become a test which measures soil strength. It can be shown that the undrained strength of a clay soil at its liquid limit is about 2 kPa (Wood, 1985; see §5.2.2), and that the British Standard and the Scandinavian standard procedures both seek the water content at which the clayey soil has this strength.

Results of site investigation are frequently presented in terms of the profiles of liquid limit, plastic limit and natural water content w with depth (Fig 1.31) because these profiles can reveal a lot about the nature of the soils (and the internal consistency of the site investigation). There are many correlations of soil mass properties with plasticity index, I_P

$$I_P = w_L - w_P \tag{1.34}$$

If the actual water content, w, of the soil is known then a liquidity index, I_L, can be defined which is somewhat equivalent to relative density[3]

$$I_L = \frac{w - w_P}{w_L - w_P} \tag{1.35}$$

---

[3]Relative density increases with decreasing void ratio, but liquidity index increases with increasing void ratio.

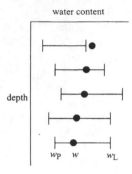

Figure 1.31: Profiles of index properties and natural water content

and correlations of certain properties with liquidity index can be obtained (Muir Wood, 1990).

These classification index tests are simple standard tests which can be rapidly performed with low cost equipment but whose results can help to slot the soils into an explicit or implicit bank of past experience. They make up a *classification model* which can be useful for sharing information across different sites and can provide a basis for moderation of other estimates of soil properties. If soils from two sites have similar index properties and similar particle characteristics (and similar geological histories) then it is to be expected that other mechanical properties will fit into a consistent pattern across the sites.

1.9   Conclusion

Various modelling issues have been aired in this introductory chapter and examples have been given of some of the different types of modelling activities that are in regular use by geotechnical engineers. The remainder of this book will concentrate on theoretical, numerical and physical modelling.

The equations of equilibrium and of strain compatibility for a continuous material are well established. In order to analyse the deformations of a geotechnical system it is necessary to provide a link between stresses and strains in the form of a constitutive model. In Chapter 2 we will discuss elements of the mechanical behaviour of soils and deduce that simple linear elastic or perfectly plastic models are inadequate in detail—though they may be appropriate in certain circumstances. Chapter 3 will then explore some of the alternative possibilities for forms of constitutive model which may be more generally applicable to soils. This is something of an open field and the intention is to open the eyes of geotechnical engineers to the possibilities of constitutive modelling without suggesting that work has come to a conclusion in this area. In fact it will be shown that there are often several ways in which the same experimental observation can be modelled.

The extreme nonlinearity of most plausible constitutive models makes it essential to use numerical procedures to obtain solutions to boundary value problems—the behaviour of complete geotechnical systems of interest to geotechnical engineers. The essence of numerical approximation is discussed in Chapter 4 but numerical analysis is seen as a tool to be used and the emphasis is certainly not on the theoretical basis of numerical analysis. However, the power of the computers that are available to all geotechnical engineers has increased so much that it is quite reasonable to suggest that numerical analysis tools should be used much more as part of the routine of geotechnical design, incorporating the constitutive models of today and recognising the inadequacy of some of the simplifying assumptions that have been imposed in the past for reasons of calculational expediency. However, increased use of numerical tools cannot obviate the need to ensure the reliability of the results.

The importance of scaling laws in the design and interpretation of physical models will be described in Chapter 5. It will be shown in Chapter 2 that soils are nonlinear history dependent materials. The understanding of scaling laws for stress related quantities for such materials is not necessarily straightforward and the extrapolation of observations made in small scale physical models to the full scale prototype response is simplified if the stress level of the physical models is similar to that of the prototype. This can be achieved by subjecting the physical model to an artificial gravitational field on a geotechnical centrifuge. Chapter 6 provides an introduction to geotechnical centrifuge modelling.

Modelling should always be of only adequate complexity. Numerical modelling will not be required or appropriate in all circumstances. Elastic analyses are regularly used as part of geotechnical design and for validation of computational tools that are to be used for more elaborate calculations. The use of perfect plasticity underpins much of the ultimate limit state design in geotechnical engineering. Some of the possibilities for simple calculations using plasticity models and other theoretical models of aspects of geotechnical behaviour are described in Chapter 7.

One of the prime applications of geotechnical modelling, whether theoretical/numerical or physical, is to assess the consequences of soil-structure interaction. Soil-structure interaction problems tend to be driven by stiffness or deformation properties of soils. Constitutive modelling of prefailure deformation properties is thus vital. The importance of soil-structure interaction will be demonstrated in Chapter 8.

2

Characteristics of soil behaviour

2.1   Introduction

It has been shown briefly in Chapter 1 (and will be shown more extensively in Chapter 5) that application of techniques of physical geotechnical modelling requires correct application of scaling laws in order to be able to extrapolate behaviour observed in (usually) small physical models to the behaviour that can be expected in a prototype geotechnical structure. Correct development of these scaling laws requires some understanding of the factors that influence the behaviour of the materials that are being modelled.

Numerical geotechnical modelling combines uncontroversial laws of equilibrium and of compatibility—continuity of displacement fields—through so-called constitutive relations which relate the changes in loads applied to elements of geotechnical materials to the deformations or gradients of displacement that develop in those elements. In this chapter we are concerned to describe some of the characteristics of the mechanical behaviour of geotechnical materials, primarily soils, which make them interesting and challenging materials to model. A hierarchy of possibilities for constitutive models that attempt to reproduce *some* of these characteristics is presented in Chapter 3 (see §3.4.1, for example). An extensive discussion of soil behaviour in the context of so-called 'critical state soil mechanics' is given by Muir Wood (1990) and that book will provide a complement to the descriptions of soil behaviour in this chapter.

We will include a very brief discussion of the influence of strain rate on the mechanical behaviour of soils in section §3.7. However, our concern will primarily be with slow deformation of soils. Several phenomena that are important in understanding certain applications of soils or soil-like materials will therefore be excluded. Many granular materials are stored in silos and are subsequently discharged as a more or less rapid flow through a hopper. Rock avalanches and rapid landslides involve inertial effects and the mechanics of collisions between individual blocks. Sediment transport by flow of water in rivers, lakes and oceans is obviously extremely important in its influence on the structure of the

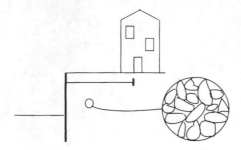

Figure 2.1: Particle-continuum duality for geotechnical system

soils that are eventually formed by sedimentation. Large changes in geometry will not be specifically addressed although they certainly occur in each of the applications just cited. Numerical procedures and constitutive models need to be specially formulated and adapted to cope with such large deformations.

2.2   Particle-continuum duality

If you pick up a handful of dry sand it can be poured, flowing almost like a liquid. Yet we know that, contrary to biblical expectations, we can safely construct major buildings sitting on the sand. We know that most soils have reached their present location by being transported by wind, water or ice. While being transported they were, at least if transported by air or water, present as dilute particle suspensions with little interaction between individual particles. We are faced with the problem of describing materials which clearly are composed of individual particles but which clearly also exist in forms in which the particles interact beneficially so that they appear to be strong materials which can be relied upon for engineering applications which are usually so much larger than the individual particles that we have to smear out the properties and create an equivalent continuum for any analysis. In deciding how we should describe and model the mechanical behaviour of soils we have to come to terms with this particle-continuum duality (Fig 2.1).

As we have seen in the brief presentation of the classification model (§1.8), soils can exist in a wide range of densities or volumetric packings and it is certain that the interesting properties of soils require us to develop models which are able to describe and accommodate large changes in volume as the particles rearrange themselves and displace some of the fluid which fills the surrounding voids. There may in general be more than one fluid in the voids—typically water and air at near surface depths in temperate climates—but there could be water, and liquid and gaseous hydrocarbon in certain circumstances. It will suffice here to think merely of soils which are saturated with a single pore fluid. There are two approaches which might be used to attempt to construct constitutive models of soil behaviour.

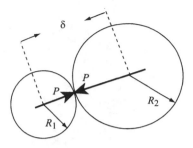

Figure 2.2: Hertzian contact of two spherical particles

Computing power exists today to be able to describe assemblies of very large numbers of individual particles interacting in three dimensions. Computation of the response of such an assembly requires two things. A house-keeping operation is required to check on the relative position of neighbouring particles and note when they begin to interact. If they do interact then an assumption is required concerning the physical law governing their interaction: basically a law relating approach of two particles in contact with the forces generated at the contact. The summation of all the effects at the particle contact level then produces the response of the entire assembly.

Textbook solutions (eg Johnson, 1985) can be obtained as a first attempt at describing the interaction of pairs of particles. The elastic contact of two spherical particles of radii R_1 and R_2 (Fig 2.2) is described by Hertzian contact theory which suggests that the load P should vary with a power of the relative approach of their centres δ:

$$P = \frac{4}{3}E^*(R^*\delta^3)^\alpha \qquad (2.1)$$

The exponent $\alpha = 1/2$ and R^* and E^* are given by

$$\frac{1}{R^*} = \frac{1}{R_1} + \frac{1}{R_2} \qquad (2.2)$$

$$\frac{1}{E^*} = \frac{(1-\nu_1^2)}{E_1} + \frac{(1-\nu_2^2)}{E_2} \qquad (2.3)$$

where E_1 and ν_1, and E_2 and ν_2 are Young's modulus and Poisson's ratio for the material of the two spheres.

This is of interest because it demonstrates that even for elastic *particle* material response the contact law is nonlinear. The behaviour of a contact under combined normal and tangential forces can also be analysed theoretically for elastic spheres and this can be extended to allow for limiting friction (local ratio of tangential to normal stress) across the region of contact.

Actual sand particles are usually rather unspherical (the shapes which influence the particle interaction will be linked with the crystalline structure of the material around the contact points) and the difficulties of geometric characterisation have rather limited the analyses of assemblies of particles of more realistic

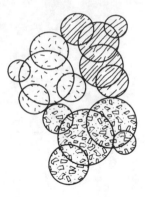

Figure 2.3: Agglomerated spherical particles

shapes in either two or three dimensions. An approximation to non-spherical particles can be created by glueing together two or more spherical particles of different diameters (Fig 2.3), but the spherical nature of the contact remains. Analytical results are also available for the contact between a conical and a spherical object. In real soils it is likely that abrasion will also occur at the contacts, where the level of local stress is often rather high, and in principle particle fracture should therefore be included. Such analyses have their attraction for studying the behaviour of assemblies of sand- or gravel-like particles where the particles interact by (apparently) rather straightforward mechanical laws at their contacts.

For clay-like materials on the other hand electron micrographs indicate that the 'particles' consist of packets of clay molecules which are distinctly non-spherical (Fig 2.4). The particles are sufficiently small that surface electrical forces between particles are significant and it is no longer sufficient to describe particle interactions in purely mechanical contact terms. Being formed of packets of molecules one should also consider the possibility of deformations occurring *within* the packets as well as *between* the packets. The shape of such particles makes it likely that particle bending will have a significant influence on the deformation response of a system of particles. The analysis of assemblies of such particles has been much more rarely attempted.

Classic experiments were conducted by Drescher and De Josselin de Jong (1972) on the shearing of a two-dimensional random assembly of photoelastic discs having six different diameters from 8 to 20 mm. The photoelastic property of the material of the discs and the size of the discs allowed the way in which stresses were transmitted through particle contacts to be very clearly observed— and the actual stresses in individual discs to be determined. Fig 2.5 is very illuminating: from it can be deduced a diagram such as Fig 2.6 which shows the network of contact forces passing through the particles. The lines in Fig 2.6 have a thickness which indicates the magnitude of the contact force; the line is drawn linking the centres of the contacting particles. It is found that some contacts are loaded much more heavily than others—and some particles

Figure 2.4: Scanning electron micrograph of Bothkennar clay (reproduced by kind permission of Professor MA Paul and Dr BF Barras)

hardly loaded at all—so that 'force chains' appear, roughly aligned with the direction of the imposed major principal stress. Soil particles are born free but are everywhere in chains.

Numerical calculations with random assemblies of (usually) circular particles have managed to introduce much larger numbers of particles—and can tackle three-dimensional assemblies too. These have confirmed the character of the load carrying networks revealed in Figs 2.5 and 2.6.

The arrangement of the soil particles adjusts itself to accommodate these force chains which form a network with a typical dimension much larger than that of individual particles—perhaps of the order of ten particle sizes in these simple analyses. The mechanical response of the assembly is intimately linked with the creation and adjustment of this fabric of more heavily loaded particle contacts. As loads are increased in such analyses deformation is seen to occur by buckling of chains of particles and the establishment of new sets of contacts. In particular, rotation of principal axes requires realignment of the load carrying contacts and is likely to imply softer material response than a loading which retains the current direction of principal stresses. Buckling remains a helpful analogy: the application of even a small lateral perturbation to the top of a structural column has a dramatic effect on the load at which a buckling instability will occur. Physicists see sand piles as examples of self organised criticality and have called the material 'fragile matter' because the structure is so sensitive to small changes in the nature of the loading (eg Cates et al, 1998).

If one were confident about one's ability to describe the mechanical behaviour at particle contacts then one could envisage modelling real geotechnical proto-type situations as boundary value problems containing assemblies of particles. The numbers of particles involved are, however, daunting. Suppose the typical particle size is about 5 mm and a typical prototype dimension 10 m. For a plane strain problem a block of soil of the order of $10 \times 10 \times 1$ m will need to be

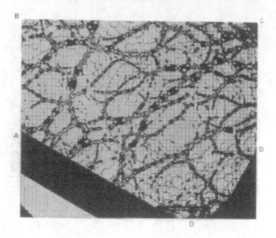

Figure 2.5: Photoelastic picture of random assembly of circular discs (from Drescher and De Josselin de Jong, 1972)

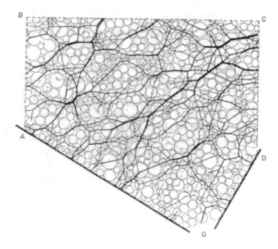

Figure 2.6: Network of chains of contact forces determined from photoelastic experiments on assembly of circular discs (from Drescher and De Josselin de Jong, 1972)

analysed: effects of geotechnical structures are expected to extend away from the structure to distances at least of the same order as the size of the structure—the height of a retaining wall for example. The width of the block being studied needs to be large enough to accommodate any effects of patterned soil fabric or particle structure that may develop as described in the previous paragraphs. The number of particles contained in this volume is then of the order of 8×10^8 which is not completely out of reach but would probably require quite long running analyses. (Of course, if the problem of concern actually involves rockfill with individual particles having dimensions of the order of 1 m then the possibly of analysing every particle in the boundary value problem is entirely reasonable, provided that the detail of the contact laws is fully understood.)

The number of particles to be included in the analysis could perhaps be reduced by replacing the actual particles by 'macroparticles' which were small enough in comparison with typical problem dimensions for the overall response not to be dominated by individual particles but sufficiently large to keep the numbers of particles to a reasonable level. This has the apparent attraction of retaining the particulate nature of the material. It is evident that it is the fabric of the network of force chains rather than the size of individual particles that must be small by comparison with the problem dimension. There remains the problem of tuning the mechanical properties of the macroparticle contacts to ensure that the overall response is satisfactory: these macroparticles have somehow to scale up the properties of the smaller particles.

The alternative to the particulate approach is the continuum hypothesis. Here we propose that the material is continuous at all scales that interest us—all quantities are infinitely differentiable. Instead of working in terms of forces and relative displacements at particle contacts we now work in terms of continuum concepts such as stress and strain. Stress is only relevant at a scale considerably larger than the individual particles and the network of force chains between particles. Strain is defined in terms of gradient of a field of displacement. Analyses and observations of particle assemblies show that individual particles rotate as well as slide at particle contacts—in fact for some assemblies of circular particles rotation seems to be the dominant mechanism (consistent with the buckling of force chains) and interparticle friction has a lower effect than might have been anticipated (see, for example, Thornton, 2000). Conventional definitions of strain do not admit rotation as a field variable. Particle rotation can be seen as a consequence of out of balance moments being imposed on the particles. These too cannot be incorporated in conventional definitions of stress: we assume that only normal and shear tractions (and not moments) can be transmitted across any surface in the *continuum* and the need for moment equilibrium forces the symmetry of the stress tensor.

There are some features of response of granular materials—particularly those associated with high gradients of displacement and particle rotation in narrow failure zones—which cannot be satisfactorily modelled unless enriched continuum approaches are adopted (see, for example, Oda and Iwashita (eds) (1999)). For example, we might permit 'couple stresses' to apply moments across the boundaries of our element and then correspondingly include rotations in our description of displacement gradient. This is well beyond the scope of this book

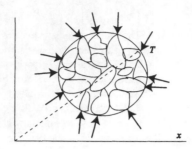

Figure 2.7: Volume element used for estimation of average stress state

but is mentioned to indicate that we may need to be prepared to question some of our starting hypotheses in order to make progress in modelling strange or difficult phenomena.

Practically, we have to work in terms of the continuum quantities stress and strain in order to be able to estimate the behaviour of geotechnical systems. It is also inevitable that our understanding of the behaviour of soils as assemblies of individual particles should in general be mediated through observation of the behaviour of samples of soils in the laboratory—each sample containing a very large number of particles. The constitutive models that we present in Chapter 3 are therefore all constructed in terms of components of stress and strain. The most appropriate use of analyses of particulate assemblies seems at present to be to provide inspiration for the continuum constitutive models.

Given such an inhomogeneous numerical (or photoelastic) assembly we have to go through a homogenisation process in order to derive an average continuum behaviour from the local observations (Drescher and De Josselin de Jong, 1972). With a finite volume V of a continuum, the stress state σ_{ij} wil be everywhere in equilibrium but variable. An average stress state $\bar{\sigma}_{ij}$ is

$$\bar{\sigma}_{ij} = \frac{1}{V} \int_V \sigma_{ij} \mathrm{d}V \qquad (2.4)$$

which, by application of Gauss' divergence theorem, can be converted to a surface integral of tractions t_j over the surface S of the volume

$$\bar{\sigma}_{ij} = \frac{1}{V} \int_S x_i t_j \mathrm{d}S \qquad (2.5)$$

where x_i is the coordinate of a point on S. For a boundary intersecting a physical or numerical assembly of particles, the integral becomes a summation of discrete forces T, with components T_j, over all the particle contacts on the boundary (Fig 2.7)

$$\bar{\sigma}_{ij} \approx \frac{1}{V} \sum_{m=1}^{N} x_i^m T_j^m \qquad (2.6)$$

and moment equilibrium of the assembly of particles should ensure that $\bar{\sigma}_{ij} = \bar{\sigma}_{ji}$.

An exactly similar argument can be used to develop an average strain tensor. Strain is calculated from gradients of displacement so we need to consider the construction of the average, over the volume, of typical displacement gradients $u_{i,j}$, adopting this shorthand notation for $\partial u_i / \partial x_j$. As before:

$$\bar{u}_{i,j} = \frac{1}{V} \int_V u_{i,j} \mathrm{d}V \tag{2.7}$$

and from the divergence theorem this can again be converted to an integral over the surface S of the volume V:

$$\bar{u}_{i,j} = \frac{1}{V} \int_S u_i n_j \mathrm{d}S \tag{2.8}$$

where n_j is the normal to the elemental surface area $\mathrm{d}S$. For a system of particles we convert this to a summation

$$\bar{u}_{i,j} \approx \sum_{m=1}^{N} \Delta u_i^m d_j^m \tag{2.9}$$

where Δu_i is the relative translation of the centres of two particles on the edge of the volume and d, with components d_j, is a 'complementary area vector' which assigns an area and a direction to each contact on the boundary (Fig 2.8). The macroscopic average strain $\bar{\epsilon}_{ij}$ is then

$$\bar{\epsilon}_{ij} = \frac{\bar{u}_{i,j} + \bar{u}_{j,i}}{2} \tag{2.10}$$

and the average rigid body motion $\bar{\omega}_{ij}$ is given by the skew symmetric part of $\bar{u}_{i,j}$;

$$\bar{\omega}_{ij} = \frac{\bar{u}_{i,j} - \bar{u}_{j,i}}{2} \tag{2.11}$$

There is thus a clear route for moving from particle analysis to continuum interpretation—although, of course, much richness of the actual particle response and nature of the transmission of loads is lost in the transformation.

As well as providing information about particle movements and the transmission of forces through the granular medium, numerical analyses of particle assemblies can give information about the evolving 'fabric' of the material. The fabric includes various elements (Oda and Iwashita, 1999):

- the orientation fabric, describing the orientation of non-spherical particles;

- the void fabric, describing the size and orientation of voids; and

- the multigrain fabric, describing the interaction betwen neighbouring particles.

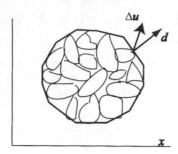

Figure 2.8: Definition of complementary area vector d for boundary of element used to define average strain state

There are two elements to the multigrain fabric: the geometric fabric which simply describes the orientation of contacts, and the kinetic fabric (Chen *et al.*, 1988) which describes how these contacts are actually being used to carry forces through the soil. Evidently the same orientation and geometric fabrics can carry many different external loads and, as a corollary, one might propose that kinetic fabric can change very much more rapidly—through formation and elimination of particle contacts—than orientation fabric which requires significant particle rotation and relative movement.

2.3   Laboratory element testing

The possibilities for performing real physical experiments on homogeneous soil samples (even ignoring for the moment the actual particulate nature of the soil) are limited by the ingenuity of the designers of soil testing devices; the possibilities for performing and deconstructing numerical experiments on assemblies of soil particles are much greater. Any general element of soil in a geotechnical system will experience changes in all of the six components of stress to which it is subjected (Fig 2.9)[1]. Any constitutive model (Chapter 3) that is used in numerical analysis (Chapter 4) will be expected to make reasonable predictions of the soil behaviour under such general stress changes. The reliability of the constitutive model can best be checked by pitting it against carefully conducted laboratory experiments which expose uniform soil samples to similarly general stress or strain changes.

Most data from laboratory tests on soil elements have come from tests performed in the standard triaxial apparatus (Fig 2.10). A sample is contained in a membrane and subjected to lateral pressure through this membrane. It is also subjected to axial deformation through rigid end platens. The loading is axisymmetric and (neglecting end effects and possible problems associated with

---

[1]The sign convention adopted throughout this book *except where specifically noted* assumes that compressive stresses and strains are positive. A shear stress τ_{ij} is positive if it acts on a plane facing in the positive i direction but is directed in the *negative j* direction.

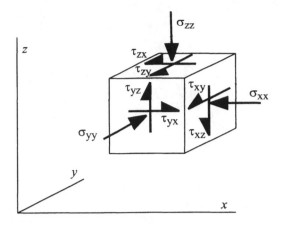

Figure 2.9: Soil element subjected to general state of stress

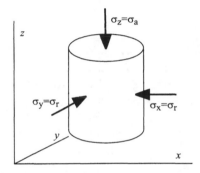

Figure 2.10: Axisymmetric (triaxial) testing configuration

initial anisotropy of the sample) we have two lateral principal stresses equal to the cell pressure σ_r

$$\sigma_x = \sigma_y = \sigma_r$$

and an axial principal stress $\sigma_z = \sigma_a$. Depending on the way in which the test is conducted the axial stress may be the major principal stress ('compression')—so that the intermediate and minor principal stresses are equal—or the minor principal stress ('extension')—so that the intermediate and major principal stresses are equal. The test has two degrees of freedom and is more strictly described as a confined uniaxial test than a triaxial test. However, it is likely to remain the most widely used soil testing device. The other test apparatus described here are, on the whole, either purely research devices or will enter commercial application only for particularly subtle or sensitive geotechnical projects.

Axial symmetry may not seem particularly relevant to many geotechnical systems: it matches exactly the conditions on the centreline beneath a circular

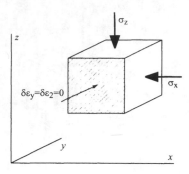

Figure 2.11: Plane strain element test

load (such as an oil tank), but not the stresses at any point off the centre-line. Plane strain may seem more generally applicable (Fig 2.11). The relevant element test can be achieved either by inserting fixed lateral boundaries in a triaxial cell or by development of a dedicated biaxial apparatus. The latter may be advantageous in permitting either σ_x or σ_z to be the major principal stress. In either, the intermediate strain increment, which from symmetry is also a principal strain increment, is zero and the corresponding stress σ_y will be the intermediate principal stress (and will be a *dependent* stress quantity taking whatever value is required to maintain the plane strain condition):

$$\delta\epsilon_y = 0; \quad \sigma_y = \sigma_2$$

Such a device also has two degrees of freedom.

True triaxial apparatus or cubical cells permit the application and control of three independent principal stresses, and corresponding principal strain increments, with fixed coincident principal axes (Fig 2.12). Loads can be imposed either by means of flexible cushions (Ko and Scott, 1967)—imposing direct control of stresses—or by a cunning arrangement of nested rigid platens (Hambly, 1969)—imposing direct control of strains. Suitable control systems can permit either device to be used, in principle, for stress or strain controlled testing with three degrees of freedom.

The simple shear apparatus (Fig 2.13) permits some control of rotation of principal axes and has been somewhat widely used for commercial testing—especially for cyclic testing linked with offshore and seismic applications. It is a plane strain device, $\delta\epsilon_y = 0$, but also prevents any direct strain in the x direction, $\delta\epsilon_x = 0$. The only two degrees of freedom are therefore the shear strain, γ_{zx}, and the vertical strain, ϵ_z, which is therefore also the volumetric strain. There are stresses τ_{zx} and σ_{zz} associated with each of these strain components. The intermediate principal stress σ_y is not an independent quantity because it has to take an appropriate value to maintain the plane strain condition. The final stress component, σ_{xx}, is also dependent on the response of the soil. The simple shear apparatus was developed as an improvement on the shear box (Fig 2.14) (Roscoe, 1953) which clearly makes no pretence of imposing uniform

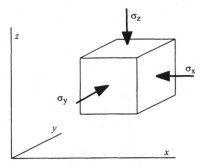

Figure 2.12: True triaxial apparatus

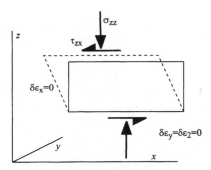

Figure 2.13: Simple shear apparatus

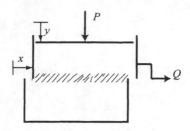

Figure 2.14: Direct shear apparatus (shear box)

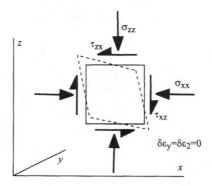

Figure 2.15: Directional shear cell

conditions on the soil being tested. Unfortunately, the need for the soil to be able to change in volume, and the kinematic need for the ends of the sample to become longer as the sample changes section from rectangle to parallelogram (even at constant volume), mean that there is a difficulty in providing the necessary complementary shear stress τ_{xz} on the ends of the sample. Stresses and strains within a simple shear sample are inevitably nonuniform and it has to be interpreted as a boundary value problem rather than a single homogeneous element (Airey, Budhu and Wood, 1985).

The directional shear cell (Fig 2.15) (Arthur *et al.*, 1977) is another plane strain device ($\delta\epsilon_y = 0$ which aims to impose controlled shear and normal stresses on two sets of initially orthogonal flexible boundaries of a sample which is free to undergo all the associated deformations, in principle homogeneously. This single element test has three degrees of freedom (σ_{xx}, σ_{zz}, τ_{zx}) but the complexities of the loading arrangements have severely limited its use.

The torsional hollow cylinder apparatus (Fig 2.16) (Saada and Baah, 1967), on the other hand, has been much more widely adopted. A hollow cylindrical sample is subjected to axial load F, and axial torque T through rigid end platens, and internal and external pressures (p_i, p_o) through containing membranes. This gives us four degrees of freedom, which is nice, but one should quickly understand that this is at the expense of inevitable internal variations of stress

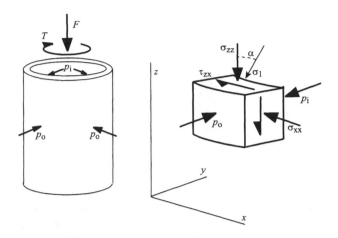

Figure 2.16: Torsional hollow cylinder apparatus

and strain. We are testing a closely controlled boundary value problem not a single homogeneous soil element. Any discussion of stress:strain response has to be deduced from average quantities.

For example, with internal and external radii r_i, r_o, we could propose an average axial stress

$$\bar{\sigma}_{zz} = F/\pi \left(r_o^2 - r_i^2\right) \qquad (2.12)$$

and shear stress

$$\bar{\tau}_{zx} = 3T/\left[2\pi \left(r_o^3 - r_i^3\right)\right] \qquad (2.13)$$

But this has assumed a uniform shear stress over the section whereas, for a rotation θ of the top of a sample of height h, the imposed shear strain varies linearly from $r_i\theta/h$ at the inner edge to $r_o\theta/h$ at the outer edge of the hollow cylinder. Should we assume some corresponding linear variation of shear stress with radius? This will certainly change the average in (2.13)—in fact the volume average for the shear stress then becomes

$$\bar{\tau}_{zx} = \frac{4T}{3\pi} \frac{\left(r_o^3 - r_i^3\right)}{\left(r_o^2 - r_i^2\right)\left(r_o^4 - r_i^4\right)} \qquad (2.14)$$

And if the stress:strain response is not in fact linear—elastic-plastic perhaps?— then the radial variation (and volumetric average) of shear stress will be different again.

If the internal and external pressures are different then the obvious average radial stress is

$$\bar{\sigma}_r = \bar{\sigma}_{yy} = (p_i + p_o)/2 \qquad (2.15)$$

(although this is not actually the volume average of the radial stress). The corresponding average circumferential stress, from radial equilibrium of a small element of the hollow cylinder (Fig 2.16), is

$$\bar{\sigma}_\theta = \bar{\sigma}_{xx} = (p_o r_o - p_i r_i)/(r_o - r_i) \qquad (2.16)$$

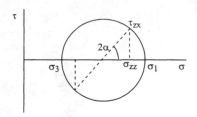

Figure 2.17: Mohr's circle of stress for torsional hollow cylinder

In all proposed expressions for average stress quantities the departure from uniformity is reduced as the ratio of wall thickness to average radius reduces—large thin samples are preferred. There are evidently compromises to be made between apparatus size, ease of sample preparation and internal stress uniformity.

If we sacrifice one degree of freedom and keep $p_i = p_o$ then $\bar{\sigma}_r = \bar{\sigma}_\theta = \bar{\sigma}_{xx} = \bar{\sigma}_{yy} = \sigma_2$ but now, as we vary the axial load and torque, the geometry of Mohr's circle (Fig 2.17) tells us that

$$\sigma_2 = \sigma_1 \sin^2 \alpha + \sigma_3 \cos^2 \alpha \tag{2.17}$$

or that the quantity b which indicates the value of the intermediate principal stress relative to the other two principal stresses is

$$b = \frac{\sigma_2 - \sigma_3}{\sigma_1 - \sigma_3} = \sin^2 \alpha \tag{2.18}$$

where α is the angle made by the major principal stress to the horizontal.

This relationship of the intermediate stress to the major and minor principal stresses is now a function of the value of α and cannot be chosen independently. This sacrifice of a degree of freedom, while leading to reduced uncertainty of the values of some of the averaged stress components, somewhat constrains the stress paths that can be followed in the torsional hollow cylinder apparatus.

Interpretation of the results of hollow cylinder tests is something of an exercise in deconvolution—working back from system response to underlying contributory elemental behaviour. Four degrees of freedom seem to be the maximum that we can achieve in laboratory element testing and we are always going to have to extrapolate using our constitutive model in order to describe the final two degrees of freedom (and we have to use a postulated model to deconvolve the hollow cylinder response itself).

2.4   Stress and strain variables

Having established that concepts of stress are likely to be helpful we need to consider the ways in which we can most usefully characterise the stresses to which

Figure 2.18: 'Wavy' plane through granular material

elements of soil are subjected. Triaxial testing provides the bulk of the available data on the mechanical behaviour of soil elements and therefore provides a useful background against which to introduce ideas of constitutive modelling of soils in Chapter 3—even though the bulk of the applications in geotechnical engineering do not have the axial symmetry that is implied by the triaxial apparatus.

The triaxial apparatus provides the possibility for confined uniaxial testing of soils or other materials and evidently provides two degrees of freedom in control of externally applied stress states. There are, however, many different ways in which these two degrees of freedom can be chosen for presentation and interpretation of test results: there are many different national traditions. However, when it comes to the development of constitutive models then it is necessary to start to impose some constraints on the choice of variables.

Development of constitutive models describing the link between changes in stress and changes in strain requires correct choice of strain increment and stress variables. Soils consist of more or less rigid particles separated by voids. Volume changes are recognised to be an important feature of the mechanical response of soils. Volume changes of saturated soil require pore water movement which is controlled and limited by the permeability of the soil. Our experience tells us that undrained response of soils is often important where the permeability of the soil prevents flow of water. Undrained deformation implies constant volume deformation (the compressibility of pore fluid is usually negligible in most civil engineering applications) and hence pure distortion, change in shape at constant size. It is convenient, then, to divide soil deformations into compression (change of volume) and distortion (change of shape) and to choose the strain increment variables correspondingly.

The principle of effective stress proposes that it is the effective stresses that control the deformation behaviour of the soil—all the constitutive models that are discussed in Chapter 3 will take this principle as axiomatic. Various attempts have been made to prove the validity of the principle of effective stress: here it will be taken simply as a hypothesis or conjecture only weakly non-falsified which has been found to work well within the context of understanding of saturated soil behaviour. Drawing a 'wavy plane' through a granular material (Fig 2.18) (see, for example, Lambe and Whitman (1979)) to support an argument of partition of total stress between pore pressure and intergranular stress (intergranular normal forces averaged over the whole area) can only support the principle of effective stress if the area of contact—over which the intergranular stress but not the pore pressure acts—is small by comparison with the total area. Classic experiments by Laughton (quoted by Bishop (1959)) on lead shot, subjected to such high pressures that the observed flats on the surface of the originally spherical particles indicated a contact area of 60%, showed that the principle of effective stress still applied under these extreme conditions.

Formally, given a total stress tensor σ_{ij} (compression positive) and pore pressure u, the effective stress tensor σ'_{ij} is given by

$$\sigma'_{ij} = \sigma_{ij} - u\delta_{ij} \tag{2.19}$$

where δ_{ij} is the Kronecker delta

$$\delta_{ij} = 1, \quad i = j; \quad \delta_{ij} = 0, \quad i \neq j \tag{2.20}$$

Expression (2.19) simply tells us that the pore pressure affects only the normal stresses and not the shear stresses supported by the soil element: this does not surprise us because we know that, in the context of soil mechanics, water has negligible shear stiffness and negligible ability to carry shear stresses.

In presentation of constitutive models in Chapter 3, we will concentrate on conditions which are attainable in conventional triaxial tests, where the states of stress and strain are assumed to be axisymmetric. In any test we can obviously identify the axial and radial strain increments $\delta\epsilon_a$ and $\delta\epsilon_r$ respectively, and corresponding axial and radial effective stresses σ'_a and σ'_r. However, since volumetric effects are recognised to be important let us choose as our first strain increment variable for constitutive modelling the volumetric strain increment $\delta\epsilon_p$:

$$\delta\epsilon_p = \delta\epsilon_a + 2\delta\epsilon_r \tag{2.21}$$

At many stages in numerical and constitutive modelling we need to make statements about increments of work done in deforming soil elements. This leads to the idea of work conjugacy of strain increment and stress quantities. In the context of volumetric deformations we need to choose a 'volumetric' effective stress p' such that the work done in changing the volume of a unit element of soil is given by

$$\delta W_v = p'\delta\epsilon_p \tag{2.22}$$

This requires that this volumetric stress should be the mean effective stress:

$$p' = \frac{1}{3}\left(\sigma'_a + 2\sigma'_r\right) \tag{2.23}$$

The subscript p for the volumetric strain indicates that this p strain is linked with the p stress.

Our two degrees of deformational freedom for soil samples in the triaxial apparatus are compression (change of size) and distortion (change of shape). We need next to choose a pair of strain increment and stress variables which can describe distortional processes. We are now bound more by convenience than by any direct theoretical constraint. When we perform triaxial tests we directly measure the 'deviator stress' q which is the amount by which the axial stress exceeds the radial stress:

$$q = \sigma_a - \sigma_r = F/A \tag{2.24}$$

where F is the axial force and A is the cross sectional area of the sample. If we choose q as our distortional or shear stress then we have no remaining choice in the definition of the distortional strain increment $\delta\epsilon_q$

$$\delta\epsilon_q = \frac{2}{3}\left(\delta\epsilon_a - \delta\epsilon_r\right) \tag{2.25}$$

and the increment of work required to change the shape of a unit element of soil is given by

$$\delta W_d = q\delta\epsilon_q \tag{2.26}$$

and again the subscript q reminds us of the link between the distortional q strain increment and the q stress.

The total work done per unit volume during any strain increment is the sum of the volumetric and distortional terms:

$$\delta W = \delta W_v + \delta W_d = p'\delta\epsilon_p + q\delta\epsilon_q = \sigma_a'\delta\epsilon_a + 2\sigma_r'\delta\epsilon_r \tag{2.27}$$

The relationships between stress variables and between strain increment variables can be summarised in matrix form

$$\begin{pmatrix} p' \\ q \end{pmatrix} = \begin{pmatrix} \frac{1}{3} & \frac{2}{3} \\ 1 & -1 \end{pmatrix} \begin{pmatrix} \sigma_a' \\ \sigma_r' \end{pmatrix} \tag{2.28}$$

$$\begin{pmatrix} \sigma_a' \\ \sigma_r' \end{pmatrix} = \begin{pmatrix} 1 & \frac{2}{3} \\ 1 & -\frac{1}{3} \end{pmatrix} \begin{pmatrix} p' \\ q \end{pmatrix} \tag{2.29}$$

$$\begin{pmatrix} \delta\epsilon_p \\ \delta\epsilon_q \end{pmatrix} = \begin{pmatrix} 1 & 2 \\ \frac{2}{3} & -\frac{2}{3} \end{pmatrix} \begin{pmatrix} \delta\epsilon_a \\ \delta\epsilon_r \end{pmatrix} \tag{2.30}$$

$$\begin{pmatrix} \delta\epsilon_a \\ \delta\epsilon_r \end{pmatrix} = \begin{pmatrix} \frac{1}{3} & 1 \\ \frac{1}{3} & -\frac{1}{2} \end{pmatrix} \begin{pmatrix} \delta\epsilon_p \\ \delta\epsilon_q \end{pmatrix} \tag{2.31}$$

$$\frac{q}{p'} = \frac{3\left[(\sigma_a'/\sigma_r') - 1\right]}{(\sigma_a'/\sigma_r') + 2} \tag{2.32}$$

$$\frac{\sigma_a'}{\sigma_r'} = \frac{1 + (2/3)\,(q/p')}{1 - (1/3)\,(q/p')} \tag{2.33}$$

$$\frac{\delta\epsilon_p}{\delta\epsilon_q} = \frac{3 + (3/2)\,(\delta\epsilon_a/\delta\epsilon_r)}{(\delta\epsilon_a/\delta\epsilon_r) - 1} \tag{2.34}$$

$$\frac{\delta\epsilon_a}{\delta\epsilon_r} = \frac{3 + (\delta\epsilon_p/\delta\epsilon_q)}{(\delta\epsilon_p/\delta\epsilon_q) - (3/2)} \tag{2.35}$$

It is also convenient to define a stress ratio η

$$\eta = \frac{q}{p'} \tag{2.36}$$

which is equivalent to a mobilised friction ϕ_m'. Under conditions of triaxial compression, in which $q > 0$ and the axial stress is greater than the radial stress,

$$\frac{\sigma_a'}{\sigma_r'} = \frac{1 + \sin\phi_m'}{1 - \sin\phi_m'} = \frac{3 + 2\eta}{3 - \eta} \tag{2.37}$$

and

$$\sin \phi'_m = \frac{\sigma'_a - \sigma'_r}{\sigma'_a + \sigma'_r} = \frac{3\eta}{6 + \eta} \tag{2.38}$$

so that

$$\eta = \frac{6 \sin \phi'_m}{3 - \sin \phi'_m} \tag{2.39}$$

Under conditions of triaxial extension, in which $q < 0$ and the axial stress is less than the radial stress,

$$\frac{\sigma'_a}{\sigma'_r} = \frac{1 - \sin \phi'_m}{1 + \sin \phi'_m} = \frac{3 + 2\eta}{3 - \eta} \tag{2.40}$$

and

$$\sin \phi'_m = \frac{\sigma'_r - \sigma'_a}{\sigma'_a + \sigma'_r} = -\frac{3\eta}{6 + \eta} \tag{2.41}$$

so that

$$\eta = -\frac{6 \sin \phi'_m}{3 + \sin \phi'_m} \tag{2.42}$$

It is unfortunate that the symbols p' and q have frequently also been used for the quantities $(\sigma'_a + \sigma'_r)/2$ and $(\sigma'_a - \sigma'_r)/2$ respectively. These definitions are not helpful in the context of constitutive modelling of response in triaxial tests because it is not possible to use them to develop correct work statements—there are no corresponding work-conjugate strain increment quantities. The engineer must be careful, in using quoted experimental data, to ensure that the definitions of p' and q are indeed as anticipated.

For axisymmetric states of stress there are only two degrees of freedom and the introduction of the stress variables q and p' clearly implies no loss of information. However, the complete stress state at any location (a symmetric second order tensor) has six independent components. We can choose these to be the normal and shear stresses on three mutually orthogonal planes (Fig 2.9) but for every stress state there is a set of three mutually orthogonal planes on which the shear stresses vanish (Fig 2.19). The normals to these three planes define a set of three principal directions or principal axes and the corresponding normal stresses are the three principal stresses $\sigma_1, \sigma_2, \sigma_3$. The two stress states in Figs 2.9 and 2.19 are exactly equivalent and related to each other by the rules of tensor transformation—or stress resolution.

If we know that the directions of the principal axes are not changing or if we choose to ignore the effects of changing the directions of the principal axes (this may be unwise) then the principal stresses alone are sufficient to describe the stress state. If we reckon that even for these more general stress states it is still important to think of the volumetric effects separately from the distortional effects then we can propose that the mean stress p will still be a useful stress variable:

$$p = \frac{1}{3} (\sigma_1 + \sigma_2 + \sigma_3) = \frac{1}{3} (\sigma_{xx} + \sigma_{yy} + \sigma_{zz}) \tag{2.43}$$

This mean stress can also be defined as

$$p = \frac{1}{3} \text{tr} (\sigma_{ij}) \tag{2.44}$$

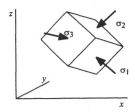

Figure 2.19: Principal axes and principal stresses for general state of stress

where tr signifies the trace, or the sum of the leading diagonal terms of the tensor.

For a given state of stress at a point in a continuum the elements of the stress tensor $\sigma_{xx}, \sigma_{yy}, \sigma_{zz}, \tau_{yz}, \tau_{zx}, \tau_{xy}$ will change if the directions of the coordinate axes x, y, z are changed. However, the values of the principal stresses will not change: these quantities are said to be 'invariant' to choice of coordinate axes. *Any* function of the three principal stresses will also be invariant to choice of axes but, by convention, *the* three invariants of a stress tensor are defined as

$$I_1 = \sigma_1 + \sigma_2 + \sigma_3 = \text{tr}(\sigma_{ij}) = 3p \qquad (2.45)$$

$$I_2 = \frac{1}{2}\left\{[\text{tr}(\sigma_{ij})]^2 - \text{tr}(\sigma_{ij}^2)\right\} = \sigma_2\sigma_3 + \sigma_3\sigma_1 + \sigma_1\sigma_2 \qquad (2.46)$$

$$I_3 = \det\sigma_{ij} = \sigma_1\sigma_2\sigma_3 \qquad (2.47)$$

If we subtract the volumetric stress then we can create a stress deviator tensor s_{ij} which describes solely the distortional components of stress

$$s_{ij} = \sigma_{ij} - p\delta_{ij} \qquad (2.48)$$

It is easy to see that

$$\text{tr}(s_{ij}) = 0 \qquad (2.49)$$

The second invariant of the stress deviator tensor is defined (as in (2.46)) as

$$J_2 = \frac{1}{2}\text{tr}(s_{ij}^2) = \frac{I_1^2}{3} - I_2 \qquad (2.50)$$

Alternative expressions for J_2 are

$$J_2 = \frac{1}{2}\left[(\sigma_1 - p)^2 + (\sigma_2 - p)^2 + (\sigma_3 - p)^2\right] \qquad (2.51)$$

and

$$J_2 = \frac{1}{6}\left[(\sigma_{yy} - \sigma_{zz})^2 + (\sigma_{zz} - \sigma_{xx})^2 + (\sigma_{xx} - \sigma_{yy})^2 + \tau_{yz}^2 + \tau_{zx}^2 + \tau_{xy}^2\right] \qquad (2.52)$$

Noting that, for an axisymmetric system of stresses

$$J_2 = \frac{1}{3}(\sigma_{zz} - \sigma_{xx})^2 \qquad (2.53)$$

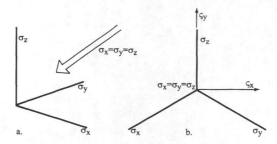

Figure 2.20: (b) π-plane deviatoric view of (a) principal stress space down space diagonal $\sigma_x = \sigma_y = \sigma_z$

we deduce that a generalised definition of our distortional stress q (2.24) will be

$$q = (3J_2)^{1/2} \tag{2.54}$$

There is a class of theoretical constitutive models that can be constructed within the framework of so-called J_2 plasticity for which the stress:strain response is controlled only by the first and second invariants of the stress tensor, p and J_2. This might also be called the Von Mises generalisation from the axisymmetric condition. It will be shown subsequently (§2.7) that this generalisation is not obviously directly appropriate for soils, though it does confer some mathematical simplification.

The generalised distortional stresses for true triaxial tests—in which all three principal stresses are changed without changing the orientations of principal axes—can be displayed in a two-dimensional diagram which is a view of the so-called π-plane—a view of principal stress space down the line $\sigma_x = \sigma_y = \sigma_z$ (Fig 2.20). For plotting purposes this is equivalent to projecting onto orthogonal axes ς_x and ς_y (Fig 2.20):

$$\varsigma_x = \frac{\sigma_y - \sigma_x}{\sqrt{2}}; \quad \varsigma_y = \frac{(\sigma_z - \sigma_x) - (\sigma_y - \sigma_z)}{\sqrt{6}} \tag{2.55}$$

which are both clearly functions of independent stress differences. Under conditions of axial symmetry, $\sigma_x = \sigma_y$, and $\varsigma_x = 0$ and $\varsigma_y = \sqrt{(2/3)}q$. Contours of constant generalised q (2.54) are circles in the π-plane.

For test apparatus which permit rotation of principal axes we have to include the imposed shear stresses in our selection of stress variables. For example, for simple shear (Fig 2.13) or directional shear cell (Fig 2.15) tests we can display the two distortional degrees of freedom in a diagram with axes $\beta = (\sigma_{zz} - \sigma_{xx})/2$; and τ_{zx} (Fig 2.21) with corresponding work-conjugate strain increments $\delta\varsigma = \delta\epsilon_{zz} - \delta\epsilon_{xx}$ and $\delta\gamma_{zx}$. In this form of plotting of deviatoric stress information the length of the stress vector is $(\sigma_1 - \sigma_3)/2$ where σ_1 and σ_3 are major and minor principal stresses respectively. The inclination of the stress vector to the β axis is 2α where α is the inclination of the major principal stress to the z

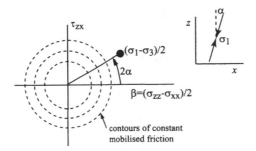

Figure 2.21: Deviatoric stress plane for rotation of principal axes

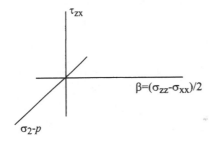

Figure 2.22: Additional deviatoric stress axis for intermediate principal stress

direction. Mobilised friction only involves major and minor principal stresses and, provided the out-of-plane stress σ_2 is indeed the intermediate principal stress, contours of constant mobilised friction for given values of mean stress become circles in the $\beta:\tau_{zx}$ stress plot (Fig 2.21).

For the torsional hollow cylinder apparatus (Fig 2.16) we need to display three degrees of freedom of distortional information: the third axis could logically be $\sigma_2 - p = \sigma_{yy} - p$ (from (2.48)) (Fig 2.22).

As an example of the application of these stress variables we can consider the stress path followed in a simple shear test (Fig 2.13). The results relate to tests on dry sand so that total and effective stresses are identical. It is found experimentally (Airey, Budhu and Wood, 1985) that data from simple shear tests on sands and clays, in which complete information about the stress tensor is available from boundary stress measurements, fit quite closely a relationship (Fig 2.23):

$$\frac{\tau_{zx}}{\sigma_{zz}} = k \tan \alpha \qquad (2.56)$$

where k is a soil constant. In plane strain tests such as this the intermediate principal stress σ_2 is a dependent quantity, taking whatever value is required in order to maintain the plane strain condition. It is consequently helpful to think

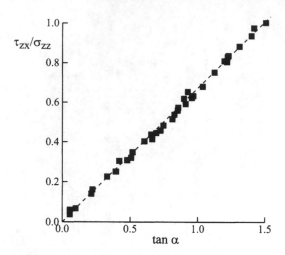

Figure 2.23: Relationship between shear stress ratio and principal stress axis direction for simple shear tests on Leighton Buzzard sand (after Airey *et al.*, 1985)

of stress changes using a plane strain mean stress s:

$$s = \frac{\sigma_{xx} + \sigma_{zz}}{2} = \frac{\sigma_1 + \sigma_3}{2} \qquad (2.57)$$

Since it is found that, in simple shear tests on sand, $\sigma_2 \approx 0.8s$ (Stroud, 1971) or more generally $\sigma_2 \approx k_2 s$, we have

$$\frac{p}{s} \approx \frac{(2 + k_2)}{3} \qquad (2.58)$$

Equation (2.56) can be converted into a stress path (Fig 2.24)

$$\tau_{zx}^2 = k^2 \sigma_{zz}^2 - 2k\beta\sigma_{zz} = 2ks\sigma_{zz} - k\left(2 - k\right)\sigma_{zz}^2 \qquad (2.59)$$

$$s = \sigma_{zz} - \beta \qquad (2.60)$$

and we see that the kinematic constraints of the simple shear test empirically constrain the distortional stress path to take a parabolic form (Fig 2.24a).

The mobilised friction is given by

$$\sin \phi_m = \frac{k\sigma_{zz} - \beta}{\sigma_{zz} - \beta} \qquad (2.61)$$

and a link can be obtained between the orientation α of the major principal stress and the mobilised friction:

$$\tan^2 \alpha = \frac{2 \sin \phi_m - k\left(1 + \sin \phi_m\right)}{k\left(1 - \sin \phi_m\right)} \qquad (2.62)$$

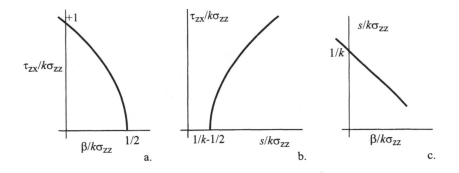

Figure 2.24: Empirically deduced stress paths for simple shear tests (after Muir Wood *et al.*, 1998)

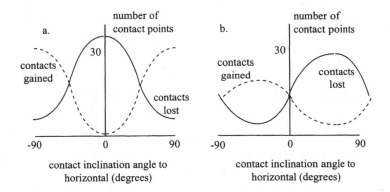

Figure 2.25: Random assembly of rods: variation of distribution of contact normals: (a) one-dimensional compression to 12% vertical strain and (b) simple shear to 12% shear strain (after Oda and Iwashita, 1999)

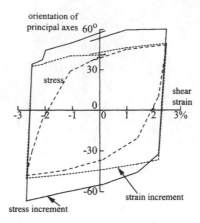

Figure 2.26: Orientations of principal axes of stress, stress increment and strain increment in a cycle of simple shear on Leighton Buzzard sand (after Wood and Budhu, 1980)

The dramatic effect of simple shear deformation on the fabric of a granular material is shown in Fig 2.25 (Oda and Iwashita, 1999) for a test on a two-dimensional random assembly of cylindrical rods of three different diameters. The geometric fabric of this pseudo-granular material is described in terms of the numbers of particle contacts in different directions. The effect of compression in the z direction to 12 % strain is to generate contacts in this vertical direction and to lose contacts in the orthogonal horizontal direction (Fig 2.25a). The effect of subsequent simple shear to $\gamma_{zx} = 12\%$ is to generate contacts at some inclination to the horizontal (Fig 2.25b): the fabric is changing dynamically during the shearing process.

Another indication of a related effect can be detected from observations of the way in which the principal axes of strain increment vary during a simple shear test. Typical data for a single simple shear cycle on a sample of sand are shown in Fig 2.26). The principal axes of stress increment are in one sense not independent but are linked with the slope of the stress path shown in Fig 2.24a—that path was an empirical deduction and not theoretically ordained. When studied directly, it is found that, after each reversal of straining, the principal axes of strain increment initially coincide with those of stress increment but progressively tend towards those of stress as the strain is monotonically increased. From a constitutive point of view this could be interpreted as a progressive transition from an elastic to a perfectly plastic model of response.

If the simple shear test eventually reaches a condition where shearing continues without further change in volume then the Mohr circle of strain increment is centred on the origin (Fig 2.27) and with coincident axes of principal strain increment and of principal stress the measured shear stress τ_{zx} must lie at the top of the Mohr circle of stress (Fig 2.27), with $\alpha = \pi/4$, so that $\tau_{zx}/\sigma_{zz} = \sin\phi_{cv}$

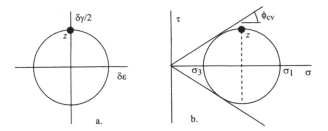

Figure 2.27: Mohr's circle of (a) strain increment and (b) stress for simple shear sample undergoing constant volume (critical state) deformation

where ϕ_{cv} is the angle of friction for constant volume (or critical state) shearing (§2.6.1). We can then deduce that $k = \sin \phi_{cv}$ in (2.56).

2.5 Stiffness

We perform laboratory tests in order to assemble observations which can help us to develop and calibrate constitutive models of soil response which provide the link between strain increments ϵ and stress increments σ which will be required for performance of numerical analysis of geotechnical systems. Formally we require to develop and populate a stiffness matrix $\boldsymbol{D}$

$$\delta\boldsymbol{\sigma} = \boldsymbol{D}\delta\boldsymbol{\epsilon} \qquad (2.63)$$

and this *incremental* link between stress and strain is probably the most useful definition of stiffness.

2.5.1 Nonlinearity: secant and tangent stiffness

Typical stress:strain relationships for soil are not linear. This nonlinearity has to be characterised and then modelled. One way in which the nonlinearity can be described is by showing how the stiffness varies with strain. Stiffness for nonlinear materials can be defined in two quite different ways (Fig 2.28), using either secant stiffness

$$G_s = \frac{\tau}{\gamma} \qquad (2.64)$$

or tangent stiffness

$$G_t = \frac{\delta\tau}{\delta\gamma} \qquad (2.65)$$

The use of the term 'stiffness' can quickly lead to false expectations because of its general association with 'elasticity'—and even more general association with 'linear elasticity'. This confusion is greatest when secant stiffness is being used because this stiffness merely defines an average stiffness over a chosen range of strain from an arbitrary zero. Tangent stiffness is more obviously useful

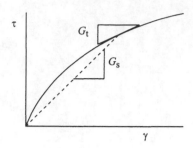

Figure 2.28: Tangent and secant shear stiffness

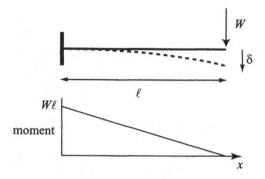

Figure 2.29: Cantilever under tip loading

because it is describing the way in which the soil will respond in generating change in stress as a result of a small imposed deformation from the *current* state of the soil—and the elastic-plastic models to be developed in Chapter 3 will generate just such general stiffness relationships.

One of the dangers of adopting any elements of the language of elasticity is that one can be rapidly seduced into adopting the entire accompanying framework of elastic analysis. Let us give a simple example, using a structural analogy: a steel cantilever beam, built in at one end and subjected to a tip point load W (Fig 2.29). This structural problem is capable of exact analysis whereas for most geotechnical systems we have to resort to numerical procedures or approximations and the direct physical insight is obscured.

The beam is of length ℓ, with flexural rigidity EI, rectangular cross section ($b \times 2d$) and full plastic moment M_p. A familiar problem might be to determine the limiting value of W to cause plastic collapse of the beam. A slightly more subtle problem is to compute the relationship between the tip deflection δ of the beam and the force W necessary to produce this deflection.

While the beam is behaving purely elastically the link between W and δ is

$$\frac{W}{\delta} = \frac{3EI}{\ell^3} \tag{2.66}$$

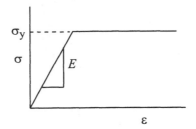

Figure 2.30: Idealised elastic-perfectly plastic stress:strain response for steel

As the deflection and the load increase there comes a point at which the moment at the root of the cantilever is sufficiently large that the stress in the extreme fibres of the beam just reaches the yield stress σ_y of the steel. The corresponding moment is the yield moment M_y which, for a beam of rectangular section, is:

$$M_y = \frac{2}{3}M_p \tag{2.67}$$

The yield value of the tip load is

$$W_y = \frac{M_y}{\ell} = \frac{2M_p}{3\ell} \tag{2.68}$$

and the corresponding tip deflection is

$$\delta_y = \frac{2M_p\ell^2}{9EI} \tag{2.69}$$

The compression and extension behaviour of the steel is assumed to be given by the bilinear relationship shown in Fig 2.30. Once the steel starts to yield then it is no longer able to accept any additional stress. As the deflection is increased beyond the yield value the stiffness of the cantilever must fall. Following Baker and Heyman (1969) we can obtain an explicit solution as follows.

Let the moment at the root of the cantilever be λM_p $(2/3 \le \lambda \le 1)$ so that the tip load for this statically determinate cantilever is

$$W = \frac{\lambda M_p}{\ell} \tag{2.70}$$

There will be a zone of partial plasticity extending into the cantilever from the root. At any particular location the elastic region occupies a fraction α of the section $(1 \ge \alpha \ge 0)$ (Fig 2.31) and the moment in the partially plastic state is

$$M = \left(1 - \frac{\alpha^2}{3}\right) M_p \tag{2.71}$$

At the root of the cantilever

$$\alpha = \alpha_o = \sqrt{3(1 - \lambda)} \tag{2.72}$$

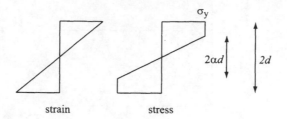

strain   stress

Figure 2.31: Strain and stress distributions across section of beam with partial plasticity

Moment equilibrium then tells us that, within this partially plastic region (Fig 2.29), measuring x from the root of the cantilever,

$$1 - \frac{\alpha^2}{3} = \lambda \left(1 - \frac{x}{\ell}\right) \tag{2.73}$$

and hence, substituting $\alpha = 1$, that the partially plastic section extends a distance x_o from the root:

$$x_o = \left(1 - \frac{2}{3\lambda}\right)\ell \tag{2.74}$$

The elastic core of the beam controls the curvature in just the same way as for a fully elastic beam so that we can write

$$\frac{d^2y}{dx^2} = \frac{\sigma_y}{\alpha E d} = \frac{2}{3}\frac{M_p}{\alpha E I} \tag{2.75}$$

Integrating this equation we deduce that the slope of the cantilever at position x_o is:

$$\left.\frac{dy}{dx}\right|_{x=x_o} = \frac{4}{9}\frac{W\ell^2}{EI\lambda^2}\left(1 - \alpha_o\right) \tag{2.76}$$

and the deflection is

$$y|_{x=x_o} = \frac{4}{81}\frac{W\ell^3}{EI\lambda^3}\left(2 - 3\alpha_o + \alpha_o^3\right) \tag{2.77}$$

Combining these values with the behaviour of the elastic cantilever of length $2\ell/3\lambda$ we find that the deflection of the tip of the cantilever is

$$\delta = \frac{4}{81}\frac{W\ell^3}{EI\lambda^3}\left(10 - 9\alpha_o + \alpha_o^3\right) \tag{2.78}$$

where α_o and λ are parametrically linked.

The resulting elastic-plastic relationship between load and deflexion is shown in Fig 2.32. The limiting load is $W_p = M_p/\ell$ corresponding to $\lambda = 1$, $\alpha_o = 0$ and the corresponding (finite) tip deflection is

$$\delta_p = \frac{40}{81}\frac{W_p\ell^3}{EI} = \frac{20}{9}\delta_y \tag{2.79}$$

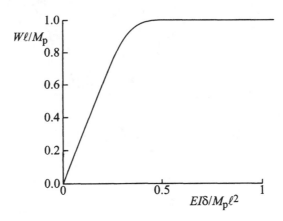

Figure 2.32: Load:deflection relationship for progressively yielding cantilever

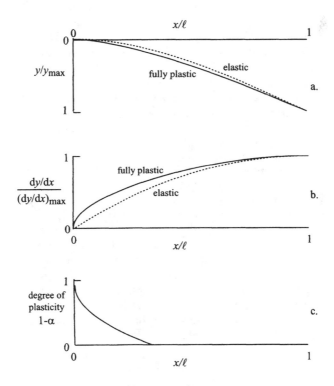

Figure 2.33: Deflected shape of cantilever

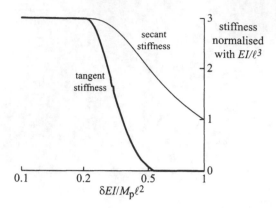

Figure 2.34: Incremental (tangent) and secant stiffness for cantilever

Now at the moment that this fully plastic collapse tip deflection is reached the shape and slope of the deflected cantilever are as shown in Fig 2.33a, b (and the variation in degree of plasticity is shown in Fig 2.33c). For comparison, the shape of an elastic cantilever with the same tip deflection is also shown. So far as the tip response is concerned we could happily describe an equivalent elastic secant stiffness which would have to vary with deflection as shown in Fig 2.34. However, while this would describe the overall system response in some respects it would mislead us into deducing that the distribution of deflection along the cantilever had still the elastic form. Figs 2.33a and b show this deduction to be false.

This simple structural system provides an illustration of two other points. Tangent stiffness falls much more rapidly than secant stiffness because it effects the way in which the cantilever wants to behave *now*. Once the cantilever has a fully developed plastic hinge at its root the tangent stiffness (in terms of increment of tip load/increment of tip deflection) falls to zero—the cantilever has no further ability to accept additional load. The secant stiffness continues falling gently and only reaches zero at infinite deflection.

If we consider a one-dimensional normalised loading diagram (Fig 2.35a) we find that there are three identifiable regimes. For $0 \leq \lambda \leq 2/3$ the response is elastic; for $2/3 \leq \lambda \leq 1$ the response is elastic-plastic. The value $\lambda = 1$ indicates the collapse load. The regime $\lambda > 1$ is inaccessible.

We can now consider a corresponding one-dimensional deflection diagram (Fig 2.35b), using a normalised deflection δ^*

$$\delta^* = \frac{EI\delta}{M_p\ell^2} \tag{2.80}$$

we find that the elastic region corresponds to $0 \leq \delta^* \leq 2/9$; the elastic-plastic region corresponds to $2/9 \leq \delta^* \leq 40/81$; and that the collapse load corresponds to all deflections $\delta^* > 40/81$. There is a many to one mapping from deflection to load.

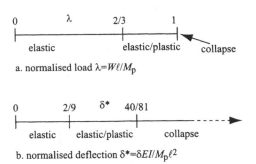

Figure 2.35: Regimes of response of cantilever terms of (a) load and (b) deflection

More importantly, whereas the whole of the deflection diagram is accessible (there is no limit to the deflections that can be imposed on the tip of the cantilever—we are not concerned with changes in geometry and the practical details), the accessible part of the load diagram is restricted to the region below the collapse load. This gives us some hints about problems of experimental control (if we want to explore the whole of the load:deflection relationship we will be well advised to do this by steadily increasing the deflection rather than progressively increasing the load) and numerical control (we will see subsequently (§3.3, §3.4) that it is always more secure in using constitutive models to compute soil response to work from strain increment to stress increment, than from stress increment to strain increment, because it is quite likely that some stress increments will try to take the soil into a forbidden area—beyond the collapse condition).

2.5.2 Stiffness and strain measurement

Perceptions of stiffness are intimately linked to one's ability to measure strains from changes in length of a known initial gauge length. Stiffness variation in a monotonic test—for example, a triaxial compression test—is typically presented (Fig 2.36) in a plot of shear modulus (usually, regrettably, secant modulus) against strain, with the strain plotted on a logarithmic scale because much of the initial variation of stiffness occurs at very small strains. Tangent stiffness varies with strain much more rapidly than secant stiffness (Fig 2.36)—and, of course, if the stress:strain response reveals strain softening after some peak then the tangent modulus will actually become negative. This indicates negative incremental stiffness which sounds, rightly, as though it might be rather important for the response of a soil element or an entire geotechnical system but is concealed in the secant stiffness which remains resolutely positive.

Technology for measurement of small deformations of soil samples has developed tremendously over the past two decades or so and, although, for routine testing, the resolution of strain measurement remains somewhat coarse (Fig

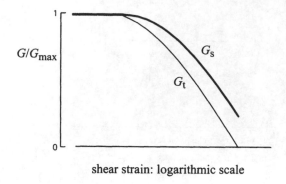

shear strain: logarithmic scale

Figure 2.36: Variation of tangent and secant shear stiffness of soils in monotonic shearing

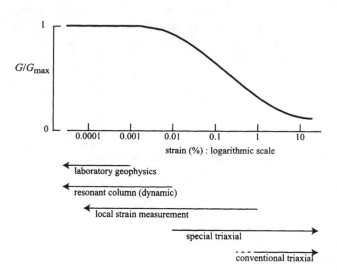

Figure 2.37: Stiffness and resolution of measuring equipment (adapted from Atkinson and Sällfors, 1991)

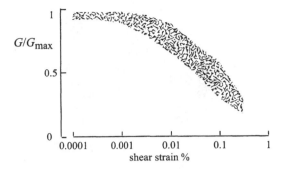

Figure 2.38: Range of secant shear stiffness degradation data for Quiou sand from resonant column and torsional shear tests. Values of shear stiffness G normalised with shear stiffness at very small strain G_{max} (after LoPresti *et al.*, 1997)

2.37), high quality research has been able to bridge the gap between the 'low' stiffnesses typically measured in traditional laboratory tests and the high stiffnesses measured using geophysical techniques in the field or the laboratory (Fig 2.38).

2.5.3   Stiffness and history: stress response envelopes

The S-shaped curves of Figs 2.36, 2.37, 2.38 are typically presented for monotonically increasing strain amplitudes and this is evidently a first stage towards the description of the evolution of the stiffness matrix D in (2.63). However, soil elements in geotechnical systems will be subjected to nonmonotonic paths following long term geological and shorter term construction histories and we will expect to use the laboratory testing possibilities that are available to us to explore the incremental effects of stress changes in a very general way. We need to have a coherent strategy for the conduct of this testing—one which is not too much prejudiced by our existing ideas of the way in which we want our soil to behave.

One way of illustrating the link between strain increments and stress increments which can be useful both for planning and interpreting programmes of testing is through the generation of stress response envelopes. Stress response envelopes were introduced by Gudehus (1979) as a way of illustrating the nature of the characters of response predicted by different classes of constitutive model. Inevitably, such envelopes will usually be presented as two-dimensional curves but these curves are sections through stress response hypersurfaces. If we restrict ourselves to more limited spaces then this will usually be because of the limitations of the testing apparatus which are available to us. Thus, for axially symmetric states of stress attainable in the conventional triaxial apparatus, envelopes can be shown in terms of volumetric (mean effective) stress and distortional stress changes resulting from the application of increments of

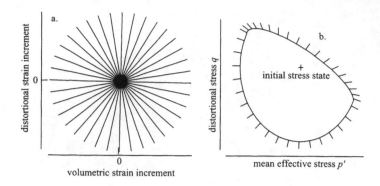

Figure 2.39: (a) Rosette of strain probes and (b) resulting stress response envelope for axisymmetric state of stress

volumetric and distortional strain. For the true triaxial tests to be shown here the envelopes will be presented in terms of deviatoric stress and strain components, neglecting a third dimension, the volumetric (or isotropic) stress and strain, which will, however, also be relevant.

If, from a given initial stress state, a series of *strain* probes of identical normalised magnitude is imposed, then the resulting envelope of *stress* responses provides a visual indication of the generalised stiffness of the soil. An example is shown in Fig 2.39 for an axisymmetric (triaxial) state of stress. The strain increments are defined in terms of volumetric strain $\delta\epsilon_p$ (2.21) and distortional strain $\delta\epsilon_q$ (2.25). The rosette of strain increments of standard length is shown in Fig 2.39a. A solid curve joins the resulting stress increments from the common initial stress, presented in terms of mean effective stress p' and distortional stress q, in Fig 2.39b). In this figure, at each point on the stress response envelope a little line indicates the direction of the corresponding strain probe.

The shape of the stress response envelope is expected to depend on the stress history of the soil. Simply, near failure we expect the stiffness for continued loading—increased distortional strain—to be considerably lower than for unloading—reversal of distortional strain—and the stress response envelope will be flattened towards the loading direction (A in Fig 2.40). At lower stress ratios the envelope is likely to be more rounded (B in Fig 2.40) with the generalised stiffness less dramatically influenced by the direction of the probe though still indicating lower stiffness for continued loading, higher stiffness for reversal of loading. There may be some partially unloaded states for which the response is much more independent of loading direction (C in Fig 2.40). For soil in a state which tends to lead to strain softening (probably at high stress ratios) the initial stress state may lie outwith the response envelope (D in Fig 2.40)—this is an indication that *all* strain increments imposed on our soil element will lead to reduction of stress ratio. We will return to this perplexing response subsequently (§3.4.1, §3.4.2).

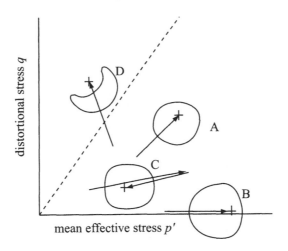

Figure 2.40: Schematic expected history dependence of stress response envelopes (points + indicate initial stress states)

relative to the response envelope will vary with history. As presented in (2.63) the stiffness matrix D is independent of the magnitudes of the strain increment and stress increment that it links—strictly it applies only to infinitesimal increments. However, we have seen that soil stiffness falls with monotonically increasing strain. Instrumentation does not permit us to determine response envelopes for 'zero' amplitude of strain increments. However, it is instructive, in gathering data to inspire our constitutive modelling, to look at response envelopes determined for different, finite, magnitudes of strain increment from a given initial stress state. This then provides a more general presentation of the stiffness variation with strain shown in Fig 2.38.

Results of triaxial stress probes on natural Pisa clay, interpreted as stress response envelopes, are shown in Fig 2.41[2]. At small strain magnitudes (0.1-0.2%) the stress response envelopes are strongly linked in position and shape to the starting stress state. As the magnitude of strain increases the importance of the starting point apparently reduces. The envelopes are closely bunched for *increasing* distortional stress, much more widely separated for *reducing* stress. We can present this series of envelopes as a series of variations of generalised secant stiffness

$$S = \sqrt{\Delta p'^2 + \Delta q^2} / \sqrt{\Delta \epsilon_p^2 + \Delta \epsilon_q^2}$$

with strain $\epsilon = \sqrt{\Delta \epsilon_p^2 + \Delta \epsilon_q^2}$—showing the effect of stress path direction (Fig 2.42).

Stress response envelopes from true triaxial tests on kaolin are shown in a π-plane deviatoric view of stress space in Figs 2.43 for two different initial histories—one (Fig 2.43a) has isotropic compression to O followed by shearing

---

[2]Data kindly replotted by Luigi Callisto.

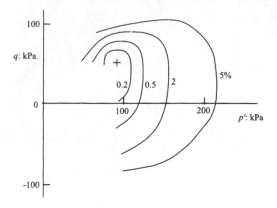

Figure 2.41: Schematic stress response envelopes for natural Pisa clay (contour values for $(\Delta\epsilon^2{}_p + \Delta\epsilon_q^2)^{1/2}$ from initial stress state +)(data from Callisto, 1996)

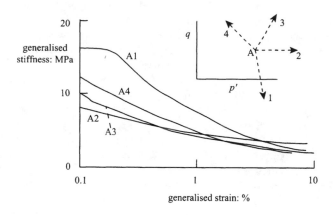

Figure 2.42: Schematic response of natural Pisa clay shown as generalised variation of stiffness S with strain (ϵ) for different stress probe directions (data from Callisto, 1996)

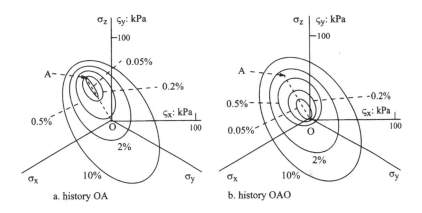

Figure 2.43: Schematic deviatoric stress response envelopes from true triaxial probing of kaolin with deviatoric history (a) OA and (b) OAO (inspired by Muir Wood, 2004)

(at constant mean effective stress) to A; the other (Fig 2.43b) has history OA followed by unloading back to O. The response envelopes are drawn for different values of a distortional strain

$$\epsilon = \frac{1}{\sqrt{3}}\sqrt{\left[(\epsilon_y - \epsilon_z)^2 + (\epsilon_z - \epsilon_x)^2 + (\epsilon_x - \epsilon_y)^2\right]}$$

which is proportional to the second invariant of the strain deviator tensor (compare 2.52). Again, the small strain envelopes in each case are closely linked to the recent stress history, OA or OAO respectively. However, as the strain magnitudes increase, the detail of the starting stress state seems to become largely irrelevant—by the 10% envelope, memory of what has gone before has been somewhat swept out.

Stress response envelopes from true triaxial tests on Leighton Buzzard sand are shown in Fig 2.44 (data from Sture *et al.*, 1988). These tests were performed in a cubical cell true triaxial apparatus in which the stresses were imposed through flexible boundary cushions: the magnitudes of strains that can be imposed while still retaining deformational uniformity are limited. Comparing deviatoric histories (imposed at constant mean stress) ABC, ABD the comments made previously are reinforced. Failure is lurking in the π-plane at some finite distance from the isotropic stress axis, A, so it is to be expected that the several stress response envelopes will be closely packed together there.

Finally, to demonstrate that the principle of generation of stress response envelopes is quite generally useful and applicable, envelopes are shown in Fig 2.45 for two stress histories imposed on Leighton Buzzard sand in a directional shear cell (Fig 2.15) (original data again from Sture *et al.*, 1988). The data are somewhat sparse but the pattern is familiar. Principal stress rotation is occurring but, with appropriate choice of deviatoric stress variables, the required conduct of the tests and interpretation of results is straightforward.

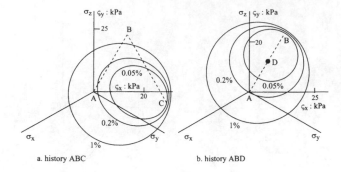

Figure 2.44: Schematic stress response envelopes from true triaxial probing of Leighton Buzzard sand with deviatoric history (a) ABC and (b) ABD (inspired by data from Sture *et al.*, 1988)

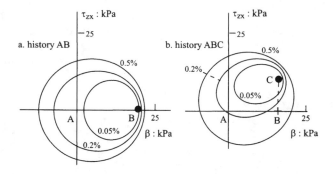

Figure 2.45: Schematic stress response envelopes from directional shear cell probing of Leighton Buzzard sand with deviatoric history (a) AB and (b) ABC (inspired by data from Sture *et al.*, 1988)

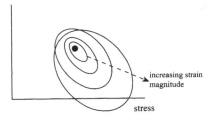

Figure 2.46: Schematic stress response envelopes for strain softening soil

In all these examples the picture is similar. There is a strongly kinematic element to the small strain response envelopes—they are carried around intimately with the most recent stress history. The nonlinearity of stress:strain response—the way in which incremental stiffness decays with continuing monotonic strain—is encapsulated in the relative positions of the response envelopes for increasing magnitudes of strain. The larger the strain magnitude the less the location of the envelope seems to care about the recent stress history and the more it is aware of other constraints on soil response—such as the limitations that failure criteria might impose. Failure will be discussed subsequently but, if failure is, simplistically, interpreted as the occurrence of zero incremental stiffness, then obviously the stress response envelopes for increasing strain magnitudes must pack closely together as they approach the failure boundary. The strains needed to identify failure may be large but this character of response is discernible in Figs 2.41 to 2.45. Post peak softening of material response would reveal itself in intersection of response envelopes at larger strains with those corresponding to smaller strains (Fig 2.46)—but the results shown have not extended this far.

2.5.4 Anisotropy of stiffness

Soils are isotropic materials which find themselves in anisotropic circumstances as a result of their history of deposition and past loading. This anisotropy manifests itself in anisotropic arrangements of particles and in the forces carried by the contacts between the particles. Since the soil particles are in general neither spherical nor even sub-spherical, the anisotropy of geometric fabric contains the layout of the centres of the particles, and the orientations of the particles. In principle, knowing this geometric fabric together with the information about orientation and activation of contacts between the particles and contact forces—the kinetic fabric (Chen *et al.*, 1988)—and the characteristics of the interparticle actions, the mechanical response of the soil system could be anticipated. Progress is being made on such descriptions of evolving fabric but it is presently more convenient to work at a larger scale and observe consequences of evolving fabric anisotropy on mechanical response in terms of continuum concepts such as stress and strain.

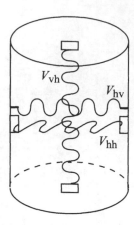

Figure 2.47: Bender elements used to discover anisotropy of small strain stiffness in triaxial test (after Pennington *et al.*, 1997)

Experimental techniques exist by which detailed information about the stiffness of soils at very small strain levels can be determined in the laboratory. Stiffness can be determined from static measurements using sensitive deformation measuring devices to record strain response to applied stress probes. Stiffness can also be determined from dynamic measurements using laboratory geophysics—typically, in element tests, using piezoceramic bender elements which can be used to generate and detect low amplitude shear waves in soils (see §6.7).

Generally accepted empirical relationships describe the variation of very small strain stiffness with stress state (kinetic fabric) for (implicitly) unchanging geometric fabric. This implies a particular form of the variation of stiffness anisotropy with stress state represented, for example, through the ratio of axial to radial stress or mobilised friction. A combination of static and dynamic data can be used to deduce all elements of a cross anisotropic stiffness matrix (Lings *et al.*, 2000) (§3.2.4).

Once the changes in stress (or imposed deformation) from any initial reference state become sufficient to disturb the geometric fabric then the simple relationship breaks down—observations of stiffness combine both geometric and kinetic effects. Experimental observations of soil response in triaxial and other testing apparatus, as well as numerical analysis of assemblies of regular or irregular particles, suggest that kinematic hardening constitutive models are likely to be required in order to simulate observed behaviour. However, theoretical studies with such models indicate that anisotropy of stiffness influences not only the pattern of deformation that develops in a geotechnical system but also the potential for bifurcation of material response and the development of localised deformation and rupture surfaces.

The techniques for bender testing are well established. The arrangement used for the tests reported here is shown in Fig 2.47 (Pennington *et al.*, 1997). Bender elements have been placed both in end platens and through the flexible membrane for cylindrical triaxial specimens so that shear waves can be

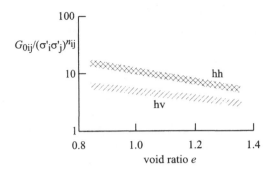

Figure 2.48: Typical anisotropy of stiffness of Gault clay as function of void ratio (inspired by Pennington *et al.*, 1997)

propagated through the sample either from end to end or from side to side, in each case with two possible polarisations. While various input waveforms for the driver elements have been explored, the data here have come from tests in which a single sine wave pulse has been sent. Appropriate interpretation of the received signals can then be used to estimate shear wave velocities through the soil V_{vh} (for vertically propagated, horizontally polarised waves), and V_{hv} and V_{hh} (for horizontally propagated waves with either vertical or horizontal polarisation). If the medium through which the waves are travelling is elastic then corresponding 'zero strain' shear moduli G_{0vh}, G_{0hv}, G_{0hh} can be estimated:

$$G_{0ij} = \rho V_{ij}^2 \qquad (2.81)$$

where ρ is the bulk density of the medium through which the waves are travelling and subscripts $_i$ and $_j$ take the values $_v$ or $_h$ as appropriate. The stiffness of soils decreases with strain on any monotonic excursion. The strain amplitude applied to the soil by a bender element is not zero but sufficiently small that the resulting deformation—for most of the passage through the specimen—can be treated as elastic and essentially at 'zero strain' amplitude. If the medium through which the waves are passing is elastic, and strains are small, then it is axiomatic that $G_{0vh} = G_{0hv}$[3].

It is found empirically that this very small strain shear stiffness of soils is influenced in a systematic way by the stress state in the soil, by the volumetric packing of the soil (through void ratio e), and by the current geometric fabric through an expression of the form (Roesler, 1979)

$$\frac{G_{0ij}}{p_r} = S_{ij} F_{ij}(e) \left(\frac{\sigma_i' \sigma_j'}{p_r^2} \right)^{n_{ij}} \qquad (2.82)$$

---

[3]Pennington *et al.* (2001) suggest that reported differences between these deduced moduli may result from roughness of the end platens in which bender elements are typically mounted in order to estimate V_{vh} and hence G_{0vh}. Arroyo (2001) notes that there may be effects linked to the way in which shear waves are transmitted in cylindrical samples in addition to effects of the non-point-like nature of the shear wave source which will tend to lead to an apparent asymmetry of the *deduced* stiffness matrix in anisotropic elastic soil.

where S_{ij} are elements of a fabric tensor and $F_{ij}(e)$ are functions of void ratio e. The first two terms in this expression, S_{ij} and $F_{ij}(e)$, are two facets of the tensor geometric fabric, with e related to its first invariant. A reasonable fit to data can be obtained using

$$F_{ij}(e) = e^{m_{ij}} \qquad (2.83)$$

where m_{ij} is a constant for any pair of values of $_i$ and $_j$. The term in (2.82) involving the principal effective stresses in the directions of propagation and of polarisation, σ'_i and σ'_j respectively, encapsulates the kinetic fabric contribution. The stress p_r is a reference stress (taken by Nash et al. (1999) as 1 kPa for simplicity but logically linked with some property of the soil mineral). Typically the exponent $m_{ij} \approx -2$ (though in detail the value depends on $_i$ and $_j$), and $n_{ij} \approx 0.25$. This expression is written for directions of propagation and polarisation coincident with principal axes of stress and anisotropy. The analysis of wave propagation through anisotropic elastic media in which these axes do not coincide is complex (Crampin, 1981).

Typical data from one-dimensionally compressed and 'isotropically' compressed reconstituted Gault clay in Fig 2.48 demonstrate that expression (2.82) matches the data of shear stiffness G_{0hh} and G_{0hv} satisfactorily. The 'isotropically' compressed material has been consolidated one-dimensionally from slurry and then subjected to isotropic stresses. Even consolidation stresses five times higher than those imposed during the initial one-dimensional preparation do not erase the initial anisotropic geometric fabric of the clay. Consequently the stiffness characterisation in Fig 2.48 does not greatly distinguish between these two consolidation histories once the effects of void ratio and magnitudes of principal stresses have been taken into account.

Bender elements mounted in triaxial samples give limited information about the anisotropic stiffness properties of the soil. In combination with high resolution measurement of strains full characterisation is possible assuming a symmetry of anisotropy matching the symmetry of loading of the specimen: cross anisotropy. The complete anisotropic description of the Gault clay is presented by Lings et al. (2000) (see §3.2.4). Here we will use the simpler treatment of cross anisotropy presented by Graham and Houlsby (1983) in which just one extra parameter α (instead of the theoretical three extra parameters) is introduced. This model implies that the ratio of the two shear stiffnesses under consideration is

$$\frac{G_{hh}}{G_{hv}} = \alpha \qquad (2.84)$$

and the ratio of direct horizontal and vertical drained stiffnesses

$$\frac{E_h}{E_v} = \alpha^2 \qquad (2.85)$$

By subsuming a five parameter material into three parameters some information about the material is lost and implicit relationships between elastic parameters are imposed. However, Lings et al. found the simplified three parameter representation serendipitously successful for matching the Gault clay data. Ignoring effects of geometric fabric change, (2.82), with $n_{ij} = 0.25$, and (2.84) and (2.85) can be used to link α and stress ratio η (2.36).

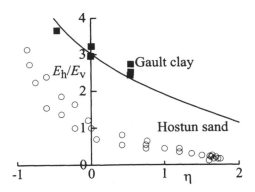

Figure 2.49: Dependence of anisotropy parameter α on stress ratio for reconstituted Gault clay (data from Pennington *et al.*, 1997) and dense Hostun sand (data from Gajo *et al.*, 2001): solid line is equation (2.86)

$$\frac{E_h}{E_v} = \alpha^2 = \left(\frac{G_{hh}}{G_{hv}}\right)^2 = \left(\frac{S_{hh}}{S_{hv}}\right)^2 \left(\frac{3-\eta}{3+2\eta}\right)^{0.5} \tag{2.86}$$

This is shown in Fig 2.49 with data for the Gault clay, taking the value of S_{hh}/S_{hv} from the data in Fig 2.48. The value of $E_h/E_v = \alpha^2$ falls as the stress ratio increases: the degree of anisotropy is decreasing.

Elastic anisotropy can also be estimated by performing small cycles of undrained unloading and reloading during a drained test. The slope of the effective stress path depends only on measurements of changes in stress and of pore pressure (Fig 2.50). The slope of the effective stress path can be deduced from the observed changes in pore pressure and known changes in applied total stresses (2.102) and this is independent of resolution of strain measurement. For the Graham and Houlsby (1983) description of anistropic elasticity (§3.2.4) the slope of the stress path in an undrained unloading-reloading cycle is

$$\frac{\delta q}{\delta p'} = \frac{3}{2} \frac{2 - 2\nu^* - 4\alpha\nu^* + \alpha^2}{1 - \nu^* + \alpha\nu^* - \alpha^2} \tag{2.87}$$

where a value of Poisson's ratio $\nu^* = 0.2$ has been assumed in order to convert stress path slopes to values of α.

Data from a triaxial test on Hostun sand are shown in Fig 2.49 (Gajo *et al.*, 2001). Samples were prepared by dry pluviation: one test was performed with constant cell pressure the other with constant axial stress. The implied stiffness ratio (2.85) changes significantly as stress ratio increases: a typical history of variation is shown in Fig 2.51. Under initial isotropic stresses the material has only slight depositional anisotropy, but α and E_h/E_v fall as stress ratio increases. The stress-strain response for this dense sand shows a peak stress ratio and Fig 2.49 shows that E_h/E_v is still falling. Stress ratio alone is not

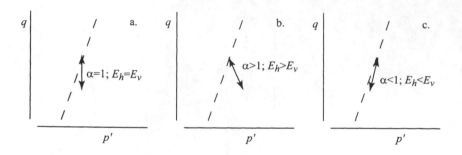

Figure 2.50: Stiffness anisotropy deduced from slope of effective stress path for undrained unload/reload cycle within drained test

sufficient to deduce the evolution of anisotropy. The continuing irrecoverable strain produces particle rearrangement (and also changes in void ratio) and an associated change in geometric fabric.

Evolving anisotropy of small strain stiffness—which provides the anchor for data such as those shown in Fig 2.38—is a significant characteristic of soil response which has only begun to be understood in the laboratory as experimental techniques have improved. Anisotropy of small strain stiffness may potentially have a significant influence on the response of geotechnical systems—especially systems where it is intended to keep deformations to a minimum (for example, tunnels in urban areas) so that the nonlinear (irrecoverable, plastic) deformation of the soil does not so easily swamp the small strain (elastic?) deformation.

We have supposed—reasonably—that the anisotropy that we are investigating and describing has a symmetry that coincides with the symmetry of the triaxial test. We are dealing with soil samples which have been deposited in a gravitational field by consolidation from slurry (clay) or by pluviation (sand) and then subjected to radial stresses and axial strains in the triaxial apparatus. This is cross anisotropy or transverse isotropy: horizontal stiffness differs from vertical stiffness but every horizontal direction is identical. If the soil has only ever experienced stresses and strains with this symmetry then it is appropriate that the anisotropy of stiffness should be of this form. The anisotropy reflects the current fabric of the soil and this has obviously resulted from the history of that soil. Sedimentary soils that have been deposited over areas of large lateral extent, and have always known a horizontal ground surface, know only this axial symmetry.

There are plenty of other histories which depart from this symmetry: any soil in a slope, any soil which has been pushed around by ice or by man or by local tectonic action. Every geotechnical construction will start imposing its own local asymmetry on the fabric of the soil and hence on the evolving stiffness characteristics. If we place in the triaxial apparatus a sample which possesses either cross anisotropy—but with axes that do not happen to coincide with the axes of the testing apparatus—or anisotropy of some more general type, then, subjected to the only loadings that the triaxial apparatus is able to impose,

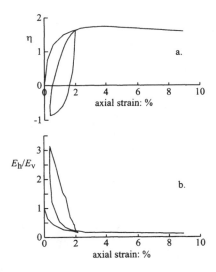

Figure 2.51: Evolution of anisotropy with strain during cycle of drained triaxial compression of dense Hostun sand ($e_o = 0.63 - 0.67$) (data from Gajo *et al.*, 2001)

the soil will have to break out in some way and respond in a non-axisymmetric manner. Confined between rough ends, the sample will have to flex (Fig 2.52b); confined between smooth ends it will skew as it is compressed (Fig 2.52c). Either way the sample will do things that we will not normally discover unless we are conscious that they may be there. It is tempting to assume that things that we choose not to observe do not exist.

Various ways have been proposed to characterise the anisotropy of soils. The terms inherent anisotropy and induced anisotropy (sometimes qualified as stress-induced anisotropy) have been used implying that there is some difference in quality between two types of anisotropy: the anisotropy of a soil as it is discovered in the ground contrasted with the anisotropy that develops as a result of some subsequent perturbation. Alternatively one might use the terms initial and subsequent anisotropy to indicate the sequence of events. The process of formation of a soil (by sedimentation or by glacial transport or by *in-situ* weathering) will imply certain stress changes (and other effects) which will leave the soil in an anisotropic state. At the simplest level, one-dimensional deposition implies an anisotropy of stress state and of deformation and hence an anisotropy of particle arrangement (and orientation) and fabric that is expected to lead to cross-anisotropic deformation properties. Truly isotropic fabrics are likely to be rare—every soil is aware of a reference direction that is the direction of gravitational acceleration.

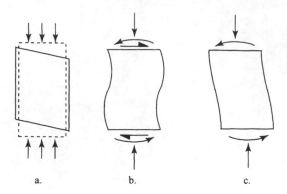

Figure 2.52: Effect of nonalignment of axes of material anisotropy with axis of symmetry of triaxial apparatus: (a) response under uniform state of imposed stress; (b) loading between rough platens; (b) loading between smooth platens (after Saada and Bianchini, 1977)

2.6   Dilatancy

The chief characteristic which distinguishes most soils from other engineering materials (such as metals and plastics) is the high proportion of the volume of the material which is made up of void filled with single or multi-phase fluid. For a typical medium dense sand about a third of the volume is void; for a normally consolidated clay voids might make up towards half of the volume. Naturally, if anything is done to disturb the arrangement of the soil particles by distorting the boundary of the soil sample then it is to be expected that rearrangement will be accompanied by some change in the volumetric packing: this is dilatancy.

A simple illustration of the 'need' for, or inevitability of, dilatancy, is provided by the thought experiment in Fig 2.53a. A loose, two-dimensional packing of circular particles is sheared. This shearing implies that the particles in each row move sideways over the particles in the row below—as they do so they fall into the gaps between those particles and the volume occupied by the soil reduces. The relationship between horizontal movement (shear displacement) and vertical movement (volume change) is shown in Fig 2.53c.

A complementary result is obtained if the two dimensional set of circular particles is initially in its densest possible packing (Fig 2.53b). Now as the particles in one layer are displaced sideways they are forced to climb over the particles in the underlying layer and the volume occupied by the soil increases (Fig 2.53c). We note that the nature of the volume change that occurs is strongly influenced by the density of the packing.

Consider a classic shear box in which a soil sample is sheared by the relative movement of the top and bottom halves of the box (Fig 2.54). Most of the deformation of the soil occurs in a thin zone around the interface between the two halves of the box. When sands are sheared in a shear box they change in volume so that a typical set of data obtained from a shear box test might

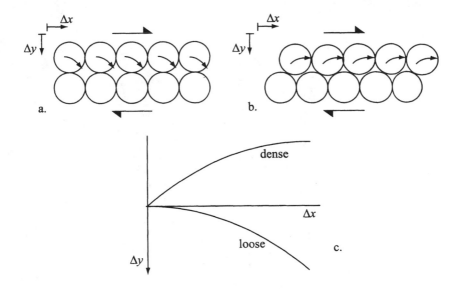

Figure 2.53: (a) Shearing of loosely packed layers of circular discs; (b) shearing of densely packed layers of circular discs; (c) volume change in shearing of loosely and densely packed layers of circular discs

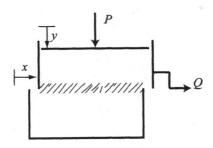

Figure 2.54: Direct shear box

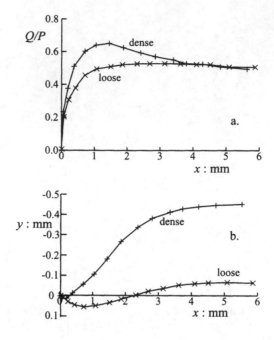

Figure 2.55: Shear load:displacement response and volume changes in direct shear test on Ottawa sand (data from Taylor (1948))

look like Fig 2.55. The incremental work done in a shear box in shearing a soil sample which is currently supporting vertical load P and horizontal load Q and is undergoing horizontal (shearing) displacement δx and vertical (volumetric) displacement δy is:

$$\delta W = P\delta y + Q\delta x \tag{2.88}$$

Taylor (1948) proposed that this work was entirely dissipated in friction at all stages of a shear test so that:

$$P\delta y + Q\delta x = \mu P\delta x \tag{2.89}$$

or

$$\frac{\delta y}{\delta x} = \mu - \frac{Q}{P} \tag{2.90}$$

The ratio of vertical to horizontal movements indicates the rate at which volumetric expansion occurs with continuing shearing

$$\frac{\delta y}{\delta x} = -\tan\psi \tag{2.91}$$

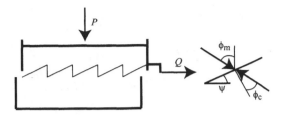

Figure 2.56: Inclined shear surfaces causing dilatancy ψ and consequential mobilised friction ϕ_m

where ψ is the angle of dilation. The ratio of shear load Q to vertical load P is an indication of the currently mobilised angle of shearing resistance ϕ_m

$$\frac{Q}{P} = \tan \phi_m \tag{2.92}$$

So expression (2.90) can be written alternatively as

$$\tan \psi = \tan \phi_m - \tan \phi_c \tag{2.93}$$

to indicate that the angle of dilation varies with the mobilised angle of shearing resistance and has the value zero when the mobilised angle of shearing resistance has the special value $\phi_c = \tan^{-1} \mu$. This value, corresponding to constant volume shearing, is called the critical state angle of shearing resistance.

A similar result can be obtained by thinking of the relative deformation of the two halves of the shear box occurring between soil particles on a sort of sawtooth interface (Fig 2.56). The available friction on the inclined surfaces is ϕ_c but, because of the inclination of the sliding surfaces, at angle ψ to the horizontal, the friction that is generated on horizontal surfaces is actually ϕ_m where

$$\phi_m = \phi_c + \psi \tag{2.94}$$

Both expressions (2.93) and (2.94) suggest that there should be some link between dilatancy and mobilised friction. Either expression is able to provide at least a first approximation to the observed response. When the mobilised friction Q/P is less than μ the sand is contracting; when the mobilised friction is greater than μ the sand is expanding. The ratio $-\delta y/\delta x$ gives an indication of the tendency to volume increase for the sand: the dilatancy. Expressions such as (2.93) and (2.94) which link dilatancy with mobilised friction are called stress-dilatancy relationships or flow rules and they describe the link between mobilised friction and mobilised dilatancy.

Data from Taylor's shear box tests on samples of Ottawa sand prepared either dense (initial specific volume 1.562) or loose (initial specific volume 1.652) can be interpreted in this way using the observed vertical movements of the top half of the shear box to estimate the critical state angle of shearing resistance ϕ_c which, according to these simple flow rules, should be a soil property and hence constant throughout the test. Results are shown in Fig 2.57. The simple

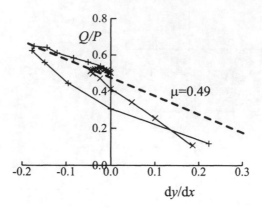

Figure 2.57: Ottawa sand: link between dilatancy and mobilised friction in direct shear tests (data from Taylor, 1948)

model appears to work reasonably well indicating a value of $\mu \approx 0.49$ in (2.90) corresponding to a critical state angle of friction $\phi_c \approx 26°$[4].

A direct analogy can be drawn between the behaviour observed in the shear box and the behaviour to be expected in the triaxial apparatus. The shear and normal displacement increments in the shear box become the distortional and volumetric strain increments and the shear and normal loads become the distortional and volumetric effective stresses:

$$\delta x \quad \rightarrow \quad \delta\epsilon_q \tag{2.95}$$

$$\delta y \quad \rightarrow \quad \delta\epsilon_p \tag{2.96}$$

$$Q \quad \rightarrow \quad q \tag{2.97}$$

$$P \quad \rightarrow \quad p' \tag{2.98}$$

and the stress-dilatancy relationship or flow rule (2.90) becomes

$$\frac{\delta\epsilon_p}{\delta\epsilon_q} = M - \frac{q}{p'} = M - \eta \tag{2.99}$$

where M is the critical state stress ratio at which constant volume shearing can occur.

2.6.1   Critical states: state variable

The concept of 'critical states' has been mentioned (§2.6). These are asymptotic states in which shearing of the soil can continue without further change in effective stress or density. The exact nature of the fabric of the soil at a critical state is not clear. It is certainly intended that any initial interparticle bonding

---

[4]There are reasons why one might expect the experimental data to lie below this line in the early stages of the test: see section §3.4.1.

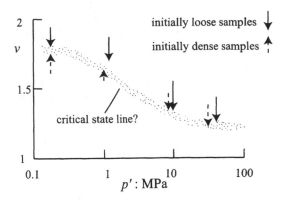

Figure 2.58: Schematic critical state line deduced from triaxial tests on Chattahoochee River sand (data from Vesić and Clough, 1968)

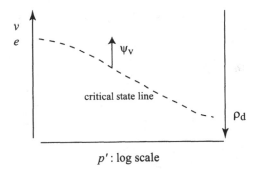

Figure 2.59: Definition of state variable ψ_v

should have been broken down so that the particles are all individually free to move and rotate. However, it is not clear whether or not the particle orientations should all be random and the structure isotropic—in fact the evolving anisotropy shown in Fig 2.51 rather suggests that the soil reaches a limiting but non-zero degree of anisotropy of stiffness (and hence, by implication, of fabric) as shearing proceeds. From an experimental point of view, there may well be good reasons why dense soils with strength dependent on density or bonded soils (in other words soils which might be expected to show some post-peak strain softening of strength) will not reach homogeneous critical states in laboratory tests so that accurate deduction of critical state conditions may be difficult—especially if internal deformations of the sample can only be deduced from external measurements.

Experimental evidence suggests that the density of soils which have reached a critical state is dependent on stress level: the higher the stress level the higher the density and lower the void ratio or specific volume. The idea of a critical state line emerges (Fig 2.58) and, no matter what the detailed shape of the critical state line, we can define a state variable ψ_v (Fig 2.59)

$$\psi_v = v - v_{cs} \tag{2.100}$$

as an indication of the distance of the current specific volume v of our soil away from the critical state specific volume v_{cs} at the current mean stress. The spread of the data in Fig 2.57 suggests that perhaps the link between mobilised friction and dilatancy in (2.93) or (2.94) should be somehow moderated by the current value of state variable which changes continuously throughout each test.

2.6.2 Pore pressure parameter

Let us conduct another thought experiment. We have a sample of soil contained in a special testing apparatus which allows us complete freedom to change all the components of stress or strain. There are some stress perturbations that we can impose which imply purely distortional strain responses. The link between the stress changes and the imposed strain changes is the subject of constitutive modelling. However, there are obviously many stress perturbations which imply a strain response containing a compression component. The volume of the sample subjected to any of these would have to change in order to accommodate the new stress state. If the volume is prevented from changing, either because we have physically closed the drainage lines from the sample, or because the sample consists of a soil which has such low permeability that it is not possible for the water to move around and out of the pores and the sample during the time interval in which the external stresses were changed, then the prevention of movement of the pore water implies that the pore water is meeting a resistance and hence that a pore water pressure develops (the pressure may be positive if the water is trying to escape because the soil wants to compress, or negative because the sample is trying to expand and draw in water). This pressure acts to partition the externally imposed stresses: part is carried by the pore pressure and part by the interparticle forces within the soil (§2.4). The resulting stresses carried by the soil particles are the *effective* stresses whereas the externally applied stresses are the *total* stresses.

Returning to axisymmetric triaxial conditions, for an increment in external, total, stresses $\delta p, \delta q$ leading to corresponding changes in effective stresses $\delta p', \delta q$, in order to maintain the undrained, constant volume condition, we know from the principle of effective stress (2.19) that the change in pore pressure δu is

$$\delta u = \delta p - \delta p' \tag{2.101}$$

The change in mean effective stress $\delta p'$ is a material response to the change in distortional stress δq[5] so we can write

$$\delta p' = -a\delta q \tag{2.102}$$

and

$$\delta u = \delta p + a\delta q \tag{2.103}$$

---

[5]See §3.3.4, §3.4.1, §3.4.2 for exploration of the undrained response of some candidate constitutive models.

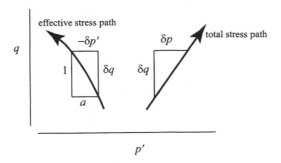

Figure 2.60: Total and effective stress paths: pore pressure parameter a

linking change of pore pressure with change in applied *total* stresses through a pore pressure parameter a (§1.2.2)[6] which indicates the current slope of the undrained effective stress path (Fig 2.60). We have already seen in the earlier discussion of anisotropy (§2.5.4) that $a = 0$ for isotropic elastic soil: the effective stress path for an undrained test on such a soil is vertical in the p', q effective stress plane. Although we do not expect undrained effective stress paths to be straight, we could use an average value of a to characterise soils: $a > 0$ implies a contractive soil which will tend to develop positive pore pressures when sheared; and $a < 0$ implies an expanding soil which will tend to develop negative pore pressures.

However, (2.103) divides the observed pore pressure change into two parts: one, δp, over which we have full control, choosing the total stress change to which the soil will be subjected; the other, $a\delta q$, over which we have no control—it is an indication of the way in which the soil chooses to keep its volume constant. The pore pressure that we actually observe is the sum of these two elements and it is only if we take away the part that we control (δp—which can arbitrarily be positive or negative according to *our* whim) that we can understand what the soil is trying to tell us about its volume change proclivities.

2.7 Strength

We can define *strength* as the ability of soils to carry stress. Usually we are concerned with carrying *shear* stress, although for cemented (or interlocked) soils (and rocks) there will also be some actual more or less reliable tensile strength. In general, the stress:strain response of a soil (Fig 2.61) will show a rapid climb to a peak shear stress followed by some softening to a large strain shear stress. (In some soils the softening may be very slight or absent.)

If we try to follow such a softening stress:strain response then it is clear that, if we control the test by successively adding shear load, then the sample will 'run away', following the dashed line in Fig 2.61, when the peak stress is

---

[6]a is actually a *variable* rather than a parameter since it will not usually be a soil constant.

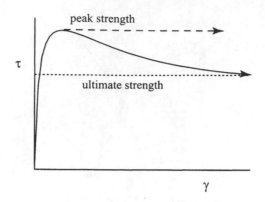

Figure 2.61: Typical shear stress:shear strain relationship for soils

reached. Even if the test is controlled by progressively increasing the boundary shear strain γ (in whatever test device we are using), it is likely that there will be imperfections in the sample which will start softening ahead of other slightly stronger regions, leading to concentration of shear deformation into a localised shear band. (The detailed analysis of the onset of localisation need not concern us here.)

We have already seen that the term 'critical state' is used to describe the states of soils which are able to continue shearing to large strains without changing stresses or density, and have seen that the critical state specific volume—or density of packing—is expected to be a function of stress level. The critical state is thought of as a state in which the soil is being continuously reworked at a scale which involves many particles. We could imagine this applying, at least, to the soil in the shear band even if the entire sample does not remain homogeneous. Observations show that shear bands in rather uniform sands have a typical thickness of $10 - 15d_{50}$ where d_{50} is the mean particle size (Muir Wood, 2002). Where it has been possible to estimate the deformations within the shear band itself it seems that conditions have tended towards a plausible critical state (Fig 2.62).

However, in a soil which is formed of platey, clay mineral particles, localisation leads to reorientation of the particles parallel to the shear band which ends up as a sliding surface between slickensided, polished layers of soil, generating a much lower residual frictional resistance than would be expected for a shearing process that was continuously churning up the soil particles. It is thought that the possibility that such residual strength may be important will depend on whether it is possible to form a sub-planar failure surface through the soil. If a soil is predominantly formed of rotund particles (a sand or gravel) or contains many such particles in a clay matrix (a glacial till or residual soil) the residual strength is not an issue: Lupini *et al.* (1981) link this to the granular specific volume of the soil (Fig 2.63)—the volume occupied by unit volume of granular material, treating the clay mineral present together with the actual

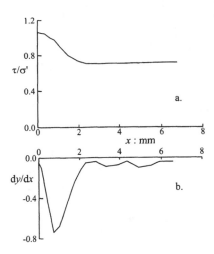

Figure 2.62: (a) Mobilised friction and (b) dilation to critical state in shear band within biaxial test on Karlsruhe sand (data from Vardoulakis, 1978)

voids. If the granular specific volume is high (or even infinite for a clay) then failure planes through the non-granular material are expected and the concept of residual strength is important.

Peak strength is more dependent on initial conditions in the soil. We expect that, at a given stress level, the denser the soil the higher will be the strength. Data from sands suggest that it is useful to link peak strength with a state variable (2.100) (Fig 2.59) in order to introduce a link with stress level as well as with density. As the state variable ψ_v becomes more negative the peak strength increases (Fig 2.64)—but equally, if we shear the soil to an eventual critical state, then its strength tends to the critical state strength as the state variable rises to zero. There is a hint here at a route to constitutive modelling which will be picked up in section §3.4.1. It should not surprise us, given our discussion of stress-dilatancy relationships in section §section:dilatancy, that the peak angle of dilation also correlates well with state variable (Fig 2.65).

Because most of the soils with which we are dealing are—at least eventually—unbonded, we have presented strength as a purely *frictional* phenomenon. A link between peak frictional strength and state variable implies that for soils of a given density or specific volume the strength will increase as the stress level falls and the state variable becomes more negative (Fig 2.66). If we assume a critical state line which is locally straight in a semi-logarithmic compression plane ($v : \ln p'$) with slope λ (though the principle of the argument is not dependent on the specific form of this relationship), then a linear relationship between peak friction ϕ'_p and state variable ψ_v (Fig 2.64, Fig 2.67b: solid line) can be converted into a failure relationship in the $p' : q$ effective stress plane (Fig 2.67a: solid lines). If interpreted without thought about its origin—particularly if the data do not cover a sufficiently wide range to reveal significant curvature—we

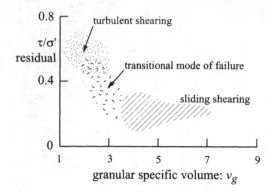

Figure 2.63: Potential for reduced residual strength dependent on granular specific volume (after Lupini *et al.*, 1981)

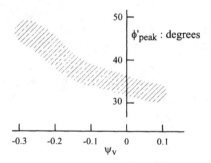

Figure 2.64: Dependence of peak strength of sands on state variable (inspired by Been and Jefferies, 1986)

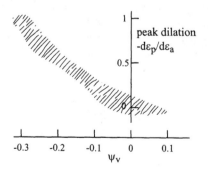

Figure 2.65: Dependence of peak dilatancy of sands on state variable (inspired by Been and Jefferies, 1986)

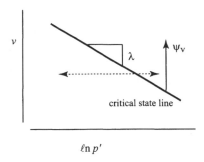

Figure 2.66: States of samples with identical state variable but different stress levels

might be tempted to assume that the soil is telling us that it possesses some cohesion.

The Hvorslev strength relationship invoked to describe the peak strengths of clays (Schofield and Wroth, 1968) comes to an essentially similar result. A locally linear strength relationship for soil *of a given density* is assumed in the $p' : q$ plane—implying a nonlinear dependence of peak strength on state variable ψ_v (Fig 2.67b: dotted line), and again giving an apparent cohesion (Fig 2.67a: dotted lines). Each line in Fig 2.67a is dependent on the density for which it is determined: peak strength data for soils cannot be properly understood unless they are corrected for the values of state variable at the moment of failure.

The language of friction in the context of soil strength implies a limiting ratio of shear stress to normal stress and defines a limiting Mohr circle (Fig 2.68). For a purely frictional soil, we can write the failure criterion as

$$\frac{\tau}{\sigma'} \leq \tan \phi' \tag{2.104}$$

or, in terms of major and minor principal effective stresses σ'_1 and σ'_3 respectively

$$\frac{\sigma'_1}{\sigma'_3} \leq \frac{1 + \sin \phi'}{1 - \sin \phi'} \quad \text{or} \quad \frac{\sigma'_1 - \sigma'_3}{\sigma'_1 + \sigma'_3} \leq \sin \phi' \tag{2.105}$$

The intermediate principal stress σ'_2 does not enter this equation.

For conventional axisymmetric triaxial tests the definitions of our stress variables p' (2.23) and q (2.24), the former of which *does* include the intermediate principal stress, imply that the lines of constant ϕ' will have different slopes (Fig 2.69) for triaxial compression ($\sigma'_a = \sigma'_1$; $\sigma'_r = \sigma'_2 = \sigma'_3$) and for triaxial extension ($\sigma'_a = \sigma'_3$; $\sigma'_r = \sigma'_1 = \sigma'_2$). In compression (2.39)

$$\eta = \frac{6 \sin \phi'}{3 - \sin \phi'} \tag{2.106}$$

and in extension (2.42)

$$\eta = -\frac{6 \sin \phi'}{3 + \sin \phi'} \tag{2.107}$$

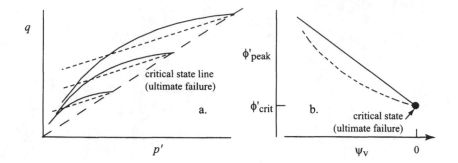

Figure 2.67: Linear dependence of strength on state variable (solid lines) and Hvorslev strength relationship (dotted lines): (a) failure relationships in $p\prime$: q plane (each line or curve describes strengths of soils with the same density or specific volume); (b) implied or assumed link between peak strength and state variable

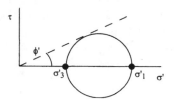

Figure 2.68: Mohr circle and Mohr-Coulomb frictional failure criterion

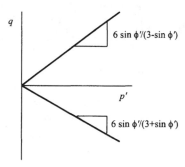

Figure 2.69: Mohr-Coulomb failure criterion in $p\prime$: q effective stress plane

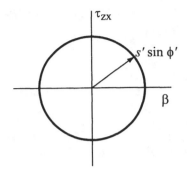

Figure 2.70: Mohr-Coulomb failure criterion in $\beta : \tau_{zx}$ deviator stress plane

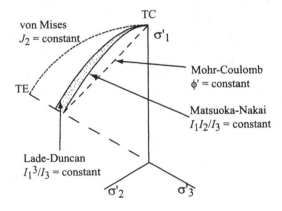

Figure 2.71: Failure criteria in π-plane deviatoric view of principal stress space (TC: triaxial compression; TE: triaxial extension)

In the stress plane $\beta = (\sigma'_z - \sigma'_x)/2 : \tau_{zx}$, which was introduced (Fig 2.21) to display stress paths for simple shear, directional shear and torsional hollow cylinder tests, for a given mean stress a line of constant angle of friction is a circle of radius $s' \sin \phi'$ (Fig 2.70), where $s' = (\sigma'_1 + \sigma'_3)/2$ is the mean effective stress in the plane of shearing. The radius of this circle, the length of the stress vector from the origin of this plot, is $(\sigma'_1 - \sigma'_3)/2$.

In the π-plane view of stress space, used for display of deviatoric stress paths for true triaxial tests, the Mohr-Coulomb failure criterion plots as an irregular hexagon for a constant mean stress section (one 60° segment is shown in Fig 2.71). This shows us immediately why a model generalisation based on the second stress invariant J_2 (2.54) will not be particularly satisfactory for soils. In fact, failure data for sand tend to lie somewhere between the Mohr-Coulomb hexagon and the J_2 circle. Two alternative failure criteria have been quite widely used to better describe the deviatoric failure conditions.

The Matsuoka-Nakai criterion (Matsuoka and Nakai, 1982) states that

$$\frac{I_1 I_2}{I_3} = \text{constant} \tag{2.108}$$

where I_1, I_2 and I_3 are the three invariants of the stress tensor defined in (2.45), (2.46), (2.47). This gives a curved failure locus which circumscribes the Mohr-Coulomb hexagon, passing through the vertices of that hexagon.

On the other hand, some experimental data suggest that the frictional strength of sands in triaxial extension is slightly higher than that in triaxial compression. Lade's (Lade and Duncan, 1975) failure criterion is expressed as

$$\frac{I_1^3}{I_3} = \text{constant} \tag{2.109}$$

This is also plotted in Fig 2.71. Evidently the common characteristic of both these failure criteria is that, unlike the Mohr-Coulomb criterion, they include the intermediate principal stress. Different sands, tested in different laboratory devices, show strength data which favour one or other of these criteria. Both are in agreement in proposing that, for states of stress lying between triaxial compression and triaxial extension, the available friction is somewhat higher—perhaps at its maximum as much as 10% higher—than that in triaxial compression. Plane strain conditions tend to fall in this intermediate region: plane strain frictional strengths will usually be underestimated if the angle of shearing resistance is determined using triaxial compression tests.

3

Constitutive modelling

3.1   Introduction

It will be seen in discussing numerical modelling in Chapter 4 that analysis of any problem requires noncontroversial statements of equilibrium and of kinematics or compatibility (the definition of strain) but that the link between these is provided by some statement of the much more uncertain link between stress change and strain change: the constitutive response.

Geotechnical journals and conferences abound with constitutive models. The choice of model is to some extent a matter of mathematical aesthetics and subjective judgement. We cannot hope to describe all possible constitutive modelling proposals here. However, there are some models which have been so widely used that they are rather generally available in all numerical analysis programs that are intended for application to geotechnical problems: isotropic elasticity; elastic-perfectly plastic Mohr-Coulomb; and Cam clay. We will present not only these models but also modest developments from these models. Our thesis is that engineers are more likely to make use of models which can be clearly seen as incrementally different from models with which they have some familiarity than to make use of models which adopt a completely different language. Thus a hardening plasticity model will be presented which is an obvious and logical extension of the perfectly plastic Mohr-Coulomb model.

The choice of model to be used for analysis is in the hands of the modeller. The second, rational proposal is that the modeller should develop some awareness of the particular features of soil history and soil response that are likely to be important in a particular application and ensure that the constitutive model that is adopted is indeed able to reproduce these features. As in all modelling, *adequate complexity* should be sought. It is too easy to discover that key elements of response are obscured by unnecessary and detachable elements of the constitutive model.

Much of the presentation and description of constitutive models in this chapter will concentrate on conditions that are accessible in the conventional triaxial apparatus. This is, and is likely to remain, the most commonly used soil testing apparatus and hence the source of data against which constitutive models

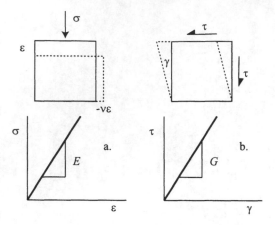

Figure 3.1: Linear relationship between stress and strain for (a) compression and (b) shearing of elastic element

have to be tuned. Geotechnical engineers have a familiarity with the triaxial apparatus and with triaxial test results so perhaps by presenting the models in this way there is a chance that some confidence in the concepts of constitutive modelling can be created. Once again we choose compressive stresses and compressive strains to be positive.

3.2   Elastic models

A linear relationship between stress and strain (Fig 3.1) is the simplest link that can be proposed, implying a constant proportionality between general stress increments and strain increments. For an isotropic, linear elastic material the full link between general stress increments and strain increments can be written as a compliance relationship

$$
\begin{pmatrix} \delta\epsilon_{xx} \\ \delta\epsilon_{yy} \\ \delta\epsilon_{zz} \\ \delta\gamma_{yz} \\ \delta\gamma_{zx} \\ \delta\gamma_{xy} \end{pmatrix}
= \frac{1}{E}
\begin{pmatrix}
1 & -\nu & -\nu & 0 & 0 & 0 \\
-\nu & 1 & -\nu & 0 & 0 & 0 \\
-\nu & -\nu & 1 & 0 & 0 & 0 \\
0 & 0 & 0 & 2(1+\nu) & 0 & 0 \\
0 & 0 & 0 & 0 & 2(1+\nu) & 0 \\
0 & 0 & 0 & 0 & 0 & 2(1+\nu)
\end{pmatrix}
\begin{pmatrix} \delta\sigma'_{xx} \\ \delta\sigma'_{yy} \\ \delta\sigma'_{zz} \\ \delta\tau_{yz} \\ \delta\tau_{zx} \\ \delta\tau_{xy} \end{pmatrix}
\qquad (3.1)
$$

indicating a dependency of strain increments on stress increments. Alternatively, this can be written as a stiffness relationship, showing stress increments as a function of strain increments[1].

$$
\begin{pmatrix} \delta\sigma'_{xx} \\ \delta\sigma'_{yy} \\ \delta\sigma'_{zz} \\ \delta\tau_{yz} \\ \delta\tau_{zx} \\ \delta\tau_{xy} \end{pmatrix} = \frac{E}{(1+\nu)(1-2\nu)} \times
$$

$$
\begin{pmatrix} 1-\nu & \nu & \nu & 0 & 0 & 0 \\ \nu & 1-\nu & \nu & 0 & 0 & 0 \\ \nu & \nu & 1-\nu & 0 & 0 & 0 \\ 0 & 0 & 0 & \frac{1-2\nu}{2} & 0 & 0 \\ 0 & 0 & 0 & 0 & \frac{1-2\nu}{2} & 0 \\ 0 & 0 & 0 & 0 & 0 & \frac{1-2\nu}{2} \end{pmatrix} \begin{pmatrix} \delta\epsilon_{xx} \\ \delta\epsilon_{yy} \\ \delta\epsilon_{zz} \\ \delta\gamma_{yz} \\ \delta\gamma_{zx} \\ \delta\gamma_{xy} \end{pmatrix} \quad (3.2)
$$

If we compress an element of isotropic elastic material without providing any lateral constraints (Fig 3.1a) there will in general be some lateral strain. At the same time that we can determine a direct stiffness, Young's modulus E, we can also discover a strain ratio, Poisson's ratio ν. It turns out that for the isotropic elastic material there are only two constitutive degrees of freedom which, as shown in (3.1) and (3.2), could be Young's modulus and Poisson's ratio.

These expressions describe Hooke's law of elasticity for general states of stress. We can specialise these expressions for principal stresses and strains, referred to orthogonal axes x, y and z, in compliance form:

$$
\begin{pmatrix} \delta\epsilon_x \\ \delta\epsilon_y \\ \delta\epsilon_z \end{pmatrix} = \frac{1}{E} \begin{pmatrix} 1 & -\nu & -\nu \\ -\nu & 1 & -\nu \\ -\nu & -\nu & 1 \end{pmatrix} \begin{pmatrix} \delta\sigma'_x \\ \delta\sigma'_y \\ \delta\sigma'_z \end{pmatrix} \quad (3.3)
$$

and in stiffness form:

$$
\begin{pmatrix} \delta\sigma'_x \\ \delta\sigma'_y \\ \delta\sigma'_z \end{pmatrix} = \frac{E}{(1+\nu)(1-2\nu)} \begin{pmatrix} 1-\nu & \nu & \nu \\ \nu & 1-\nu & \nu \\ \nu & \nu & 1-\nu \end{pmatrix} \begin{pmatrix} \delta\epsilon_x \\ \delta\epsilon_y \\ \delta\epsilon_z \end{pmatrix} \quad (3.4)
$$

It may be remarked that, when expressed in terms of all 6 general stress and strain components, (3.1) or (3.2), or in terms of the principal stress and strain components, (3.3) or (3.4), the elastic compliance and stiffness matrices are symmetric: this is an inevitable property of the elastic material. In fact we can attach thermodynamic requirements to an elastic (strictly, a hyperelastic) system. The system is conservative or path independent, which implies that the strains obtained are independent of the sequence in which the stresses are applied or removed and that we can superpose independent systems of stresses in order to deduce the result of applying combinations of stresses. There exists

---

[1]Recall that it was seen in §2.5.1 that it appeared to be more secure to move from strain to stress than from stress to strain—and this will be seen to be a feature of several of the models that are described in this chapter.

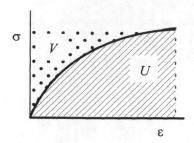

Figure 3.2: Definition of strain energy density (U) and complementary energy density (V)

a strain energy density function $U(\epsilon)$ which describes the strain energy per unit volume as a function of a general strain state, referred to some origin of strain (Fig 3.2). There is no generation or loss of energy in any closed stress path.

By definition, an increment of strain energy U is the sum of all products of stress components and corresponding work-conjugate strain increment components. Restricting ourselves to principal stress conditions for simplicity, we can write:

$$\delta U = \sum (\sigma_i \delta \epsilon_i) \tag{3.5}$$

and thus

$$\sigma_i = \frac{\partial U}{\partial \epsilon_i} \tag{3.6}$$

where the index i takes the values x, y and z in turn. Differentiating this one more time, we recover the components D_{ij} of the stiffness matrix:

$$D_{ij} = \frac{\partial^2 U}{\partial \epsilon_i \partial \epsilon_j} \tag{3.7}$$

and the symmetry of the stiffness matrix is anticipated.

For the isotropic linear elastic material, and working in terms of principal stresses and strains, the strain energy density function is:

$$U = \frac{E}{2(1 - 2\nu)(1 + \nu)} \left[(1 - \nu) \left(\epsilon_x^2 + \epsilon_y^2 + \epsilon_z^2 \right) + 2\nu \left(\epsilon_y \epsilon_z + \epsilon_z \epsilon_x + \epsilon_x \epsilon_y \right) \right] \tag{3.8}$$

There is a parallel between compliance and stiffness formulations, interchanging stresses and strains. We can similarly define a complementary energy density function V (Fig 3.2) such that

$$\delta V = \sum (\epsilon_i \delta \sigma_i) \tag{3.9}$$

and thus

$$\epsilon_i = \frac{\partial V}{\partial \sigma_i} \tag{3.10}$$

Differentiating this one more time, we recover the components C_{ij} of the symmetric compliance matrix:

$$C_{ij} = \frac{\partial^2 V}{\partial \sigma_i \partial \sigma_j} \tag{3.11}$$

For the isotropic linear elastic material

$$V = \frac{1}{2E} \left[(\sigma_x^2 + \sigma_y^2 + \sigma_z^2) - 2\nu (\sigma_y \sigma_z + \sigma_z \sigma_x + \sigma_x \sigma_y) \right] \tag{3.12}$$

We can specialise (3.3) and (3.4) further for the axisymmetric conditions of the triaxial test, so that the x and y axes are interchangeable horizontal or radial axes and z is the vertical, axial direction:

$$\begin{pmatrix} \delta \epsilon_a \\ \delta \epsilon_r \end{pmatrix} = \frac{1}{E} \begin{pmatrix} 1 & -2\nu \\ -\nu & 1-\nu \end{pmatrix} \begin{pmatrix} \delta \sigma_a' \\ \delta \sigma_r' \end{pmatrix} \tag{3.13}$$

$$\begin{pmatrix} \delta \sigma_a' \\ \delta \sigma_r' \end{pmatrix} = \frac{E}{(1+\nu)(1-2\nu)} \begin{pmatrix} 1-\nu & 2\nu \\ \nu & 1 \end{pmatrix} \begin{pmatrix} \delta \epsilon_a \\ \delta \epsilon_r \end{pmatrix} \tag{3.14}$$

and, paradoxically, we have lost the symmetry that we had come to expect from elastic compliance and stiffness matrices. However, in writing these expressions in terms of axial and radial components of stress and strain we are no longer using work conjugate quantities.

In introducing appropriate quantities with which to describe conditions of stress and strain in soils in section §2.4 we suggested that there was some advantage in separating soil response into compression—change of size—and distortion—change of shape—and then choosing stress and strain variables accordingly. For the axisymmetric conditions of the triaxial test, we introduced mean effective stress (volumetric stress) $p' = (\sigma_a' + 2\sigma_r')/3$ and distortional stress $q = (\sigma_a' - \sigma_r')$ and corresponding work-conjugate strain increments: volumetric strain $\delta \epsilon_p$ and distortional strain $\delta \epsilon_q$. Transforming (3.13), from equations (2.28) to (2.31)

$$\begin{pmatrix} \delta \epsilon_p \\ \delta \epsilon_q \end{pmatrix} = \frac{1}{E} \begin{pmatrix} 1 & 2 \\ \frac{2}{3} & -\frac{2}{3} \end{pmatrix} \begin{pmatrix} 1 & -2\nu \\ -\nu & 1-\nu \end{pmatrix} \begin{pmatrix} 1 & \frac{2}{3} \\ 1 & -\frac{1}{3} \end{pmatrix} \begin{pmatrix} \delta p' \\ \delta q \end{pmatrix} \tag{3.15}$$

$$\begin{pmatrix} \delta \epsilon_p \\ \delta \epsilon_q \end{pmatrix} = \frac{1}{E} \begin{pmatrix} 3(1-2\nu) & 0 \\ 0 & \frac{2}{3}(1+\nu) \end{pmatrix} \begin{pmatrix} \delta p' \\ \delta q \end{pmatrix} \tag{3.16}$$

or

$$\begin{pmatrix} \delta \epsilon_p \\ \delta \epsilon_q \end{pmatrix} = \begin{pmatrix} \frac{1}{K} & 0 \\ 0 & \frac{1}{3G} \end{pmatrix} \begin{pmatrix} \delta p' \\ \delta q \end{pmatrix} \tag{3.17}$$

and in stiffness form

$$\begin{pmatrix} \delta p' \\ \delta q \end{pmatrix} = \begin{pmatrix} K & 0 \\ 0 & 3G \end{pmatrix} \begin{pmatrix} \delta \epsilon_p \\ \delta \epsilon_q \end{pmatrix} \tag{3.18}$$

and we have now discovered an alternative (but not independent) pair of elastic degrees of freedom: bulk modulus K

$$K = \frac{E}{3(1-2\nu)} \tag{3.19}$$

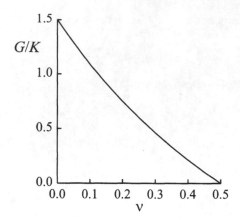

Figure 3.3: Relationship between G/K and Poisson's ratio ν for isotropic elastic material

and shear modulus G

$$G = \frac{E}{2(1+\nu)} \tag{3.20}$$

These are evidently functions of Young's modulus and Poisson's ratio and knowing any two elastic quantities we can recover any other elastic parameter.

$$E = \frac{9KG}{G+3K} \tag{3.21}$$

$$\nu = \frac{3K - 2G}{2(G+3K)} = \frac{3 - 2\frac{G}{K}}{2\left(\frac{G}{K} + 3\right)} \tag{3.22}$$

and

$$\frac{G}{K} = \frac{3(1-2\nu)}{2(1+\nu)} \tag{3.23}$$

so that Poisson's ratio is a direct indication of the ratio of shear and bulk moduli
(Fig 3.3).

Working in terms of these properly work-conjugate volumetric and distortional quantities the symmetry of the compliance (3.17) and stiffness matrices
(3.18) has been regained. The strain energy density function is:

$$U = \frac{1}{2}K\epsilon_p^2 + \frac{3}{2}G\epsilon_q^2 \tag{3.24}$$

and the complementary energy density function

$$V = \frac{1}{2}\frac{p'^2}{K} + \frac{1}{6}\frac{q^2}{G} \tag{3.25}$$

and it can be confirmed that the elements of the compliance and stiffness matrices can be obtained by appropriate differentiation.

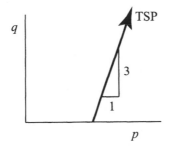

Figure 3.4: Total stress path for conventional triaxial compression test

Writing the elastic relationships in terms of p', q and $\delta\epsilon_p, \delta\epsilon_q$ makes clear the absence of any coupling between volumetric and distortional effects for the isotropic elastic model. Thus, if the mean effective stress p' is changed without change in distortional stress, the *shape* of the soil element remains unchanged. Conversely, if the soil element is subjected to a distortional strain (change in shape) without change in size or volume (undrained deformation) then there will be no tendency for the effective mean stress p' to change.

3.2.1 Conventional drained triaxial compression test

In a conventional triaxial compression test, the cell pressure (radial total stress) is kept constant and the axial strain (and hence for much of the time the axial stress) is increased. From the definitions of our stress variables we deduce that the imposed *total* stress path has gradient $\delta q/\delta p = 3$ (Fig 3.4). In a conventional *drained* triaxial compression test this will also be the *effective* stress path. If $\delta q = 3\delta p'$ then, from (3.17),

$$\frac{\delta\epsilon_p}{\delta\epsilon_q} = \frac{G}{K} \tag{3.26}$$

More generally, if $\delta q/\delta p' = \lambda$ then $\delta\epsilon_p/\delta\epsilon_q = 3G/\lambda K$.

If we plot the results of a drained triaxial compression test in terms of distortional stress q and volumetric strain ϵ_p as functions of distortional strain ϵ_q (Fig 3.5a) then we can rapidly recover the values of the two elastic soil properties. However, it might be more common to plot q and ϵ_p as functions of axial strain, ϵ_a (Fig 3.5b). Then the slope $\delta q/\delta\epsilon_a = E$ and $\delta\epsilon_p/\delta\epsilon_a = 1 - 2\nu$ and again we have sufficient information to discover any of the elastic properties. By implication, in drawing Fig 3.5 we are assuming that we might adopt an elastic treatment for the incremental response of the soil even if the overall stress:strain response is far from linearly elastic.

3.2.2 Conventional undrained triaxial compression test

We have already observed that there is complete uncoupling of volumetric and distortional effects for an isotropic elastic soil. Consequently, in an undrained

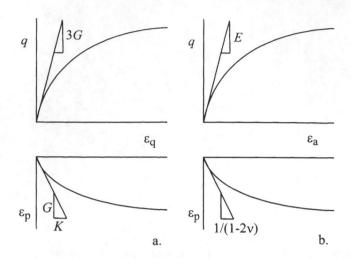

Figure 3.5: Elastic properties deduced from initial stages of conventional drained triaxial compression test: (a) plotted in terms of distortional strain ϵ_q; (b) plotted in terms of axial strain ϵ_a

test, since the volumetric strain is always zero $\delta\epsilon_p = 0$, the effective mean stress remains constant $\delta p' = 0$. The change in pore pressure δu is given by the principle of effective stress (§2.4):

$$\delta u = \delta p - \delta p' = \delta p \qquad (3.27)$$

and thus the pore pressure change merely reflects the (arbitrary) imposed change in total mean stress—and may be positive or negative depending on the chosen total stress path (Fig 3.6a). For the isotropic elastic material the soil has no intrinsic desire to change in volume as it is sheared and hence the pore pressure parameter $a = 0$ (§2.6.2). In a conventional triaxial compression test, following a *total* stress path $\delta q/\delta p = 3$ there will be a pore pressure $\Delta u = \Delta q/3$ at all stages of the test.

Under undrained conditions, distortional strain and axial strain are identical $\delta\epsilon_q = \delta\epsilon_a$ so it matters not whether we think of plotting distortional stress q against ϵ_q or ϵ_a. The slope of the stress:strain response for the elastic material is $\delta q/\delta\epsilon_q = \delta q/\delta\epsilon_a = 3G$. We are not able to determine K from an undrained test—because there is by definition no volume change.

We have noted that the stress:strain response of soils is controlled by changes in *effective* stress. That $\delta p' = 0$ in an undrained test is the response of the soil to certain imposed constraints: the drainage valve is closed. Evidently the constitutive response must be unaffected by the setting of the drainage valve. However, we could choose to take an external, total stress view of our sample and seek a total stress elastic response introducing 'undrained' elastic properties E_u, ν_u, G_u and K_u such that

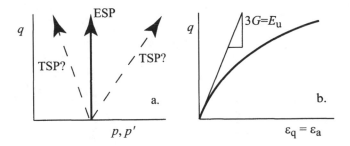

Figure 3.6: Undrained triaxial compression test on elastic soil: (a) effective (ESP) and total (TSP) stress paths; (b) elastic stiffness from initial stress:strain response

$$\begin{pmatrix} \delta\epsilon_p \\ \delta\epsilon_q \end{pmatrix} = \begin{pmatrix} 1/K_u & 0 \\ 0 & 1/3G_u \end{pmatrix} \begin{pmatrix} \delta p \\ \delta q \end{pmatrix} \tag{3.28}$$

Now, if we imagine (reasonably) that we are free to control δp and δq as we will, the fact that drainage is prevented so that $\delta\epsilon_p = 0$ can only mean that $K_u = \infty$ as expected for an incompressible material. From the definition of bulk modulus (3.19) this implies that the undrained Poisson's ratio $\nu_u = 0.5$: again, as expected from our prior knowledge of properties of incompressible materials.

The shear stiffness must be the same whether we are thinking in terms of total stresses or effective stresses: the distortion of the sample is the same, and distortional stress q, as a shear stress, or difference between two normal stresses, is the same whether thought of as a total stress or an effective stress quantity. Thus $G_u = G$ and we can deduce from (3.20) that

$$G_u = \frac{E_u}{2(1+\nu_u)} = G = \frac{E}{2(1+\nu)} \tag{3.29}$$

or

$$E_u = 3G = \frac{3E}{2(1+\nu)} \tag{3.30}$$

We conclude that the 'undrained' elastic properties of the soil are in no way independent but are directly related to the real *effective stress* stiffness characteristics of the soil.

3.2.3 Measurement of elastic parameters with different devices

We have already noted that drained triaxial tests can be interpreted to give both of the elastic parameters needed to describe the isotropic elastic soil model—provided that the imposed effective stress path involves changes in both p' and q

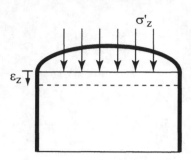

Figure 3.7: Oedometer

as indeed occurs in conventional triaxial compression with constant cell pressure. Other devices can be interpreted to give elastic properties but often these are not individual properties (E, ν or G, K) but rather some composite quantity which can only be interpreted in terms of one of these quantities if an assumption is made about, for example, ν or G/K.

For example, the oedometer (Fig 3.7) imposes one-dimensional deformation on the soil. We can manipulate (3.3) or (3.4) to discover that for this situation of zero strain in the x and y directions:

$$\frac{\delta\sigma'_x}{\delta\sigma'_z} = \frac{\nu}{1-\nu} = \frac{3K-2G}{3K+4G} \tag{3.31}$$

and that the one-dimensional, oedometric stiffness E_{oed} is

$$E_{oed} = \frac{\delta\sigma'_z}{\delta\epsilon_z} = E\frac{(1-\nu)}{(1+\nu)(1-2\nu)} = K + \frac{4}{3}G \tag{3.32}$$

We cannot deconstruct this any further to discover any of the usual individual elastic properties.

In-situ geophysics (or laboratory geophysics using bender elements §2.5.4, §6.7) can be used to determine shear modulus G directly from the shear wave velocity V_s

$$V_s = \sqrt{\frac{G}{\rho}} \tag{3.33}$$

where ρ is the density of the soil. However, the compression wave velocity V_p is concerned with the speed of propagation of a one-dimensional deformation through the soil and the corresponding link with stiffness is

$$V_p = \sqrt{\frac{K+\frac{4}{3}G}{\rho}} = \sqrt{\frac{E_{oed}}{\rho}} \tag{3.34}$$

In soils saturated with water the compression wave transmission is dominated by the effect of the pore water (which has more or less zero shear stiffness so the

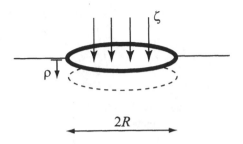

Figure 3.8: Plate loading test

compression wave transmission is governed by the bulk modulus of the water) and it may be difficult to detect the arrival of a compression wave which has travelled through the soil skeleton. Anyway, if we wished to deduce individual values of shear or bulk modulus from the compression wave velocity alone we would—as for the oedometer—have to assume a value of G/K or Poisson's ratio.

The pressuremeter (§1.2.4) generates pure shear in the surrounding ground as its cylindrical cavity is expanded (§8.8). Elastic properties, from interpretation of unload-reload cycles, give a direct indication of shear modulus G. However, it is common practice (Baguelin et al., 1978) to quote a pressuremeter modulus $E_p = G/2.6$ calculated assuming a value of Poisson's ratio $\nu = 0.3$. There is no particular reason why Poisson's ratio should have this value and it is evidently rather vital that, if values of shear modulus and bulk modulus are subsequently required, given only a value of E_p, then the same value of Poissons's ratio should be used for their calculation.

A loading test on a rigid circular plate on the surface of an elastic soil (Fig 3.8) gives a stiffness

$$\frac{\zeta}{\rho/R} = \frac{4G}{\pi(1-\nu)} \tag{3.35}$$

where ζ is the average pressure on the plate of radius R and ρ is the settlement. Again this is a composite stiffness and interpretation of any one of the conventional elastic properties requires some assumption about another one.

3.2.4 Anisotropy

An isotropic material has the same properties in all directions—we cannot distinguish any one direction from any other. Samples taken out of the ground with any orientation would behave identically. However, we know that soils have been deposited in some way—for example, sedimentary soils will know about the vertical direction of gravitational deposition. There may in addition be seasonal variations in the rate of deposition so that the soil contains more or less marked layers of slightly different grain size and/or plasticity. The scale of layering may be sufficiently small that we do not wish to try to distinguish separate materials, but the layering together with the directional deposition may nevertheless be sufficient to modify the properies of the soil in different directions—in other words to cause it to be anisotropic.

We can write the stiffness relationship between elastic strain increment $\delta\epsilon^e$ and stress increment $\delta\sigma$ compactly as

$$\delta\sigma = D\delta\epsilon^e \tag{3.36}$$

where D is the stiffness matrix and hence D^{-1} is the compliance matrix. For a completely general anisotropic elastic material

$$D^{-1} = \begin{pmatrix} a & b & c & d & e & f \\ b & g & h & i & j & k \\ c & h & l & m & n & o \\ d & i & m & p & q & r \\ e & j & n & q & s & t \\ f & k & o & r & t & u \end{pmatrix} \tag{3.37}$$

where each letter $a, b, \ldots$ is, in principle, an independent elastic property and the necessary symmetry of the stiffness matrix for the elastic material has reduced the maximum number of independent properties to 21. As soon as there are material symmetries then the number of independent elastic properties falls (Crampin, 1981).

For example, for monoclinic symmetry (z symmetry plane) the compliance matrix has the form:

$$D^{-1} = \begin{pmatrix} a & b & c & 0 & 0 & d \\ b & e & f & 0 & 0 & g \\ c & f & h & 0 & 0 & i \\ 0 & 0 & 0 & j & k & 0 \\ 0 & 0 & 0 & k & l & 0 \\ d & g & i & 0 & 0 & m \end{pmatrix} \tag{3.38}$$

and has thirteen elastic constants. Orthorhombic symmetry (distinct x, y and z symmetry planes) gives nine constants:

$$D^{-1} = \begin{pmatrix} a & b & c & 0 & 0 & 0 \\ b & d & e & 0 & 0 & 0 \\ c & e & f & 0 & 0 & 0 \\ 0 & 0 & 0 & g & 0 & 0 \\ 0 & 0 & 0 & 0 & h & 0 \\ 0 & 0 & 0 & 0 & 0 & i \end{pmatrix} \tag{3.39}$$

whereas cubic symmetry (identical x, y and z symmetry planes, together with planes joining opposite sides of a cube) gives only three constants:

$$D^{-1} = \begin{pmatrix} a & b & b & 0 & 0 & 0 \\ b & a & b & 0 & 0 & 0 \\ b & b & a & 0 & 0 & 0 \\ 0 & 0 & 0 & c & 0 & 0 \\ 0 & 0 & 0 & 0 & c & 0 \\ 0 & 0 & 0 & 0 & 0 & c \end{pmatrix} \tag{3.40}$$

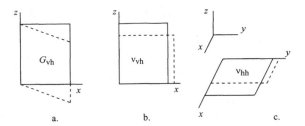

Figure 3.9: Independent modes of shearing for cross-anisotropic material

If we add the further requirement that $c = 2(a - b)$ and set $a = 1/E$ and $b = -\nu/E$ then we recover the isotropic elastic compliance matrix of (3.1).

Though it is obviously convenient if geotechnical materials have certain fabric symmetries which confer a reduction in the number of independent elastic properties, it has to be expected that in general materials which have been pushed around by tectonic forces, by ice, or by man will not possess any of these symmetries and, insofar as they have a domain of elastic response, we should expect to require the full 21 independent elastic properties. If we choose to *model* such materials as isotropic elastic or anisotropic elastic with certain restricting symmetries then we have to recognise that these are modelling decisions of which the soil or rock may be unaware.

However, many soils are deposited over areas of large lateral extent and symmetry of deposition is essentially vertical. All horizontal directions look the same but horizontal stiffness is expected to be different from vertical stiffness. The form of the compliance matrix is now:

$$D^{-1} = \begin{pmatrix} a & b & c & 0 & 0 & 0 \\ b & a & c & 0 & 0 & 0 \\ c & c & d & 0 & 0 & 0 \\ 0 & 0 & 0 & e & 0 & 0 \\ 0 & 0 & 0 & 0 & e & 0 \\ 0 & 0 & 0 & 0 & 0 & f \end{pmatrix} \tag{3.41}$$

and we can write $a = 1/E_h$, $b = -\nu_{hh}/E_h$, $c = -\nu_{vh}/E_v$, $d = 1/E_v$, $e = 1/G_{vh}$ and $f = 2(a - b) = 2(1 + \nu_{hh})/E_h$:

$$D^{-1} = $$
$$\begin{pmatrix} 1/E_h & -\nu_{hh}/E_h & -\nu_{vh}/E_v & 0 & 0 & 0 \\ -\nu_{hh}/E_h & 1/E_h & -\nu_{vh}/E_v & 0 & 0 & 0 \\ -\nu_{vh}/E_v & -\nu_{vh}/E_v & 1/E_v & 0 & 0 & 0 \\ 0 & 0 & 0 & 1/G_{vh} & 0 & 0 \\ 0 & 0 & 0 & 0 & 1/G_{vh} & 0 \\ 0 & 0 & 0 & 0 & 0 & 2(1+\nu_{hh})/E_h \end{pmatrix}$$
$$\tag{3.42}$$

This is described as transverse isotropy or cross anisotropy with hexagonal symmetry. There are 5 independent elastic properties: E_v and E_h are Young's moduli for unconfined compression in the vertical and horizontal directions respectively; G_{vh} is the shear modulus for shearing in a vertical plane (Fig 3.9a). Poisson's ratios ν_{hh} and ν_{vh} relate to the lateral strains that occur in the horizontal direction orthogonal to a horizontal direction of compression and a vertical direction of compression respectively (Fig 3.9c, b).

Testing of cross anisotropic soils in a triaxial apparatus with their axes of anisotropy aligned with the axes of the apparatus does not give us any possibility to discover $G_{vh}(= 1/e)$ since this would require controlled application of shear stresses to vertical and horizontal surfaces of the sample—and attendant rotation of principal axes. In fact we are able only to determine 3 of the 5 elastic properties. If we write (3.42) for radial and axial stresses and strains for a sample with its vertical axis of symmetry of anisotropy aligned with the axis of the triaxial apparatus, we find that

$$\begin{pmatrix} \delta\epsilon_a \\ \delta\epsilon_r \end{pmatrix} = \begin{pmatrix} 1/E_v & -2\nu_{vh}/E_v \\ -\nu_{vh}/E_v & (1-\nu_{hh})/E_h \end{pmatrix} \begin{pmatrix} \delta\sigma'_a \\ \delta\sigma'_r \end{pmatrix} \tag{3.43}$$

The compliance matrix is not symmetric because, in the context of the triaxial test, the strain increment and stress quantities are not properly work conjugate. We deduce that while we can separately determine E_v and ν_{vh} the only other elastic property that we can discover is the composite stiffness $E_h/(1-\nu_{hh})$. We are not able to separate E_h and ν_{hh} (Lings et al., 2000).

On the other hand, Graham and Houlsby (1983) have proposed a special form of (3.41) or (3.42) which uses only 3 elastic properties but forces certain interdependencies among the 5 elastic properties for this cross anisotropic material.

$$\boldsymbol{D}^{-1} = \frac{1}{E^*} \times$$
$$\begin{pmatrix}
1/\alpha^2 & -\nu^*/\alpha^2 & -\nu^*/\alpha & 0 & 0 & 0 \\
-\nu^*/\alpha^2 & 1/\alpha^2 & -\nu^*/\alpha & 0 & 0 & 0 \\
-\nu^*/\alpha & -\nu^*/\alpha & 1 & 0 & 0 & 0 \\
0 & 0 & 0 & 2(1+\nu^*)/\alpha & 0 & 0 \\
0 & 0 & 0 & 0 & 2(1+\nu^*)/\alpha & 0 \\
0 & 0 & 0 & 0 & 0 & 2(1+\nu^*)/\alpha^2
\end{pmatrix}$$
$$\tag{3.44}$$

This is written in terms of a Young's modulus $E^* = E_v$, the Young's modulus for loading in the vertical direction, a Poisson's ratio $\nu^* = \nu_{hh}$, together with a third parameter α. The ratio of stiffness in horizontal and vertical directions is $E_h/E_v = \alpha^2$ and other linkages are forced: $\nu_{vh} = \nu_{hh}/\alpha$; $G_{hv} = G_{hh}/\alpha = \alpha E^*/2(1+\nu^*)$.

For our triaxial stress and strain quantities, the compliance matrix becomes

$$\begin{pmatrix} \delta\epsilon_p \\ \delta\epsilon_q \end{pmatrix} = \frac{1}{det} \begin{pmatrix} 3G^* & -J \\ -J & K^* \end{pmatrix} \begin{pmatrix} \delta p' \\ \delta q \end{pmatrix} \tag{3.45}$$

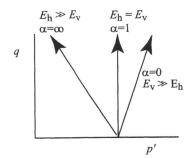

Figure 3.10: Effect of cross-anisotropy on direction of undrained effective stress path

where

$$det = 3K^*G^* - J^2 \tag{3.46}$$

and the stiffness matrix is

$$\left(\begin{array}{c} \delta p' \\ \delta q \end{array} \right) = \left(\begin{array}{cc} K^* & J \\ J & 3G^* \end{array} \right) \left(\begin{array}{c} \delta\epsilon_p \\ \delta\epsilon_q \end{array} \right) \tag{3.47}$$

where

$$K^* = E^* \frac{1 - \nu^* + 4\alpha\nu^* + 2\alpha^2}{9(1 + \nu^*)(1 - 2\nu^*)} \tag{3.48}$$

$$G^* = E^* \frac{2 - 2\nu^* - 4\alpha\nu^* + \alpha^2}{6(1 + \nu^*)(1 - 2\nu^*)} \tag{3.49}$$

$$J = E^* \frac{1 - \nu^* + \alpha\nu^* - \alpha^2}{3(1 + \nu^*)(1 - 2\nu^*)} \tag{3.50}$$

The stiffness and compliance matrices (written in terms of correctly chosen work conjugate strain increment and stress quantities) are still symmetric—the material is still elastic—but the non-zero off-diagonal terms tell us that there is now coupling between volumetric and distortional effects. There will be volumetric strain when we apply purely distortional stress, $\delta p' = 0$, distortional strain during purely isotropic compression, $\delta q = 0$, and there will be change in mean effective stress in undrained tests, $\delta\epsilon_p = 0$.

In fact the slope of the effective stress path in an undrained test is, from (3.45),

$$\frac{\delta p'}{\delta q} = \frac{J}{3G^*} = \frac{2(1 - \nu^* + \alpha\nu^* - \alpha^2)}{3(2 - 2\nu^* - 4\alpha\nu^* + \alpha^2)} \tag{3.51}$$

From our definition of pore pressure parameter a (§2.6.2) we find

$$a = -\frac{\delta p'}{\delta q} = -\frac{J}{3G^*} \tag{3.52}$$

which will, in the presence of anisotropy, not be zero.

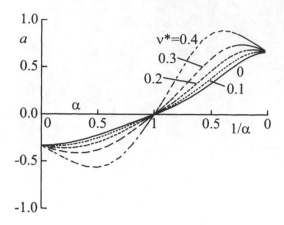

Figure 3.11: Relationship between anisotropy parameter α and pore pressure parameter a for different values of Poisson's ratio $\nu*$.

A first inspection of (3.51) merely suggests that there are limits on the pore pressure parameter of $a = 2/3$ and $a = -1/3$ for α very large ($E_h \gg E_v$) and α very small ($E_v \gg E_h$) repectively (Fig 3.10), which in turn imply effective stress paths with constant axial effective stress and constant radial effective stress respectively. The link between a and α is actually slightly more subtle. In fact, for $\nu^* \neq 0$ the relationship is not actually monotonic and the effective stress path direction overshoots the apparent limits (Fig 3.11). The deduction of a value of α (and hence $E_h/E_v = \alpha^2$) from a is not very reliable when a is around $-1/3$ or $2/3$ (recall the data presented in Figs 2.51 and 2.49, §2.5.4). For $\nu^* = 0.5$, $a = -(1+2\alpha)/[3(1-\alpha)]$ or $\alpha = (1+3a)/(3a-2)$. These relationships satisfy the expected limits for $\alpha = 0$ and $\alpha = \infty$ but there are singularities in the inversion of (3.51) for $\alpha = 1$ and $\nu^* = 0.5$.

3.2.5 Nonlinearity

We will probably expect that the dominant source of nonlinearity of stress:strain response will come from material plasticity—and we will go on to develop elastic-plastic constitutive models in the next section. However, we also have an expectation that some of the truly elastic properties of soils will vary with stress level and this can be seen as a source of elastic nonlinearity. Our thoughts about elastic materials as conservative materials—the term 'hyperelasticity' is used to describe such materials—might make us a little cautious about plucking from the air arbitrary empirical functions for variation of moduli with stresses. For example, if we were to suppose that the bulk modulus of the soil varied with mean effective stress but that Poisson's ratio (and hence the ratio of shear modulus to bulk modulus) were constant then we would find that in a closed stress

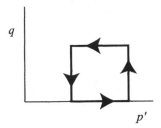

Figure 3.12: Cycle of stress changes which should give zero energy generated or dissipated for conservative material

cycle such as that shown in Fig 3.12 energy would be created (or lost) creating a perpetual motion machine in violation of the first law of thermodynamics—this would not be a conservative system. We need to find a strain energy (3.7) or complementary energy density (3.11) function which can be differentiated to give acceptable variation of moduli with stresses.

Such a complementary energy function can be deduced from the nonlinear elastic model described by Boyce (1980):

$$V = p'^{n+1} \left(\frac{1}{(n+1)K_1} + \frac{1}{6G_1} \left[\frac{q}{p'} \right]^2 \right) \qquad (3.53)$$

where K_1 and G_1 are reference values of bulk modulus and shear modulus and n is a nonlinearity parameter. The compliance matrix can then be deduced by differentiation

$$\begin{pmatrix} \delta\epsilon_p \\ \delta\epsilon_q \end{pmatrix} = p'^{n-1} \begin{pmatrix} \frac{n}{K_1} + \frac{(1-n)(2-n)}{6G_1}\eta^2 & -\frac{1-n}{3G_1}\eta \\ -\frac{1-n}{3G_1}\eta & \frac{1}{3G_1} \end{pmatrix} \begin{pmatrix} \delta p' \\ \delta q \end{pmatrix} \qquad (3.54)$$

where $\eta = q/p'$. There is again (as for the anisotropic model) coupling between volumetric and distortional effects. The stiffnesses are broadly proportional to p'^{1-n}.

Because the compliances are now varying with stress ratio η the effective stress path implied for an undrained (purely distortional) loading is no longer straight. In fact, for a reference state $p' = p'_o$, $\eta = q = 0$, the effective stress path is

$$\frac{p'_o}{p'} = \left(1 - \beta\eta^2 \right)^n \qquad (3.55)$$

where $\beta = (1-n)K_1/6G_1$. Contours of constant volumetric strain ($\delta\epsilon_p = 0$) are shown in Fig 3.13 for $n = 0.2$ and Poisson's ratio $\nu = 0.3$ implying $K_1/G_1 = 2.17$—values typical for the road sub-base materials being tested by Boyce for their small strain, resilient elastic properties.

Similarly the path followed in a purely volumetric deformation ($\delta\epsilon_q = 0$) will develop some change in distortional stress. For an initial state $p' = p'_o$, $q = q_o$ the effective stress path for such a test is

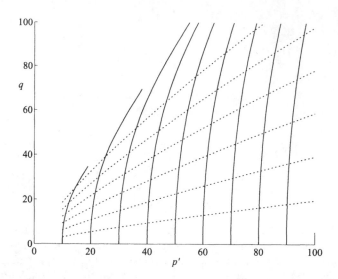

Figure 3.13: Contours of constant volumetric strain (solid lines) and constant distortional strain (dotted lines) for nonlinear elastic model of Boyce (1980)

$$\frac{q}{q_o} = \left(\frac{p'_o}{p'}\right)^{n-1}$$

(3.56)

Contours of constant distortional strain are also shown in Fig 3.13 for $n = 0.2$.

It is often proposed that the elastic volumetric stiffness—bulk modulus—of clays should be directly proportional to mean effective stress: $K = p'/\kappa$. Integration of this relationship shows that elastic unloading of clays produces a straight line response when plotted in a logarithmic compression plane ($\epsilon_p = -\ln v : \ln p'$) (Fig 3.14) where v is specific volume. But what assumption should we make about shear modulus? If we simply assume that Poisson's ratio is constant, so that the ratio of shear modulus to bulk modulus is constant, then we will emerge with a non-conservative material (Zytynski *et al.*, 1978). If we assume a constant value of shear modulus, independent of stress level, we will obtain a conservative material but may find that we have physically surprising values of implied Poisson's ratio for certain high or low stress levels. Again we need to find a strain or complementary energy function that will give us the basic modulus variation that we desire.

Houlsby (1985) suggests that an acceptable strain energy function could be

$$U = p'_r e^{\epsilon_p/\kappa} \left(\kappa + \frac{3}{2}\alpha\epsilon_q^2\right)$$

(3.57)

Incrementally this implies a stiffness matrix which, once again, contains off-diagonal terms indicating coupling between volumetric and distortional elements of deformation:

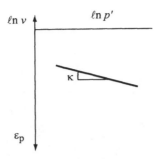

Figure 3.14: Linear logarithmic relationship between v and $p\prime$ for elastic material with bulk modulus proportional to $p\prime$

$$\begin{pmatrix} \delta p' \\ \delta q \end{pmatrix} = p' \begin{pmatrix} 1/\kappa & \eta/\kappa \\ \eta/\kappa & \eta/\epsilon_q \end{pmatrix} \begin{pmatrix} \delta\epsilon_p \\ \delta\epsilon_q \end{pmatrix} \tag{3.58}$$

It can be deduced that

$$\eta = \frac{q}{p'} = \frac{3\alpha\epsilon_q}{1 + \frac{3\alpha}{2\kappa}\epsilon_q^2} \tag{3.59}$$

so that contours of constant distortional strain are lines of constant stress ratio η (Fig 3.15). Constant volume (undrained) stress paths are found to be parabolae (Fig 3.15):

$$q^2 = 6\alpha\kappa p_i' \left(p' - p_i'\right) \tag{3.60}$$

All parabolae in this family touch the line $\eta = \sqrt{3\alpha\kappa/2}$.

The nonlinearity that has been introduced in these two models is still associated with an isotropic elasticity. The elastic properties vary with deformation but not with direction.

Although it tends to be assumed that nonlinearity in soils comes exclusively from soil plasticity—as will be discussed in the subsequent sections—we have seen that with care it may be possible to describe some elastic nonlinearity in a way which is thermodynamically acceptable. Equally, most elastic-plastic models will contain some element of elasticity—which may often be swamped by plastic deformations. It must be expected that the fabric variations which accompany any plastic shearing will themselves lead to changes in the elastic properties of the soil. The formulation of such variations of stiffness should in principle be based on the differentiation of some serendipitously discovered elastic strain energy density function in order that the elasticity should not violate the laws of thermodynamics. Evidently the development of strain energy functions which permit evolution of anisotropy of elastic stiffness is tricky. Many constitutive models adopt a pragmatic, hypoelastic approach and simply define the evolution of the moduli with stress state or with strain state without concern for the thermodynamic consequences. This may not provoke particular problems provided the stress paths or strain paths to which soil elements are subjected are not very repeatedly cyclic.

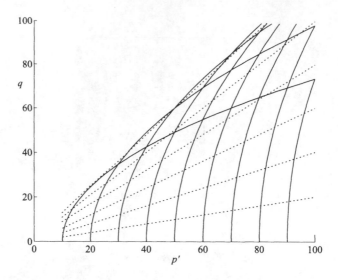

Figure 3.15: Contours of constant volumetric strain (solid lines) and constant distortional strain (dotted lines) for nonlinear elastic model of Houlsby (1985)

3.2.6   Heterogeneity

Anisotropy and nonlinearity are both possible departures from the simple assumptions of isotropic linear elasticity. A rather different departure is associated with heterogeneity. We have already noted that small scale heterogeneity—seasonal layering—may lead to anisotropy of stiffness (and other) properties at the scale of a typical sample. Many natural and man-made soils contain large ranges of particle sizes (§1.8)—glacial tills and residual soils often contain boulder-sized particles within an otherwise soil-like matrix. If the scale of our geotechnical system is large by comparison with the size and spacing of these boulders then it will be reasonable to treat the material as essentially homogeneous. However, we will still wish to determine its mechanical properties.

If we attempt to measure shear wave velocities *in situ*, using geophysical techniques, then we can expect that the fastest wave from source to receiver will take advantage of the presence of the large hard rock-like particles—which will have a much higher stiffness and hence higher shear wave velocity than the surrounding soil (Fig 3.16). The receiver will show the travel time for the fastest wave which has taken this heterogeneous route. If the hard material occupies a proportion λ of the spacing between source and receiver, and the ratio of shear wave velocities is k (and hence, neglecting density differences, the ratio of shear moduli is of the order of k^2), then the ratio of apparent shear wave velocity $\bar{V}_s$ to the shear wave velocity of the soil matrix V_s is

$$\frac{\bar{V}_s}{V_s} = \frac{k}{\lambda + k - k\lambda} \tag{3.61}$$

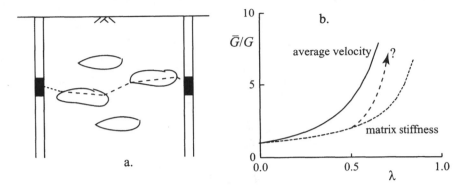

Figure 3.16: (a) Soil containing boulders between boreholes used for measurement of shear wave velocity; (b) average stiffnesses deduced from interpretation of shear wave velocity and from matrix stiffness

The deduced average shear modulus $\bar{G}$ is then greater than the shear stiffness of the soil matrix G by the ratio

$$\frac{\bar{G}}{G} = \left(\frac{k}{\lambda + k - k\lambda}\right)^2 \rightarrow \frac{1}{(1-\lambda)^2} \quad \text{as} \quad k \to \infty \qquad (3.62)$$

Laboratory testing of such heterogeneous materials is not easy because the test apparatus needs itself to be much larger than the typical maximum particle size and spacing in order that a true average property should be measured. At a small scale, Muir Wood and Kumar (2000) report tests to explore mechanical characteristics of mixtures of kaolin clay and a fine gravel ($d_{50} = 2$ mm). They found that all the properties of the clay/gravel system were controlled by the soil matrix until the volume fraction of the gravel was about 0.45-0.5. At that stage, but not before, interaction between the 'rigid' particles started rapidly to dominate. For $\lambda < 0.5$ then, this implies a ratio of equivalent shear stiffness $\bar{G}$ to soil matrix stiffness G:

$$\frac{\bar{G}}{G} = \frac{1}{1-\lambda} \qquad (3.63)$$

These two expressions, (3.62) and (3.63), are compared in Fig 3.16 for a modulus ratio $k^2 = 10000$.

3.3 Elastic-perfectly plastic models

Elastic descriptions of soil behaviour are useful for the wide range of quick analytical solutions to which they give access. If we need some idea about the stress distribution around a footing or wall or pile then at least a first estimate can be obtained using an elastic analysis (§7.2). Many of these analyses are either available as closed form results or have been previously published in

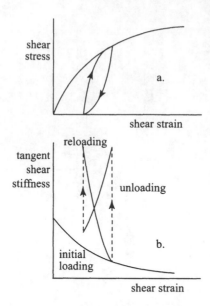

Figure 3.17: (a) Typical irreversible stress:strain response and (b) typical modulus variation for soil

the literature (for example, Poulos and Davis, 1974). Elastic analyses are also useful to give a first estimate of the deformations that may be expected for a geotechnical structure under working loads—structures which are therefore not loaded to anything approaching the failure conditions.

However, a quick comparison of the stress:strain response implied by a linear elastic description of soil behaviour with the actual stress:strain response of a typical soil shows that there are many features of soil response that the simple model is unable to capture (Fig 3.17). In particular, it is clear that most soils show nonlinear stress:strain relationships with the stiffness falling from a high initial value. If a soil is unloaded from some intermediate, prefailure condition then it will not recover its initial state but will be left with permanent, irrecoverable deformation—which we will call plastic deformation to distinguish it from the recoverable, elastic elements of deformation. During this unloading process the tangent stiffness increases initially, typically to a value higher than the initial stiffness and then falls—a similar pattern is seen on reloading (Fig 3.17).

Most soils develop significant volume changes even when they are subjected only to changes in shear stress. Most soils, if sheared to sufficiently large strains, reach a state of continuing shearing with no further change in stresses—zero incremental or tangent stiffness—at large strains. This type of behaviour, in which the tangential stiffness has fallen to zero, is described as perfect plasticity and the next stage in development of simple models is to use an elastic-perfectly plastic description of soil response (Fig 3.18). This elastic-perfectly plastic model

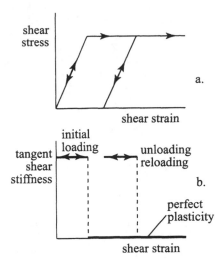

Figure 3.18: Elastic-perfectly plastic model: (a) stress:strain response and (b) modulus variation

is the first of a series of models which we will describe which gradually improve the degree with which the richness of behaviour shown in Fig 3.17 can be reproduced.

3.3.1 General elastic-perfectly plastic model

The underlying assumption of the soil models that are being developed is that the strain increments that accompany any change in stress can be divided into elastic (^{e}) (recoverable) and plastic (^{p}) (irrecoverable) parts

$$\delta\epsilon = \delta\epsilon^e + \delta\epsilon^p \qquad (3.64)$$

The strain tensor is thought of here as a six element *vector* of cartesian strain components since in this form the presentation and programming of stiffness relationships involves nothing more than straightforward matrix multiplication and manipulation. In many applications it will be a subset of this vector that will be of interest. This division of strain clearly reflects the observation that removal of loads from a sheared soil sample in general leaves the sample with some permanent changes in shape and size.

The elastic strain increment $\delta\epsilon^e$ occurs whenever there is any change in stress $\delta\sigma$ (where the stress is also thought of as a six dimensional vector of cartesian components).

$$\delta\sigma = D\delta\epsilon^e \qquad (3.65)$$

where D is the elastic stiffness matrix. The first ingredient of the model is therefore a description of the elastic behaviour which may be isotropic or anisotropic as appropriate.

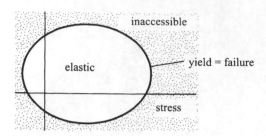

Figure 3.19: Elastic-perfectly plastic model: yield surface separating elastic and inaccessible regions of stress space

In the elastic-perfectly plastic model there is a region of stress space which can be reached elastically, without incurring any irrecoverable deformations (Fig 3.19). However, as soon as the boundary of this elastic region is reached then the material yields (or fails) at constant stress. The boundary of the elastic region is called a yield surface (Fig 3.19) and is mathematically described by a yield function: this is the second ingredient of the model.

$$f(\boldsymbol{\sigma}) = 0 \qquad\qquad (3.66)$$

The plastic strain increment $\delta\boldsymbol{\epsilon}^p$ (in (3.64)) occurs only when the stress state lies on—and remains on—the yield surface during the load increment so that

$$f(\boldsymbol{\sigma}) = 0; \quad \delta f = \frac{\partial f}{\partial \boldsymbol{\sigma}}^T \delta\boldsymbol{\sigma} = 0 \qquad\qquad (3.67)$$

where ^{T} indicates the transpose of the vector. This relation is known as the consistency condition.

The perfectly plastic soil model has been discussed so far only in terms of a limiting set of stress states which can be reached—defined by the yield function $f(\boldsymbol{\sigma})$. For the model to be useful in more extensive numerical analysis it is necessary to be able to make some statements about the nature of the deformations that occur when this limiting stress state is reached. Before we do this it may be helpful to digress in order to discuss the behaviour of a perfectly plastic structural system where the nature of the plastic deformations is rather clear.

3.3.2 A digression: collapse of portal frame

At this point a digression may be helpful in order to explore the deformations that occur in a simple perfectly plastic structural system: the collapse of a steel portal frame under combinations of vertical and horizontal loading (Fig 3.20a). The collapse of this frame is governed by three mechanisms (Fig 3.20b, c, d): beam collapse, which occurs when the vertical load is the dominant loading (Fig 3.20b); sway collapse, which occurs when the horizontal load is the dominant loading (Fig 3.20c); and a combined mechanism (Fig 3.20d).

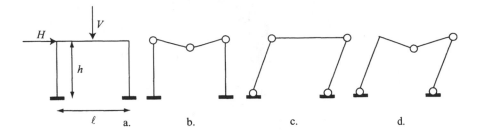

Figure 3.20: (a) Steel portal frame; (b) beam collapse; (c) sway collapse; (d) combined collapse mechanism (o indicate locations of plastic hinges)

The standard methods of plastic structural analysis can be used to demonstrate that for the beam mechanism (Fig 3.20b) the limiting load is given by

$$\frac{V\ell}{M_p} = 8 \tag{3.68}$$

For the sway mechanism (Fig 3.20c) the limiting load is given by

$$\frac{Hh}{M_p} = 4 \tag{3.69}$$

and for the combined mechanism (Fig 3.20d) the limiting loads are given by

$$\frac{Hh}{M_p} + \frac{V\ell}{2M_p} = 6 \tag{3.70}$$

These three expressions define a collapse locus—or interaction diagram—in the plot of vertical and horizontal loads (Fig 3.21). The collapse locus defines the boundary of the region of safe combinations of loads[2].

If a change in the loads applied to the structure brings the load combination to the collapse locus then collapse of the frame will occur according to one of the three mechanisms. The particular choice of mechanism will depend on the load combination at which the collapse locus is reached, according to equations (3.68)-(3.70), and not on the route by which that boundary is reached. This is obviously a difference from the response of an elastic material or elastic system for which the incremental deformation depends on the changes in loads and not on the values of the loads themselves[3].

---

[2]The collapse locus cannot be strictly described as a yield locus because, by extension from the analysis of the plastically collapsing cantilever (§2.5.1), the frame will start to yield and generate irrecoverable plastic rotations as soon as the yield moment is reached at any section. The extent of this plastic region inside the collapse locus will depend on the ratio of yield moment to full plastic moment of the structural sections from which the frame is constructed. For rectangular sections the ratio is 2/3; for more practical I sections the ratio is more typically of the order of 0.87.

[3]For a conservative nonlinear elastic material (§2.5.1) the incremental stiffness will depend on the current state of stress but consideration of (3.54) and (3.58) shows that the mechanism of elastic deformation is always dependent on the accompanying *increments* of stress.

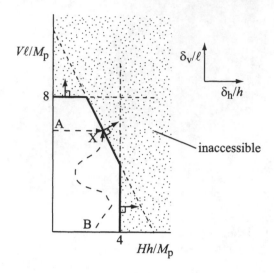

Figure 3.21: Collapse locus for steel portal frame

By appropriate choice of variables it is possible to indicate the mechanism of plastic collapse on the same diagram as the collapse locus. For the perfectly plastic system brought to collapse by changes in the applied loads, there is at collapse no possibility of determining the *magnitudes* of the deformations— failure, according to the simple model, just continues indefinitely. (The collapse could be contained if the structure were loaded using some sort of deformation control.) The three mechanisms do, however, indicate the *relative* magnitudes of the vertical movement δ_v at the centre of the beam and the horizontal movement δ_h at the top of the columns.

The work done in any deformation of the structure (δ_v, δ_h) while it carries loads V and H is

$$\delta W = V\delta_v + H\delta_h \tag{3.71}$$

The work conjugacy of (V, H) and (δ_v, δ_h) is obvious. However, for convenience the collapse locus has been plotted in terms of normalised loads $V\ell/M_p$ and Hh/M_p. A work conjugacy can be deduced between $(V\ell, Hh)$ and $(\delta_v/\ell, \delta_h/h)$.

The three mechanisms of collapse can be described in terms of the ratios of these two displacement variables. The beam mechanism gives:

$$\frac{\delta_v/\ell}{\delta_h/h} = \infty \tag{3.72}$$

The sway mechanism gives:

$$\frac{\delta_v/\ell}{\delta_h/h} = 0 \tag{3.73}$$

and the combined mechanism gives:

$$\frac{\delta_v/\ell}{\delta_h/h} = \frac{1}{2} \tag{3.74}$$

These three expressions define the directions of three vectors which can be plotted on the corresponding segments of the collapse locus to indicate the resulting collapse mechanisms (Fig 3.21). It is found in this case that these collapse vectors are normal to the corresponding segments of the collapse locus.

The important result is that the mechanism of collapse (by which is meant the ratio of the several components of plastic deformation) is in some way linked with the shape of the collapse locus (which indicates whether or not collapse is taking place) and is independent of the route by which the collapse locus was reached. Thus the quite different paths AX and BX in Fig 3.21 both end up at the same point on the collapse locus and will both lead to the same mechanism of collapse—by the combination of sway and beam modes—even though one is reaching collapse by applying only increments of horizontal load and the other by applying increments of vertical load.

3.3.3 General elastic-perfectly plastic model (continued)

In order to be able to calculate the plastic deformations we make the assumption that there exists a plastic potential function $g(\boldsymbol{\sigma})$ which can be evaluated at the current stress state such that the plastic strain increment is given by

$$\delta\boldsymbol{\epsilon}^p = \mu\frac{\partial g}{\partial \boldsymbol{\sigma}} \tag{3.75}$$

where μ is a scalar multiplier whose magnitude is essentially arbitrary since this expression merely defines the *mechanism* of plastic deformation—the ratio of the several components of plastic deformation. It is thus only the gradient of the plastic potential function $g(\boldsymbol{\sigma})$ that is required, the actual value of the function is not relevant.

Combination of (3.64), (3.65) and (3.75) gives

$$\delta\boldsymbol{\sigma} = \boldsymbol{D}\delta\boldsymbol{\epsilon} - \mu\boldsymbol{D}\frac{\partial g}{\partial \boldsymbol{\sigma}} \tag{3.76}$$

and combination of (3.76) with (3.67) allows us to determine μ

$$\mu = \frac{\frac{\partial f}{\partial \boldsymbol{\sigma}}^T \boldsymbol{D}\delta\boldsymbol{\epsilon}}{\frac{\partial f}{\partial \boldsymbol{\sigma}}^T \boldsymbol{D}\frac{\partial g}{\partial \boldsymbol{\sigma}}} \tag{3.77}$$

and hence generate an expression for the elastic-plastic stiffness matrix $\boldsymbol{D}^{ep}$ giving $\delta\boldsymbol{\sigma}$ as a function of $\delta\boldsymbol{\epsilon}$:

$$\delta\boldsymbol{\sigma} = \left[\boldsymbol{D} - \frac{\boldsymbol{D}\frac{\partial g}{\partial \boldsymbol{\sigma}}\frac{\partial f}{\partial \boldsymbol{\sigma}}^T \boldsymbol{D}}{\frac{\partial f}{\partial \boldsymbol{\sigma}}^T \boldsymbol{D}\frac{\partial g}{\partial \boldsymbol{\sigma}}}\right]\delta\boldsymbol{\epsilon} = \boldsymbol{D}^{ep}\delta\boldsymbol{\epsilon} \tag{3.78}$$

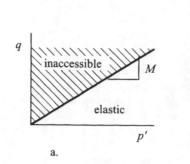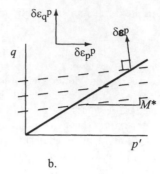

Figure 3.22: Elastic-perfectly plastic Mohr-Coulomb model (a) yield/failure locus; (b) plastic potentials

from which the stress increment can be calculated from any total strain increment that is causing yield.

Note that, as before(§2.5.1), we will expect that we can always deduce increments in stress from imposed increments in strain (the operation indicated in (3.78)) but that the reverse operation will not always be possible if our current state of stress is already on the yield/failure boundary of the elastic region (Fig 3.19). This will become clear when we consider a particular example.

3.3.4   Elastic-perfectly plastic Mohr-Coulomb model

To demonstrate how this final expression can be used we can look at the special case of the elastic-perfectly plastic Mohr-Coulomb soil model. We will apply the model to axisymmetric conditions. We introduce this model first because Mohr-Coulomb failure is something which is familiar to all undergraduate civil engineers and because elastic-perfectly plastic Mohr-Coulomb models are generally available in most finite element programs that might be used by practising civil engineers. There is a familiarity in some of the ingredients of the model.

First we define the elastic properties as usual using an isotropic elastic model:

$$\left(\begin{array}{c} \delta p' \\ \delta q \end{array} \right) = \left(\begin{array}{cc} K & 0 \\ 0 & 3G \end{array} \right) \left(\begin{array}{c} \delta\epsilon_p^e \\ \delta\epsilon_q^e \end{array} \right) \tag{3.79}$$

Next we define the yield function as (Fig 3.22a)

$$f(\boldsymbol{\sigma}) = f(p', q) = q - Mp' \tag{3.80}$$

If $f(p', q) < 0$ the soil is behaving elastically; if $f(p', q) = 0$ the soil is yielding (failing) and generating plastic deformations. To have $f(p', q) > 0$ is impossible: this defines an inaccessible region of the (p', q) stress plane (Fig 3.22a). The value of the soil property M can be related to the angle of shearing resistance ϕ of the soil in triaxial compression:

$$M = \frac{6\sin\phi}{3 - \sin\phi} \tag{3.81}$$

Finally we require some constraint on the plastic deformations in the form of a flow rule which defines the plastic deformation mechanism at the current stress state. We define a plastic potential function (Fig 3.22b)

$$g(\boldsymbol{\sigma}) = g(p', q) = q - M^*p' + k = 0 \tag{3.82}$$

where k is an arbitrary variable to allow the plastic potential function to be defined at the current state of stress and M^* is another soil property. This implies that the plastic strain increments are given by normality to the plastic potential function at the current state of stress (Fig 3.22b)

$$\begin{pmatrix} \delta\epsilon_p^p \\ \delta\epsilon_q^p \end{pmatrix} = \mu \begin{pmatrix} \partial g/\partial p' \\ \partial g/\partial q \end{pmatrix} = \mu \begin{pmatrix} -M^* \\ 1 \end{pmatrix} \tag{3.83}$$

where μ is a scalar multiplier which merely indicates the magnitude of the plastic strain increments. The ratio of the two components of plastic strain is:

$$\frac{\delta\epsilon_p^p}{\delta\epsilon_q^p} = -M^* \tag{3.84}$$

The link between M^* and angle of dilation is not so simple as the link between M and angle of shearing resistance ϕ (3.81) because, while the intermediate principal stress plays no role in the latter (Mohr-Coulomb failure is concerned only with the ratio of major and minor principal stresses), the intermediate principal strain certainly influences the former. Angle of dilation is essentially a plane strain concept (§2.6) and is thus directly relevant to many geotechnical applications—but the intermediate strain is then conveniently zero. In plane strain the angle of dilation ψ has a geometrical meaning as the tangent to Mohr's circle of strain increment (Fig 3.23a). Under conditions of triaxial compression ($\delta\epsilon_a > 0$) we can define a similar tangent angle ψ_c (Fig 3.23b). We find that the link with the triaxial strain increment ratio is:

$$\frac{\delta\epsilon_p^p}{\delta\epsilon_q^p} = \frac{3}{4}(3\sin\psi_c - 1) \tag{3.85}$$

and $\sin\psi_c < 1/3$ implies dilation. Then, while we can define an angle of dilation ψ as a material property for use in analysis from:

$$M^* = \frac{6\sin\psi}{3 - \sin\psi} \tag{3.86}$$

for triaxial compression to be exactly similar to (3.81), the direct geometrical interpretation has been lost. Angles ψ and ψ_c are linked through:

$$\sin\psi_c = \frac{1 - 3\sin\psi}{3 - \sin\psi} \tag{3.87}$$

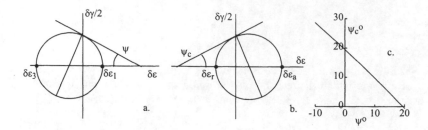

Figure 3.23: Mohr circles of strain increment and angles of dilation in (a) plane strain; (b) triaxial compression. (c) Link between ψ and ψ_c

This relationship is plotted in Fig 3.23c.

However, with this loose definition of ψ, for $M^* = 0$ plastic deformation occurs at constant volume (zero dilatancy, $\psi = 0$). Soils that contract when they are sheared plastically have negative angles of dilation: $\psi < 0$ and $M^* < 0$ (Fig 3.24c); soils that expand have positive angles of dilation: $\psi > 0$ and $M^* > 0$ (Fig 3.24c). It is generally found that for most real soils $\psi < \phi$ and $M^* < M$. A special (though physically unrealistic) case is obtained when $M^* = M$ and $\psi = \phi$.

The energy that is dissipated during an increment of plastic deformation is

$$\delta W^p = \sigma^T \delta \epsilon^p = p' \delta \epsilon_p^p + q \delta \epsilon_q^p \tag{3.88}$$

Since the soil is yielding the stresses are related by

$$q = Mp' \tag{3.89}$$

and the plastic strain increments are related by (3.84). The plastic energy thus becomes

$$\delta W^p = (M - M^*)p' \delta \epsilon_q^p \tag{3.90}$$

It is evident that if $M^* = M$ there is no plastic energy dissipation which seems likely to provide an unsatisfactory description of soil behaviour.

The complete elastic-plastic stiffness matrix (3.78) for this perfectly plastic model can now be generated:

$$D^{ep} = \left[\begin{pmatrix} K & 0 \\ 0 & 3G \end{pmatrix} - \frac{1}{KMM^* + 3G} \begin{pmatrix} MM^*K^2 & -3M^*GK \\ -3MGK & 9G^2 \end{pmatrix} \right] \tag{3.91}$$

The second term in (3.91) is only included if the soil is yielding. In stiffness form the link between stress increments and strain increments, when the soil is yielding, is

$$\begin{pmatrix} \delta p' \\ \delta q \end{pmatrix} = \frac{3GK}{KMM^* + 3G} \begin{pmatrix} 1 & M^* \\ M & MM^* \end{pmatrix} \begin{pmatrix} \delta \epsilon_p \\ \delta \epsilon_q \end{pmatrix} \tag{3.92}$$

The elastic-plastic stiffness matrix is in general asymmetric unless $M^* = M$ which, as has been shown, is physically unreasonable. However, certain

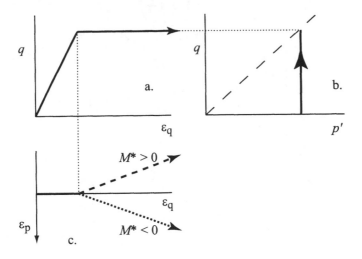

Figure 3.24: Elastic-perfectly plastic Mohr-Coulomb model: (a) stress:strain response; (b) constant $p\prime$ effective stress path; (c) volumetric strain, dependence on $M*$

numerical analysis programs require the stiffness matrix to be symmetric for solution purposes and it is for these programs that the assumption $M^* = M$ is often forced upon the user—or else some numerical subterfuge is needed to overcome the limitation of the program.

Although it is often easier to think of stress changes producing changes in strain—and physical considerations of the behaviour of soils often encourage us to move in this direction—if we look at the diagram of the (p', q) stress plane (Fig 3.22a) we can see that this will not provide a secure route for analysis because a large part of the stress plane is in fact forbidden territory. On the other hand, working from strain increments to stress increments carries no such problem: all strain increments are permitted even when the current stress state sits on the yield (failure) locus. Some of these strain increments will produce purely elastic changes in stress which take the stress state away from yield; others will force the stress state to move up or down along the yield (failure) locus in such a way that the elastic component of the strain caused by the change in stress uses up that part of the total strain increment which cannot be ascribed to the plastic strain mechanism given by (3.84) or (3.75).

We saw in section §2.5.3 that one way of illustrating the link between strain increments and stress increments is through the generation of stress response envelopes. For each of the strain increments of a rosette of increments of similar magnitude but different direction we can use the elastic-plastic stiffness form of the model ((3.79) or (3.91) depending on whether the soil is responding elastically or elastoplastically) to calculate the stress increment response (Fig 3.25). This stress response envelope consists of two parts.

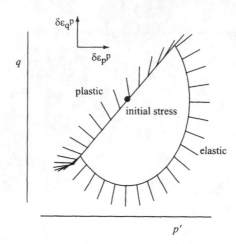

Figure 3.25: Elastic-perfectly plastic Mohr-Coulomb model: stress response envelope (calculated with $M = 1.2$, $M* = 0.2$ and initial stresses $p' = 100$, $q = Mp' = 120$)

If the strain increment can be supported by elastic unloading then the stress increment is directed away from the yield locus. For these increments the response envelope takes the form of half of an ellipse (Fig 3.26). If the strain increment requires the soil to yield then the stress state has to lie on the yield locus—all plastic stress states are, in this perfectly plastic model, confined to this one line. For these increments the stress response envelope consists of a straight line tangent to (in this case coincident with) the yield locus at the given initial stress (Fig 3.26).

Two limits may be noted. If the ratio of strain components is given by (3.84) then the stresses remain unchanged as the soil yields:

$$\begin{pmatrix} \delta\epsilon_p \\ \delta\epsilon_q \end{pmatrix} = \lambda \begin{pmatrix} -M^* \\ 1 \end{pmatrix} \quad \rightarrow \quad \begin{pmatrix} \delta p' \\ \delta q \end{pmatrix} = \begin{pmatrix} 0 \\ 0 \end{pmatrix} \qquad (3.93)$$

It is of course possible for the stress state to move along the yield locus purely elastically without incurring plastic deformation. In this case

$$\begin{pmatrix} \delta\epsilon_p \\ \delta\epsilon_q \end{pmatrix} = \lambda \begin{pmatrix} 1/K \\ M/3G \end{pmatrix} \qquad (3.94)$$

and this ratio defines the boundary of elastically attainable strain states in the corresponding strain increment plane. It should be clear from the stress response envelope in Fig 3.25 that, not only is there a part of the stress plane that is inaccessible (anywhere implying a value of $q/p' > M$), even for stress changes which lie along the boundary of the elastic region ($\delta q/\delta p' = M$) there is an infinite number of possible causative strain increments and we cannot even tell

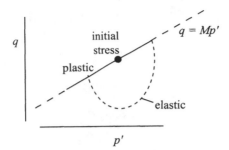

Figure 3.26: Elastic-perfectly plastic Mohr-Coulomb model: elastic and plastic sections of stress response envelope

whether the soil is behaving elastoplastically or purely elastically. The ambiguity in trying to work from stress increments to strain increments is emphasised.

A particular case is given by undrained constant volume shearing for which

$$
\begin{pmatrix} \delta\epsilon_p \\ \delta\epsilon_q \end{pmatrix} = \lambda \begin{pmatrix} 0 \\ 1 \end{pmatrix} \quad \rightarrow \quad \begin{pmatrix} \delta p' \\ \delta q \end{pmatrix} = \frac{3GKM^*\lambda}{KMM^* + 3G} \begin{pmatrix} 1 \\ M \end{pmatrix} \tag{3.95}
$$

and the stress path ascends or descends the line $q = Mp'$ for $M^* > 0$ (Fig 3.25) or $M^* < 0$ respectively (assuming that $3G + KMM^* > 0$ always).

The principal application of elastic-perfectly plastic models is to the calculation of collapse loads for geotechnical structures: such as the ultimate bearing capacity of a foundation, or the limiting stresses on a retaining structure. A discussion of the ultimate limit state analysis of such problems is given in section §7.3. In such analyses it is assumed that the soil has been sheared so much that all elements that combine to form a failure mechanism around the structure have reached the perfectly plastic failure condition. In this state the pre-failure response is of no further concern and there is a sound theoretical basis for judging the reliability of the collapse loads that are thus calculated.

The response of a geotechnical structure—for example, a footing—to increasing load or increasing deformation, calculated using an elastic-perfectly plastic soil model (§4.10.4), shows a steady transition from the initial linear elastic response to the ultimate zero stiffness perfectly plastic collapse condition (Fig 3.27). This may appear to be an entirely reasonable description of the behaviour of the system and arises because the failure of the soil propagates steadily from the stress concentrations at the edge of the footing until an overall failure mechanism can form. Once failure has started anywhere in the soil the stresses that are generated by the continuing loading must be redistributed through the remaining unfailed soil: the stiffness for continuing loading thus falls.

For each individual element of soil, however, the elastic-perfectly plastic description looks less convincing (Fig 3.28). The model can only at best describe the final failure condition together with either the initial stiffness or some average

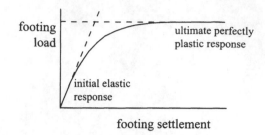

footing settlement

Figure 3.27: Numerical analysis of settlement of strip footing on elastic-perfectly plastic Mohr-Coulomb soil

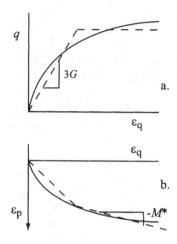

Figure 3.28: Elastic-perfectly plastic Mohr-Coulomb model compared with typical soil response in conventional drained triaxial compression test

stiffness corresponding to a stress state intermediate between the beginning and end of the test. Evidently such an average stiffness will not give an accurate description of the behaviour at any soil element—no soil element will actually experience combinations of stress and strain which fall along this assumed notional linear elastic prefailure relationship. The volumetric response is also only crudely represented (Fig 3.28b).

This deficiency of the elastic-perfectly plastic model is slightly obscured when soil behaviour is described in terms of secant stiffness. It is standard practice to show variation of shear stiffness with shear strain both for monotonic testing (Fig 3.29a)—where the stiffness falls as failure is approached—and for cyclic, or more generally non-monotonic testing—where the average stiffness in any cycle reflects the strain level at which the direction of loading was reversed (Fig 3.29a). Typical experimental data show a reduction in average cyclic (secant) stiffness with increasing strain amplitude. An elastic-perfectly plastic model

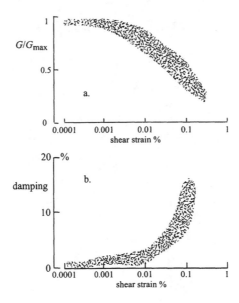

Figure 3.29: Range of (a) secant shear stiffness degradation data (normalised with shear stiffness at very small strain G_{max} and (b) damping ratio for Quiou sand from resonant column and torsional shear tests (after LoPresti *et al.*, 1997)

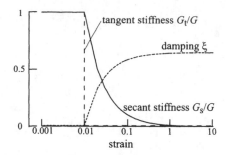

Figure 3.30: Elastic-perfectly plastic Mohr-Coulomb model: stiffness and damping variation with strain (yield strain =0.01)

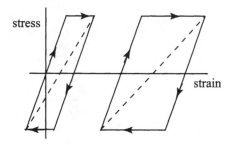

Figure 3.31: Variation of secant stiffness with strain amplitude for cycles of loading of elastic-perfectly plastic material

will also show a reduction in average stiffness with increasing strain amplitude (Fig 3.30) but when the response is considered in terms of tangent stiffness it is very clear that the secant response is the combination of a constant prefailure stiffness with varying amounts of strain imposed while the soil is at failure and hence has zero tangent stiffness (Fig 3.31).

If the shear strain to failure of an elastic-perfectly plastic soil with shear modulus G is γ_f then at a strain of γ_m the shear stress is $G\gamma_f$ and the secant stiffness G_s is

$$G_s = G\frac{\gamma_f}{\gamma_m} \tag{3.96}$$

This is plotted in Fig 3.30.

The damping ratio ξ can be calculated similarly. This is defined as

$$\xi = \frac{W}{4\pi\Delta W} \tag{3.97}$$

where W is the energy dissipated in each cycle (shown stippled in Fig 3.32), and ΔW is the maximum elastic energy stored in each cycle (shown shaded in

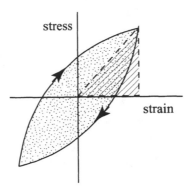

Figure 3.32: Definition of damping ratio

Fig 3.32). For the elastic-perfectly plastic material

$$W = 4G\gamma_f^2 \left(\frac{\gamma_m}{\gamma_f} - 1 \right)$$

(3.98)

and

$$\Delta W = \frac{1}{2} G \gamma_f^2 \frac{\gamma_m}{\gamma_f}$$

(3.99)

so that

$$\xi = \frac{2}{\pi} \left(1 - \frac{\gamma_f}{\gamma_m} \right)$$

(3.100)

and this is also plotted in Fig 3.30.

Elastic-perfectly plastic models are widely used because of their simplicity. They are available in every computer program that is seriously intended for numerical analysis of geotechnical problems. They require definition of elastic properties—of which there will be two for an isotropic model as in any linear elastic model; and of some failure property—for example, a limiting angle of shearing resistance for a frictional model to be used for description of drained soil conditions or a limiting shear stress for a cohesive model to be used for description of undrained soil conditions; together with some statement about the volume changes that accompany failure—for example, an angle of dilation. There is obviously need for care in the selection of the elastic properties. Not all programs give the freedom to select angles of dilation which are different from the angle of shearing resistance.

3.4 Elastic-hardening plastic models

Constitutive models form an essential link in the numerical or theoretical prediction of deformation of geotechnical structures. Perfect plasticity provides some possibilities for matching certain aspects of observed mechanical response

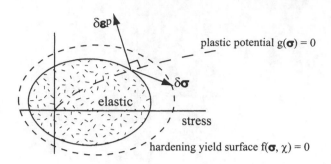

Figure 3.33: Elastic-hardening plastic model: yield surface separating elastic from plastic regions of stress space, and plastic potential for definition of plastic strain increments

of soils but in a rather limited way. Hardening plasticity opens up further modelling possibilities. Perfect plasticity enables us to reproduce the inelasticity of soil behaviour—the accumulation of irrecoverable strains. Hardening plasticity enables us in addition to describe prefailure nonlinearity.

Hardening models are natural extensions of the perfectly plastic models that have been described in section §3.3. The additional feature is that the yield function is no longer merely a function of the stresses but also introduces a hardening parameter which characterises the current size of the yield surface. An extra hardening equation is then required to define the way in which this hardening parameter changes as plastic strains occur—or in other words the penalty in permanent deformation of the material which is necessary in order to increase the size of the elastic region and *harden* the material. The general form of the ingredients of the hardening plastic model will be introduced first and then specific hardening plastic models will be developed.

There are four ingredients of the hardening plastic models—three of these are common to the perfectly plastic models.

1. *Elastic properties*: Whenever the stresses change elastic strains will occur. We may assume isotropic elastic behaviour for convenience but this is not essential.

$$\delta\boldsymbol{\sigma} = \boldsymbol{D}\delta\boldsymbol{\epsilon}^e \qquad (3.101)$$

2. *Yield criterion*: We need to define the current boundary in stress space to the region of elastic behaviour (Fig 3.33). Within this region all stress changes can be applied without incurring irrecoverable deformations. The definition of the yield function allows us to answer the question: are yield and plastic deformation occurring? For a hardening model the boundary is not fixed but will depend on the history of loading of the soil.

We write the yield criterion as a function of a hardening parameter χ[4]:

$$f(\boldsymbol{\sigma}, \chi) = 0 \qquad (3.102)$$

---

[4]There could in general be more than one hardening parameter.

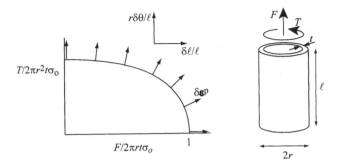

Figure 3.34: Combined tension and torsion of annealed copper tubes: yield locus and plastic strain increment vectors (σ_o is yield stress) (inspired by original data from Taylor and Quinney, 1931)

There is the usual constraint that the current stress state cannot lie outside the current yield surface but the yield surface is no longer of fixed size (as it was in the perfectly plastic model) but is able to expand in order to accommodate the imposed stress changes. The consistency condition (3.67), which states that the stress state must remain on the yield surface when plastic strains are being generated, now becomes:

$$f(\boldsymbol{\sigma}, \chi) = 0; \quad \delta f = \frac{\partial f}{\partial \boldsymbol{\sigma}}^T \delta \boldsymbol{\sigma} + \frac{\partial f}{\partial \chi} \delta \chi = 0 \qquad (3.103)$$

3. *Flow rule*: We require some way of describing the mechanism of plastic deformation. We can conveniently do this in just the same way as for the perfectly plastic model using a plastic potential to indicate the ratio of the several strain components (Fig 3.33) and to show that the plastic strains are controlled by the current stresses at yield and not by the stress increment which brought the soil to yield:

$$\delta \boldsymbol{\epsilon}^p = \mu \frac{\partial g}{\partial \boldsymbol{\sigma}} \qquad (3.104)$$

where μ is again a scalar multiplier which we have to find. It may sometimes be convenient to assume that the functions f and g are the same: the material then obeys the hypothesis of associated flow (the flow is associated with the yield criterion) or normality (the strain increment vectors are normal to the yield surface at the current stress state) but this is certainly not a necessary assumption and certainly not an assumption of which soils are aware (although the analysis of the collapse of the steel portal frame demonstrated normality and normality can also be observed in the combined tension and torsion of thin-walled annealed metal tubes: Fig 3.34).

4. *Hardening rule*: The hardening rule links the change in size of the yield surface with the magnitude of the plastic strain and hence provides a link between χ and μ.

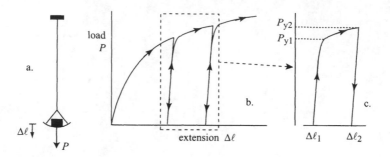

Figure 3.35: Repeated tension of annealed copper wire (inspired by original data from Taylor and Quinney, 1931)

For a perfectly plastic material, once the stress state reaches the yield surface, plastic straining can continue indefinitely: the incremental (tangent) plastic stiffness is zero. The uniaxial tension of annealed copper wire provides a simple example of a hardening plastic material (Fig 3.35). Each time the axial load is increased beyond the previous maximum load there is some further irrecoverable extension of the wire but linked with this is an increase in the yield strength of the material: the elastic region has been increased at the expense of some further irrecoverable rearrangement of the metal crystals. The elastic stiffness of the copper can be deduced from the slope of the elastic loading and unloading relationship, for changes of load and length δP and $\delta \ell$:

$$E = \frac{\delta P / A}{\delta \ell / \ell} \tag{3.105}$$

where A and ℓ are cross-sectional area and length of the wire respectively.

The plastic hardening stiffness can be defined in just the same way:

$$E^p = \frac{(P_{y2} - P_{y1})/A}{(\Delta \ell_2 - \Delta \ell_1)/\ell} \tag{3.106}$$

where $\Delta \ell_1$ and $\Delta \ell_2$ are the irrecoverable changes in length of the wire that are left when the successive maximum (yield) loads P_{y1} and P_{y2} are removed (Fig 3.35c). It is evident from Fig 3.35b that the plastic stiffness of the copper wire is not a constant but falls steadily with increasing plastic extension of the wire (Fig 3.36).

For our more general hardening plasticity model we must suppose that the hardening parameter is some general function $\chi(\epsilon^p)$ of the plastic strains. The combination of the consistency condition (3.103) and the flow rule (3.104) then gives:

$$\frac{\partial f}{\partial \boldsymbol{\sigma}}^T \delta \boldsymbol{\sigma} + \mu \frac{\partial f}{\partial \chi} \frac{\partial \chi}{\partial \epsilon^p}^T \frac{\partial g}{\partial \boldsymbol{\sigma}} = 0 \tag{3.107}$$

and if we write

$$H = -\frac{\partial f}{\partial \chi} \frac{\partial \chi}{\partial \epsilon^p}^T \frac{\partial g}{\partial \boldsymbol{\sigma}} \tag{3.108}$$

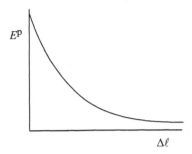

Figure 3.36: Tension of annealed copper wire: schematic variation of plastic hardening stiffness with extension

a procedure exactly similar to that used for the perfectly plastic model can be used to generate the stiffness relationship between stress increments and total strain increments:

$$\delta\boldsymbol{\sigma} = \left[\boldsymbol{D} - \frac{\boldsymbol{D}\frac{\partial g}{\partial\boldsymbol{\sigma}}\frac{\partial f}{\partial\boldsymbol{\sigma}}^{T}\boldsymbol{D}}{\frac{\partial f}{\partial\boldsymbol{\sigma}}^{T}\boldsymbol{D}\frac{\partial g}{\partial\boldsymbol{\sigma}} + H} \right] \delta\boldsymbol{\epsilon} = \boldsymbol{D}^{ep}\delta\boldsymbol{\epsilon} \qquad (3.109)$$

3.4.1 Extended Mohr-Coulomb model

The elastic-perfectly plastic Mohr-Coulomb model (§3.3.4) is widely used for geotechnical analysis. It provides a very crude match to actual shearing behaviour of soils (Fig 3.28). A natural extension is to create a hardening version of the Mohr-Coulomb model in which the size of the yield surface varies in some nonlinear way with the development of plastic strain. In the model to be described here the hardening will be linked only with distortional strain: such a distortional hardening model is found to be quite useful for the modelling of sands where it is rearrangement of the rather hard particles that dominates the response at typical engineering stress levels and irrecoverable volumetric changes are essentially linked with this rearrangement. We will present the four ingredients of the model in turn and restrict ourselves in this presentation to the stress and strain conditions that can be attained in the conventional triaxial apparatus.

1. *Elastic properties*: The elastic properties are assumed to be described by a linear isotropic elastic model which requires two stiffness properties such as shear modulus G and bulk modulus K

$$\left(\begin{array}{c} \delta p' \\ \delta q \end{array} \right) = \left(\begin{array}{cc} K & 0 \\ 0 & 3G \end{array} \right) \left(\begin{array}{c} \delta\epsilon_p^e \\ \delta\epsilon_q^e \end{array} \right) \qquad (3.110)$$

In fact for many granular materials it might be reasonable to assume that the shear stiffness is not in fact constant but varies in some way with stress level—for example:

$$G \propto p'^{\frac{1}{2}} \qquad (3.111)$$

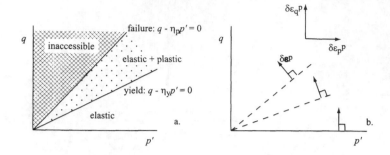

Figure 3.37: Elastic-hardening plastic Mohr-Coulomb model: (a) yield locus and failure locus separating elastic, plastic and inaccessible regions of stress plane; (b) normality of plastic strain increment vectors to yield loci

However, we have seen that cavalier introduction of nonlinear elasticity risks thermodynamic unacceptability and that if we want the shear stiffness to vary with stress level then we should really invoke an elastic strain energy function to achieve this.

2. *Yield criterion*: The yield criterion is now taken to be a generalisation of the yield criterion assumed for the perfectly plastic model

$$f(\boldsymbol{\sigma}, \chi) = f(p', q, \eta_y) = q - \eta_y p' \qquad (3.112)$$

where η_y is a hardening parameter which indicates the current size of the yield locus (Fig 3.37a). It will be seen when the hardening rule is introduced that the yield locus is allowed progressively to expand until it reaches some limiting failure size.

3. *Flow rule*: As for the perfectly plastic model it is not particularly satisfactory to assume normality of plastic strain increment vectors to the current yield locus. Normality would imply

$$\frac{\delta\epsilon_p^p}{\delta\epsilon_q^p} = -\eta_y \qquad (3.113)$$

with the directions of plastic strain increment vectors shown in Fig 3.37b. These imply that volumetric expansion accompanies shearing at all non-zero stress ratios and that the rate of volumetric expansion—possibly characterised by an angle of dilation—increases steadily as the yield stress ratio increases. A more suitable description of the plastic volume changes can be developed from the interpretation of the results of conventional direct shear tests on sand that was presented in section §2.6.

Following Taylor's (1948) proposal of a link between dilatancy and mobilised friction in a shear box test, we emerged with a stress-dilatancy equation (2.99) expressed in terms of *total* strain increments. Since we are now engaged in the development of elastic-plastic models we need to think of this rather as a flow rule which controls the ratio of *plastic* strain increments:

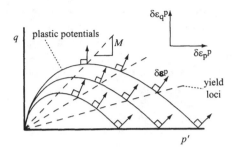

Figure 3.38: Elastic-hardening plastic Mohr-Coulomb model: plastic potential curves (solid lines) and yield loci (dashed lines)

$$\frac{\delta\epsilon_p^p}{\delta\epsilon_q^p} = M - \frac{q}{p'} = M - \eta_y \qquad (3.114)$$

where M is the critical state stress ratio at which constant volume shearing can occur. Evidently this flow rule is only invoked when the soil is yielding so that the stress ratio q/p' is then of necessity equal to η_y[5].

It will be recalled that for the perfectly plastic model we introduced the concept of a plastic potential function $g(\boldsymbol{\sigma})$ passing through the current stress state to which the plastic strain increments are normal. It can be deduced that the flow rule (3.114) corresponds to the plastic potential function

$$g(\boldsymbol{\sigma}) = q - Mp' \ln \frac{p'_r}{p'} = 0 \qquad (3.115)$$

where p'_r is an arbitrary variable introduced in order to allow us to create a member of this general class of plastic potential curves passing through the current stress state. Then

$$\begin{pmatrix} \delta\epsilon_p^p \\ \delta\epsilon_q^p \end{pmatrix} = \mu \begin{pmatrix} \frac{\partial g}{\partial p'} \\ \frac{\partial g}{\partial q} \end{pmatrix} = \mu \begin{pmatrix} M - \eta \\ 1 \end{pmatrix} \qquad (3.116)$$

which is consistent with (3.114).

These plastic potential curves are plotted in Fig 3.38 together with a set of yield loci. The directions of the plastic strain increment vectors are also shown: the difference from the directions implied from normality (Fig 3.37b) is dramatic. Now yielding at low stress ratio implies volumetric compression but the rate of

---

[5]At low stress ratios, far from failure, we can expect elastic strains to be important and hence a stress-dilatancy interpretation using *total* strain increments (§2.6) will be different from that calculated using plastic strain increments (3.114). So the discrepancy in Fig 2.57 could be anticipated if the soil does indeed know about the flow rule of (3.114). As failure is approached, plastic strains will dominate and the neglect of the difference between elastic and total strain increments becomes less important.

volumetric compression steadily decreases as the stress ratio increases. For stress ratio $q/p' = M$ plastic deformation occurs at constant volume; for stress ratio $q/p' > M$ plastic deformation is accompanied by volumetric expansion.

4. *Hardening rule*: We will assume that the soil is a distortional hardening material so that the current size of the yield locus η_y depends only on the plastic distortional strain ϵ_q^p. We are trying to describe using our model a mechanical behaviour in which the stiffness falls steadily as the soil is sheared towards failure. One of the simplest ways in which such a stiffness degradation can be described is using a hyperbolic relationship between stress ratio and distortional strain

$$\frac{\eta_y}{\eta_p} = \frac{\epsilon_q^p}{a + \epsilon_q^p} \tag{3.117}$$

or incrementally

$$\delta\eta_y = \frac{(\eta_p - \eta_y)^2}{a\eta_p}\delta\epsilon_q^p \tag{3.118}$$

or

$$\begin{pmatrix} \partial\eta_y/\partial\epsilon_p^p \\ \partial\eta_y/\partial\epsilon_q^p \end{pmatrix} = \begin{pmatrix} 0 \\ (\eta_p - \eta_y)^2/a\eta_p \end{pmatrix} \tag{3.119}$$

where η_p is a limiting value of stress ratio and a is a soil constant—which essentially just scales the plastic strain since (3.117) and (3.118) are actually functions of ϵ_q^p/a.

We now have all the information that we need to produce the complete elastic-plastic stiffness relationship (3.109):

$$\begin{pmatrix} \delta p' \\ \delta q \end{pmatrix} =$$

$$\left[\begin{pmatrix} K & 0 \\ 0 & 3G \end{pmatrix} - \frac{\begin{pmatrix} -K^2\eta_y(M - \eta_y) & 3GK(M - \eta_y) \\ -3GK\eta_y & 9G^2 \end{pmatrix}}{3G - K\eta_y(M - \eta_y) + p'(\eta_p - \eta_y)^2/(a\eta_p)} \right] \begin{pmatrix} \delta\epsilon_p \\ \delta\epsilon_q \end{pmatrix}$$

$$\tag{3.120}$$

Writing the stiffness relation in this way (compare also (3.78), (3.91), (3.109)) is convenient because it divides the stiffness into an elastic part and a plastic part. In application of the model, the elastic stiffness can be used to predict the stress change resulting from a given strain change. If this computed stress change lies outwith the current yield surface then the plastic stiffness can be applied as a corrector to bring the calculated stress state back onto the (possibly hardened) yield surface.

It can be checked that, when the yield stress ratio reaches the asymptotic value $\eta_y = \eta_p$, this stiffness relationship becomes identical with that generated for the perfectly plastic Mohr-Coulomb model(3.91) if we write $\eta_y = \eta_p = M$ and $(M - \eta_y) = (M - \eta_p) = -M^*$. As then, the stiffness matrix is not symmetric because we have assumed a nonassociated flow rule: the plastic potential function (3.115) is quite different from the yield function (3.112).

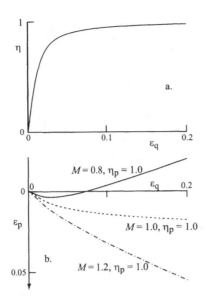

Figure 3.39: Elastic-hardening plastic Mohr-Coulomb model: triaxial compression tests with constant mean effective stress: (a) stress:strain response and (b) volumetric response for different values of M and η_p ($K = 5000$ kPa, $\nu = 0.25$, $a = 0.005$, $p' = 100$ kPa)

The information about yielding and hardening for this model can also be used to generate the plastic compliance relationship linking plastic strain increments with stress increments.

$$\begin{pmatrix} \delta\epsilon_p^p \\ \delta\epsilon_q^p \end{pmatrix} = \frac{a\eta_p}{p'(\eta_p - \eta_y)^2} \begin{pmatrix} -(M - \eta_y)\eta_y & (M - \eta_y) \\ -\eta_y & 1 \end{pmatrix} \begin{pmatrix} \delta p' \\ \delta q \end{pmatrix} \quad (3.121)$$

As before, however, this relationship is not always useful because, as the stress state nears the asymptotic stress ratio $q/p' = \eta_p$, there is once again a region of the (p', q) effective stress plane into which it is impossible for the stress increments to stray. However, as expected, we can see that as the yielding stress ratio tends to η_p so the plastic stiffness tends to zero and the compliance tends to infinity.

The volumetric response depends on the relative values of η_p and M. If $\eta_p > M$ then the model predicts compression followed by expansion (Fig 3.39b). If $\eta_p = M$ then the model predicts compression reducing until a critical state of constant volume shearing is reached (Fig 3.39b). (If $\eta_p < M$ the model predicts continuing volumetric compression as failure is approached: Fig 3.39b.) In this simple form we cannot describe strain softening with this model. In practice, pre-peak response may be adequate since working loads are not intended to produce significant amounts of failure and we are interested in pre-failure response of our geotechnical structures. If much of the soil around a structure has been

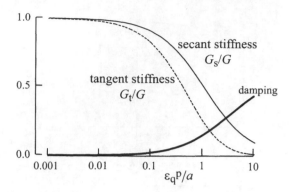

Figure 3.40: Elastic-hardening plastic Mohr-Coulomb model: variation of secant and tangent shear stiffness and damping ratio with strain

brought to failure then the overall deformations of the structure are likely to be unacceptably large.

The variation of stiffness with strain in monotonic straining and the variation of damping ratio in cyclic shearing are shown in Fig 3.40. Obviously the tangent stiffness now falls more gradually than for the elastic-perfectly plastic model (Fig 3.30).

The elastic-plastic stiffness relationship (3.120) can be used to generate the envelope of stress responses to a rosette of applied total strain increments. These are shown in Fig 3.41 for two different values of stress ratio. As the stress state approaches the peak stress ratio so the stress response envelope (which is composed of two separate elliptical sections for the elastic and elastic-plastic strain increments) becomes more and more distorted. It is evident that the stress response envelope for the elastic-perfectly plastic model (Fig 3.25, 3.26) is a degenerate version of the response envelope for the hardening model: the elastic-plastic ellipse has collapsed to a line segment.

Extended Mohr-Coulomb model: undrained effective stress path

An example of the application of this extended Mohr-Coulomb model is provided by the calculation of the effective stress path that will be followed in an undrained test. An undrained test provides a direct deformation constraint:

$$\delta\epsilon_p = \delta\epsilon_p^e + \delta\epsilon_p^p = 0 \tag{3.122}$$

The sum of the elastic and plastic volumetric strain increments is zero: any tendency of the particle structure to undergo permanent rearrangement and change in volume—for example, collapse—has to be countered by a change in mean effective stress which leads to a balancing elastic volumetric expansion. (A tendency of the volume to undergo irrecoverable expansion will correspondingly be accompanied by an elastic compression.) The shape of the effective stress path can be most easily found by requiring the elastic and plastic volumetric

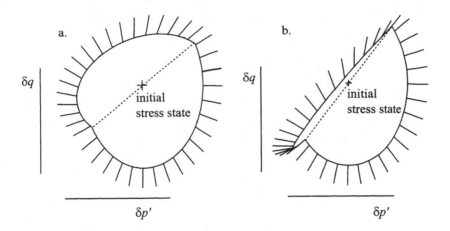

Figure 3.41: Elastic-hardening plastic Mohr-Coulomb model: stress response envelopes for (a) $\eta_y = \eta = 0.8$ and (b) $\eta_y = \eta = 1.3$ ($K = 1500$ kPa, $\nu = 0.3$, $a = 0.01$, $M = 1.2$, $\eta_p = 1.5$, $p_i' = 100$ kPa)

strain increments to be equal and opposite at all stages which from (3.110) and (3.121) implies

$$\frac{\delta p'}{K} + \frac{a\eta_p}{p'(\eta_p - \eta)^2}[-(M - \eta)\eta\delta p' + (M - \eta)\delta q] = 0 \qquad (3.123)$$

(writing η for η_y since we are assuming that the soil is yielding throughout). Noting that from the definition of stress ratio η

$$\delta q = p'\delta\eta + \eta\delta p' \qquad (3.124)$$

equation (3.123) can be rewritten

$$\frac{\delta p'}{K} + \frac{a\eta_p(M - \eta)p'\delta\eta}{p'(\eta_p - \eta)^2} = 0 \qquad (3.125)$$

and integrated to give

$$\frac{p_i' - p'}{K} = a\eta_p\left[\frac{(M - \eta_p)(\eta - \eta_i)}{(\eta_p - \eta)(\eta_p - \eta_i)} - \ln\left(\frac{\eta_p - \eta}{\eta_p - \eta_i}\right)\right] \qquad (3.126)$$

where p_i' and η_i are the initial values of mean effective stress and stress ratio. For a soil which is initially isotropically compressed with $\eta_i = 0$ this can be written

$$\frac{p' - p_i'}{aK\eta_p} = -\frac{(M/\eta_p) - 1}{(\eta_p/\eta) - 1} + \ln\left(1 - \frac{\eta}{\eta_p}\right) \qquad (3.127)$$

and this is plotted in Fig 3.42 for different values of M and η_p. As $\eta \to \eta_p$ the change in mean effective stress tends to infinity but the sign of the change

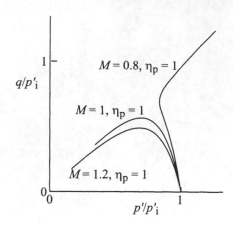

Figure 3.42: Elastic-hardening plastic Mohr-Coulomb model: undrained effective stress paths ($K = 2000$ kPa, $a = 0.02$, $p'_i = 200$ kPa)

in mean effective stress (and hence broadly the sign of the pore pressure that develops) depends on the sign of the difference between M and η_p. If $M \geq \eta_p$ then the mean stress falls steadily (and pore pressure is expected to build up). If $M < \eta_p$ then the mean stress first decreases (pore pressure build up) and then increases (pore pressure decrease). The model suggests that this increase in mean stress continues indefinitely but on the one hand the model is defective in suggesting that shearing accompanied by dilation can continue to large strains and on the other hand an undrained test would reach a physical conclusion when the pore pressure reaches a negative value of about -100 kPa and cavitation of the pore water occurs. (There is similarly a physical limit for low values of η_p in that an effective stress of zero is reached as the pore pressure continues to increase.) These limitations simply indicate some of the deficiencies of this simple model which would need to be rectified if it were to be used for analyses in which accurate representation of such response were reckoned to be essential.

Extended Mohr-Coulomb model: worked example

A sample of Mohr-Coulomb soil with hyperbolic hardening rule and with properties:

$$\phi_p = 40°, \quad a = 0.01, \quad G = 3 \text{ MPa}$$

has not been presheared. As a result the initial yield stress ratio $\eta_y = 0$. It is subjected to an initial isotropic stress state $q_i = 0$, $p'_i = 100$ kPa and then tested in triaxial compression with the mean effective stress maintained constant: $\delta p' = 0$.

   1. Calculate the shear strain required to bring the soil to a stress ratio 50% of the peak value.

   2. Calculate the secant shear stiffness at this stress ratio ($q/3\epsilon_q$).

3. Calculate the tangent shear stiffness at this stress ratio ($\delta q/3\delta\epsilon_q$).

4. Compare these stiffnesses with the initial tangent stiffness of the soil and with the unloading elastic stiffness G.

The given angle of friction must be converted to an equivalent peak stress ratio:

$$\eta_p = \frac{6\sin\phi_p}{3 - \sin\phi_p} = 1.636$$

1. From the hyperbolic hardening relationship, for $\eta_y = \eta_p/2$, the plastic shear strain is given by:

$$\frac{1}{2} = \frac{\epsilon_q^p}{a + \epsilon_q^p} \quad \rightarrow \quad \epsilon_q^p = a = 1\%$$

With $p' = p'_i$, $q = \eta_p p'_i/2$ and elastic strain is:

$$\epsilon_q^e = \frac{q}{3G} = \frac{\eta_p p'_i}{6G} = 0.91\%$$

2. The secant stiffness is:

$$G_{s50} = \frac{q}{3\epsilon_q} = \frac{\eta_p p'_i}{6\left(a + \frac{\eta_p p'_i}{6G}\right)} = 1428 \text{ kPa}$$

3. The plastic tangent stiffness can be deduced from the differentiation of the hyperbolic hardening rule with $\eta_y = \eta_p/2$:

$$\frac{\delta\eta}{\delta\epsilon_q^p} = \frac{(\eta_p - \eta_y)^2}{a\eta_p} = \frac{\eta_p}{4a} \quad \rightarrow \quad \frac{\delta q}{3\delta\epsilon_q^p} = \frac{p'_i\eta_p}{12a}$$

The elastic tangent stiffness is G. The total combined stiffness is therefore:

$$G_{t50} = \frac{\delta q}{3\left(\delta\epsilon_q^p + \delta\epsilon_q^e\right)} = \frac{\delta q}{3\left(\frac{4a\delta q}{p'_i\eta_p} + \frac{\delta q}{3G}\right)} = \frac{\eta_p p'_i}{\left(12a + \frac{\eta_p p'_i}{G}\right)} = 937 \text{ kPa}$$

4. The initial tangent stiffness can be calculated in exactly the same way. The plastic stiffness is:

$$\frac{\delta\eta}{\delta\epsilon_q^p} = \frac{(\eta_p - \eta_y)^2}{a\eta_p} = \frac{\eta_p}{a} \quad \rightarrow \quad \frac{\delta q}{3\delta\epsilon_q^p} = \frac{p'_i\eta_p}{3a}$$

and the combined stiffness:

$$G_{ti} = \frac{\delta q}{3\left(\delta\epsilon_q^p + \delta\epsilon_q^e\right)} = \frac{\delta q}{3\left(\frac{a\delta q}{p'_i\eta_p} + \frac{\delta q}{3G}\right)} = \frac{\eta_p p'_i}{\left(3a + \frac{\eta_p p'_i}{G}\right)} = 1935 \text{ kPa}$$

The collected values are:

initial tangent (and secant) stiffness	G_{ti}	1935 kPa
secant stiffness at 50% peak stress ratio	G_{s50}	1428 kPa
tangent stiffness at 50% peak stress ratio	G_{t50}	937 kPa
elastic unloading stiffness	G	3000 kPa

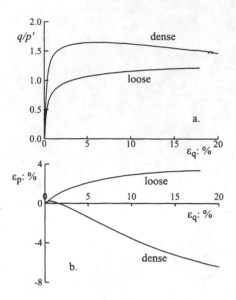

Figure 3.43: Conventional drained triaxial compression tests on Hostun sand: loose (initial void ratio = 1.0), dense (initial void ratio = 0.69), confining stress $\sigma'_r = 200$ kPa (data from Benahmed, 2001)

Note that, even at the start of the test, the occurrence of plastic strains leads to a significant decrease in stiffness from the elastic value. This elastic value is only encountered for stress changes which head back below the current yield locus on unloading.

Note also that, in the hyperbolic hardening rule, η_p is the peak stress ratio (which can only be attained at infinite strain) and a is the shear strain required to reach a stress ratio 50% of the peak. This hardening rule is evidently most useful where the soil is not expected to be loaded close to failure since real sands are expected to soften post peak towards a critical state.

Mohr-Coulomb model with post-peak softening

The Mohr-Coulomb model that has been described is able to reproduce some nonlinearity of initial response but is unable to describe the strain softening that is a familiar feature of the behaviour of dense sands (Fig 3.43). An adaptation of this model can be devised introducing a trilinear 'hardening' rule linking η_y and ϵ^p_q.

For low strains, up to the peak strength η_p, the soil is assumed to behave elastically and the size of the frictional yield locus is kept fixed:

$$\eta < \eta_p \Rightarrow \eta_y = \eta_p; \quad \delta\epsilon^p_q = 0 \tag{3.128}$$

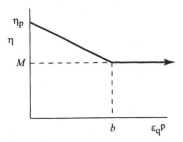

Figure 3.44: Elastic-plastic Mohr-Coulomb model with post-peak plastic softening

Once the peak strength is reached the soil starts to soften and the size of the yield locus reduces (Fig 3.44). A linear relation between the yield stress ratio and the distortional shear strain is assumed:

$$\frac{\eta_p - \eta_y}{\eta_p - M} = \frac{\epsilon_q^p}{b} \quad \text{for} \quad 0 < \epsilon_q^p < b \tag{3.129}$$

where b is a soil constant which describes the distortional strain required to bring the soil to the critical state stress ratio M. For plastic distortional strains beyond this value the soil behaves as a perfectly plastic material, and plastic distortional strains continue to occur without change in volume:

$$\eta_y = M \quad \text{for} \quad \epsilon_q^p > b \tag{3.130}$$

It is assumed that the flow rule is identical to that introduced in the previous section ((3.114) or (3.116)) so that, as the soil is sheared from the peak strength to a critical state, volumetric expansion occurs at a rate which decreases as the critical state is approached. Typical response described using this version of the Mohr-Coulomb model is shown in Fig 3.45.

Since all that has changed by comparison with the previous hyperbolic hardening model is the 'hardening' rule, the complete elastic-plastic stiffness relationship changes only through the form of the term H in (3.108).

$$H = -\frac{\partial f}{\partial \eta_y} \frac{d\eta_y}{d\epsilon_q^p} \frac{\partial g}{\partial q} = -(-p)\left(-\frac{\eta_p - M}{b}\right)(1) \tag{3.131}$$

$$\begin{pmatrix} \delta p' \\ \delta q \end{pmatrix} =$$

$$\left[\begin{pmatrix} K & 0 \\ 0 & 3G \end{pmatrix} - \frac{\begin{pmatrix} -K^2\eta_y(M - \eta_y) & 3GK(M - \eta_y) \\ -3GK\eta_y & 9G^2 \end{pmatrix}}{3G - K\eta_y(M - \eta_y) - p(\eta_p - M)/b} \right] \begin{pmatrix} \delta\epsilon_p \\ \delta\epsilon_q \end{pmatrix} \tag{3.132}$$

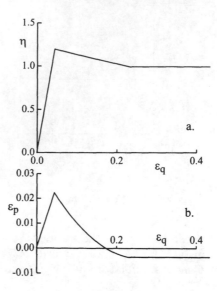

Figure 3.45: Elastic-hardening plastic Mohr-Coulomb model with post-peak softening: (a) stress:strain and (b) volumetric strain response for conventional drained triaxial compression test ($\delta\sigma_r = 0$, $p'_i = 100$ kPa) ($K = 3000$ kPa, $G = 1500$ kPa, $\eta_p = 1.2$, $M = 1$, $b = 0.2$)

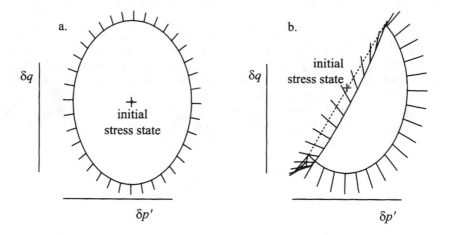

Figure 3.46: Elastic-hardening plastic Mohr-Coulomb model with post-peak softening: stress response envelopes for (a) pre-peak elastic response and (b) post-peak softening response

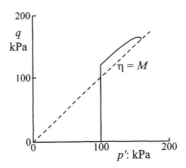

Figure 3.47: Elastic-hardening plastic Mohr-Coulomb model with post-peak softening: undrained effective stress path ($K = 3000$ kPa, $\eta_p = 1.2$, $M = 1$, $b = 0.2$, $p'_i = 100$ kPa)

The stress response envelopes for a rosette of strain increments applied before and after the peak strength are shown in Fig 3.46. Before the peak stress ratio is reached the response is purely elastic for all strain increments and the stress response envelope is a single complete ellipse (Fig 3.46a). In the softening regime (Fig 3.46b) it is evident that the same stress increment can be generated by either purely elastic unloading or by elastic plus plastic softening response and it is not just that there is a region of the stress plane which is inaccessible but there is even an ambiguity about the 'unloading' response. Such behaviour is a reality for geotechnical materials and leads to bifurcation of response and localisation of deformations in shear zones or failure surfaces. It becomes difficult to maintain uniformity of deformation of samples of such materials as they are sheared. This is a further indication of the advantage of working from strain increments to stress increments—the response is unambiguous—rather than from stress increments to strain increments—there are two choices of response, elastic unloading or plastic softening.

The undrained effective stress path followed after the peak in this model is found in the same way as for the previous hardening model by setting the plastic and elastic volumetric strain increments equal and opposite:

$$\frac{p' - p'_i}{bK\eta_p} = \frac{\left(1 - \frac{\eta}{\eta_p}\right)\left(1 + \frac{\eta}{\eta_p} - 2\frac{M}{\eta_p}\right)}{2\left(1 - \frac{M}{\eta_p}\right)} \tag{3.133}$$

with the limit, when the soil has softened to the critical state:

$$\frac{p'_f - p'_i}{bK\eta_p} = \frac{1}{2}\left(1 - \frac{M}{\eta_p}\right) \tag{3.134}$$

The path is plotted in Fig 3.47. Because the soil behaves elastically until the peak stress ratio is reached the mean effective stress does not change during this initial elastic phase of the test.

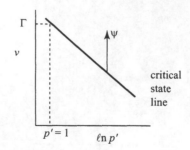

Figure 3.48: Definition of state variable ψ

Mohr Coulomb model with strength dependent on state variable

This softening model is still defective in not allowing nonlinear stress:strain response before the peak stress ratio is reached; the previous model was defective in not allowing softening. Evidently the two models could be combined, introducing a switch to softening when some designated peak stress ratio had been attained. This is rather the strategy adopted by Nova and Wood (1979). However, failure switches seem somewhat unsatisfactory from a physical point of view. A slightly more subtle way of achieving a similar result is described here.

We have described two models which can be seen as simple additions of hardening plasticity to the perfectly plastic Mohr-Coulomb model. These models can be seen as examples of some of the possibilities which the framework of hardening plasticity opens up. We can develop models which are able to reproduce those features of mechanical response that we feel are important in a particular application. This is not as arbitrary a process as it may seem: the features that we are including are certainly inspired by experimental observation. Each of the models so far described has its merits: the second model certainly introduces features of strain softening and reducing dilatancy which will be relevant at large deformations of sands but does so by incorporating a switch to turn off the softening process once the critical state stress ratio has been reached. A way in which both hardening and softening can be rather simply—and elegantly—combined in a single model, which is again clearly a development from the Mohr-Coulomb family, has been described by Muir Wood, Belkheir and Liu (1994) and by Gajo and Muir Wood (1999). A slightly simpler version of this model will be briefly presented here.

Strength of soils is linked with density (§2.7). If the density of a soil changes as it is sheared then we expect the strength to change as well. Let us make the *current* peak strength η_p in the hardening model a variable which is a function of the current density—or more appropriately a function of state variable ψ combining information of density (through specific volume v) and mean stress (Fig 3.48) (§2.6.1, §2.7). Formally we should write

$$\eta_p = M - k\psi = M - k\left(v - \Gamma + \lambda \ln p'\right) \tag{3.135}$$

where k is a soil constant linking state variable and strength. It is assumed that the critical state line can locally be described as a straight line of slope λ in a semilogarithmic compression plane (Fig 3.48); and v and p' are the current values of specific volume and mean effective stress which will in general vary during a test from their initial values v_o and p'_o. This can be written in terms of the initial specific volume (related to the initial density of the sand) and the volumetric strain ϵ_p which has occurred from the start of the test which can be divided into elastic and plastic components, ϵ_p^e and ϵ_p^p :

$$
\begin{aligned}
\eta_p &= M - k\left[v_o\left(1 - \epsilon_p\right) - \Gamma + \lambda \ln p'\right] \\
&= M - k\left[\left(v_o - \Gamma + \lambda \ln p'_o\right) + \left(\lambda \ln \frac{p'}{p'_o} - v_o \epsilon_p^e\right) - v_o \epsilon_p^p\right] \quad (3.136)
\end{aligned}
$$

We will neglect the second term in parentheses and thus assume for simplicity that the elastic volumetric strain roughly balances the effect of the change in mean effective stress. (But the model developed by Gajo and Muir Wood (1999) does not make this simplification.) Then

$$
\frac{\partial \eta_y}{\partial \epsilon^p} = \begin{pmatrix} \frac{\partial \eta_y}{\partial \epsilon_p^p} \\ \frac{\partial \eta_y}{\partial \epsilon_q^p} \end{pmatrix} = \begin{pmatrix} \frac{\partial \eta_y}{\partial \eta_p} \frac{\partial \eta_p}{\partial \epsilon_p^p} \\ \frac{\partial \eta_y}{\partial \epsilon_q^p} \end{pmatrix} = \begin{pmatrix} \frac{\eta_y}{\eta_p} k v_o \\ \frac{(\eta_p - \eta_y)^2}{a \eta_p} \end{pmatrix} \quad (3.137)
$$

and the hardening expression H (3.108) becomes

$$
H = -\frac{\partial f}{\partial \chi} \frac{\partial \chi}{\partial \epsilon^p}^T \frac{\partial g}{\partial \sigma} = p'\left[\left(M - \eta_y\right) \frac{\eta_y}{\eta_p} k v_o + \frac{(\eta_p - \eta_y)^2}{a \eta_p}\right] \quad (3.138)
$$

The yield function, flow rule and hardening rule are chosen as before ((3.112), (3.114) and (3.117)) and consequently the other elements of the elastic-plastic stiffness matrix (3.120) remain unchanged.

This model now homes in on a critical state condition, heading always towards the current peak strength following the hyperbolic hardening law but this peak strength is itself changing as the soil compresses or dilates with shearing. Thus, even though the hardening law appears to be a simple hyperbolic monotonically increasing function of strain, nevertheless the stress:strain response is able to introduce strain softening and the accompanying smooth transition between compression and dilation. The peak strength is thus a moving target which can only be attained at infinite distortional strain (it remains the asymptote of the hardening law) by which time it is identical with the critical state strength.

Typical stress-strain and volumetric strain responses calculated using this model are shown in Fig 3.49. The behaviour depends strongly on the initial value of state variable: a positive initial state variable indicates an initially loose material which tends to compress as it is sheared and shows little in the way of a peak strength; a negative initial state variable indicates an initially dense material which dilates as soon as the critical state stress ratio is exceeded on the initial loading and then shows a peak with subsequent strain softening. Evidently the stress response envelopes that are calculated (Fig 3.50) depend on whether the current state is pre-peak—in which case the response is similar to that shown for the hardening model (Fig 3.41)—or post-peak—in which case the response is similar to that shown for the softening model (Fig 3.46).

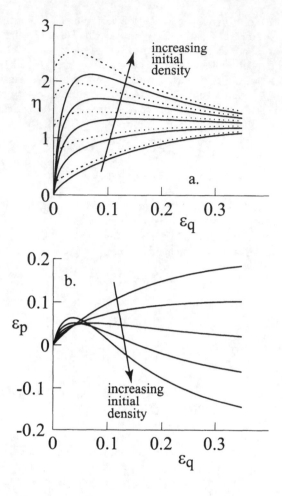

Figure 3.49: Response of elastic-hardening plastic Mohr-Coulomb model with current strength dependent on state variable in conventional drained triaxial compression tests ($\delta\sigma_r = 0$): (a) stress:strain response (dotted curves indicate variation in current peak strength; solid curves indicate mobilised strength) and (b) volumetric strain response dependent on initial density (initial value of state variable in range $-0.5 < \psi_i < 0.5$, $p'_i = 100$ kPa) ($K = 1500$ kPa, $\nu = 0.3$, $M = 1.2$, $a = 0.01$, $k = 2$)

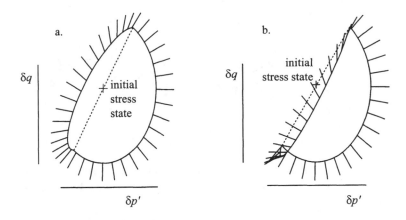

Figure 3.50: Elastic-hardening plastic Mohr-Coulomb model with strength dependent on state variable: stress response envelopes (a) pre-peak (hardening $\epsilon_q = 0.04$);(b) post-peak (softening $\epsilon_q = 0.2$) (all constitutive parameters as in Fig 3.49, $\psi_i = -0.5$)

3.4.2 Cam clay

Historically it is probably reasonable to describe Cam clay as the first *hardening* plastic model that has become generally adopted for soils. It has formed a basis for much subsequent development of soil models. Originally developed in the early 1960s, models of the Cam clay form have been widely and successfully used for analysis of problems involving the loading of soft clays. It has been less successful in describing the behaviour of sands for which models which make use of distortional hardening and nonassociated flow (§3.4.1) have generally been reckoned to be more satisfactory. A detailed description of the Cam clay model and of the behaviour of soils—especially clays—seen against the patterns of behaviour that the Cam clay model reveals is given by Muir Wood (1990); here we will present the model within the general framework of elastic-hardening plastic models that has been developed in section §3.4.

We can quickly identify a defect of the various extended Mohr-Coulomb models. For clays, an important aspect of the observed mechanical behaviour is the large change in volume that occurs during compression (Fig 3.51) when the stresses acting on a sample of soil are all increased in proportion—isotropic compression and one-dimensional compression are obvious examples. Clearly if the model is to be used to reproduce the loading of soft clays then this volumetric response must be included. However, applying such proportional stress paths to any of the Mohr-Coulomb models, as presently described, will produce solely elastic response as shown in Fig 3.51c. The irrecoverability of the volumetric response suggests that a different mechanism of plastic deformation will be required. This could be achieved by adding extra yield mechanisms to the

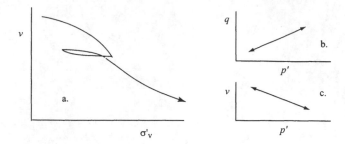

Figure 3.51: (a) Large irrecoverable volume changes in oedometer test on clay subjected to (b) typical compression stress path giving rise to (c) elastic volume changes in Mohr-Coulomb model

Mohr-Coulomb models; the Cam clay model that will be described here provides an elegant alternative route.

As before, for simplicity we will develop the model in terms of the triaxial strain increment and stress variables and work through the several ingredients of the model in turn.

1. *Elastic properties*: We will assume that the elastic behaviour of the soil is isotropic and defined by two elastic parameters, bulk modulus K and shear modulus G.

Results of oedometer tests are typically presented in semilogarithmic plots because it is found that the relationships between stress and volume change then become somewhat more linear—both during loading and during unloading. Looking at the typical loading and unloading response in an oedometer (Fig 3.51a) we can easily see the division of the volume changes into elastic and plastic parts just as for the uniaxial loading of the copper wire in Fig 3.35. It is logical then to use the average slope κ of an unload-reload line to characterise the elastic volumetric response (Fig 3.52) and to assume that κ is a soil constant:

$$v = v_\kappa - \kappa \ln p' \tag{3.139}$$

where v_κ is a reference value of specific volume on a particular unloading-reloading relationship. We can convert this to an incremental relationship

$$\delta \epsilon_p^e = -\frac{\delta v}{v} = \frac{\kappa}{v} \frac{\delta p'}{p'} \tag{3.140}$$

which implies that the bulk modulus K is not constant but is dependent on stress level (and on current packing)

$$K = \frac{\delta p'}{\delta \epsilon_p^e} = \frac{v p'}{\kappa} \tag{3.141}$$

In this form the value of κ is directly related to swelling index C_s:

$$\kappa = \frac{C_s}{\ln 10} \tag{3.142}$$

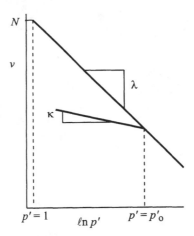

Figure 3.52: Cam clay: linear normal compression and unloading-reloading lines in semilogarithmic compression plane

An alternative possibility is to propose a linear link between volumetric strain (rather than change in volume) and log of mean stress so that then the bulk modulus depends only on p' (for example, (3.58)) which leads to slightly more elegant expressions. However, the important result is that the elastic stiffness is nonlinear and depends on the current stress level.

Having chosen one elastic property we require one more (the elastic properties of our material may be nonlinear but the material is still assumed to be isotropic). We may often find it convenient to choose a constant shear modulus G because we will see this directly from the initial behaviour in any compression test.

$$\delta\epsilon_q^e = \frac{\delta q}{3G} \tag{3.143}$$

An alternative will be to choose a constant value of Poisson's ratio ν, thus forcing a constant ratio of shear modulus and bulk modulus.

$$G = K\frac{3(1 - 2\nu)}{2(1 + \nu)} \tag{3.144}$$

Clearly if G is constant then the variation of bulk modulus K with stress will lead to a varying ν (and as the effective mean stress and hence the bulk modulus fall towards zero the value of Poisson's ratio will tend towards -1). However, if Poisson's ratio ν is assumed to be constant then G changes together with bulk modulus and we have seen that there are thermodynamic problems if we make both G and K functions of p'—it becomes possible to generate or dissipate energy on supposedly elastic cycles of stress change (Zytynski *et al.*, 1978). It is not possible to define an elastic potential which implies a constant Poisson's ratio if the bulk modulus is a function of mean stress alone (§3.2.5).

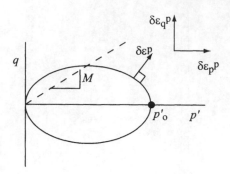

Figure 3.53: Elliptical yield locus for Cam clay model

With certain reservations then, we have the elastic stiffness and compliance relationships:

$$\begin{pmatrix} \delta p' \\ \delta q \end{pmatrix} = \begin{pmatrix} vp'/\kappa & 0 \\ 0 & 3G \end{pmatrix} \begin{pmatrix} \delta \epsilon_p^e \\ \delta \epsilon_q^e \end{pmatrix} \tag{3.145}$$

$$\begin{pmatrix} \delta \epsilon_p^e \\ \delta \epsilon_q^e \end{pmatrix} = \begin{pmatrix} \kappa/vp' & 0 \\ 0 & 1/3G \end{pmatrix} \begin{pmatrix} \delta p' \\ \delta q \end{pmatrix} \tag{3.146}$$

2. *Yield criterion*: In the triaxial stress plane (p', q) it is assumed that the yield locus has an elliptical shape passing through the origin of the stress plane (Fig 3.53). This introduces two variables: the aspect ratio of the ellipse M which controls the shape of the ellipse, the ratio of the vertical (q) axis to the horizontal (p') axis; and the size of the ellipse p'_o which is the hardening parameter χ for the Cam clay model. The equation of the ellipse can be presented in various different ways. To fit in with the general presentation of hardening plastic models we can write:

$$f(\boldsymbol{\sigma}, p'_o) = \frac{q^2}{M^2} - p'(p'_o - p') \tag{3.147}$$

so that, as usual, $f < 0$ indicates elastic behaviour, $f = 0$ indicates that yielding is occurring and $f > 0$ is not permitted.

However, the equation of the ellipse can also be written

$$\frac{p'}{p'_o} = \frac{M^2}{M^2 + \eta^2} \tag{3.148}$$

or

$$\frac{q^2}{M^2} = p'(p'_o - p') \tag{3.149}$$

Different forms of the equation are useful in different circumstances.

For stress changes $(\delta p', \delta q)$ causing yield, the change in size of the yield locus can be written:

$$\delta p'_o = (2p' - p'_o)\frac{\delta p'}{p'} + \frac{2q}{M^2}\frac{\delta q}{p'} \tag{3.150}$$

or

$$\delta p_o' = \left(\frac{M^2 - \eta^2}{M^2} \delta p' + \frac{2\eta}{M^2} \delta q \right) \qquad (3.151)$$

or

$$\frac{\delta p_o'}{p_o'} = \frac{\delta p'}{p'} + \frac{2\eta}{M^2 + \eta^2} \delta\eta \qquad (3.152)$$

and from these expressions the change in size of the yield locus required to accommodate any change in effective stress which causes yielding can be calculated.

It is often convenient to work in terms of mean stress p' and stress ratio η and expressions (3.148) and (3.152) are then obviously appropriate. As the size of the yield locus changes the shape remains the same and the locus grows from the origin. Along any line at constant stress ratio $\eta = q/p'$ the angle of intersection with any yield locus is always the same.

3. *Flow rule*: It is assumed that Cam clay obeys the hypothesis of associated flow (normality) so that the plastic strain increment vector is assumed to be normal to the yield surface at the current stress state (Fig 3.53). The plastic potential function then has the same form as the yield criterion:

$$g\left(\boldsymbol{\sigma}\right) = f\left(\boldsymbol{\sigma}, p_o'\right) = \frac{q^2}{M^2} - p'\left(p_o' - p'\right) = 0 \qquad (3.153)$$

The plastic strain increments are given by

$$\begin{pmatrix} \delta\epsilon_p^p \\ \delta\epsilon_q^p \end{pmatrix} = \mu \begin{pmatrix} \frac{\partial g}{\partial p'} \\ \frac{\partial g}{\partial q} \end{pmatrix} = \mu \begin{pmatrix} 2p' - p_o' \\ \frac{2q}{M^2} \end{pmatrix} \qquad (3.154)$$

Alternatively, using form (3.148) of the equation of the elliptical yield locus, the ratio of plastic volumetric strain to plastic distortional strain, which characterises the plastic deformation mechanism, can be written:

$$\frac{\delta\epsilon_p^p}{\delta\epsilon_q^p} = \frac{M^2 - \eta^2}{2\eta} \qquad (3.155)$$

The mechanism of plastic deformation depends only on the stress ratio at which yielding is occurring and changes continuously as the stress ratio changes. Several particular cases are of interest:

- for $\eta = 0$, $\delta\epsilon_p^p/\delta\epsilon_q^p = \infty$ which implies compression without distortion and this is appropriate for isotropic consolidation without application of distortional stresses;

- for $\eta = M$, $\delta\epsilon_p^p/\delta\epsilon_q^p = 0$ which implies distortion without compression— this is the critical state condition;

- yielding with low values of stress ratio $\eta < M$ gives $\delta\epsilon_p^p/\delta\epsilon_q^p > 0$ which implies compression plus distortion; and

- yielding with high values of stress ratio $\eta > M$ gives $\delta\epsilon_p^p/\delta\epsilon_q^p < 0$ which implies expansion plus distortion.

The implications of these different cases for the overall response of the model will be explored subsequently.

4. *Hardening rule*: The hardening rule describes the dependence of the size of the yield locus p'_o on the plastic strain. Cam clay is a volumetric hardening model in which it is assumed that the size of the yield locus depends only on the plastic volumetric strain through an expression

$$\begin{pmatrix} \partial p'_o / \partial \epsilon_p^p \\ \partial p'_o / \partial \epsilon_q^p \end{pmatrix} = \begin{pmatrix} vp'_o / (\lambda - \kappa) \\ 0 \end{pmatrix} \tag{3.156}$$

This hardening rule introduces one additional soil parameter λ. During isotropic normal compression we have change in mean stress p' with distortional stress q kept constant at zero. There will be elastic volumetric strains given by (3.140) and, because the mean stress is always at the tip of the yield surface $p' = p'_o$, there will be plastic volumetric strains given by a rearrangement of (3.156):

$$\delta \epsilon_p^p = \frac{\lambda - \kappa}{v} \frac{\delta p'_o}{p'_o} = \frac{\lambda - \kappa}{v} \frac{\delta p'}{p'} \tag{3.157}$$

The total volumetric strain is then

$$\delta \epsilon_p = \delta \epsilon_p^e + \delta \epsilon_p^p = \frac{\kappa \delta p'}{vp'} + \frac{(\lambda - \kappa) \delta p'}{vp'} = \frac{\lambda}{v} \frac{\delta p'}{p'} \tag{3.158}$$

Noting that the definition of the volumetric strain is

$$\delta \epsilon_p = \frac{\delta v}{v} \tag{3.159}$$

expression (3.158) can be integrated to give the form of the normal compression relationship linking specific volume v and mean effective stress p':

$$v = N - \lambda \ln p' \tag{3.160}$$

where N is a reference value of specific volume for unit value of mean effective stress. This is a linear normal compression relationship with slope λ in the semi-logarithmic plot (Fig 3.52). It may be noted that

$$\lambda = \frac{C_c}{\ln 10} \tag{3.161}$$

and the plastic compressibility λ can be directly related to the compression index C_c.

Now that all the ingredients of the model are in place the overall plastic compliance relationship can be deduced:

$$\begin{pmatrix} \delta \epsilon_p^p \\ \delta \epsilon_q^p \end{pmatrix} = \frac{\lambda - \kappa}{vp'(M^2 + \eta^2)} \begin{pmatrix} M^2 - \eta^2 & 2\eta \\ 2\eta & \frac{4\eta^2}{M^2 - \eta^2} \end{pmatrix} \begin{pmatrix} \delta p' \\ \delta q \end{pmatrix} \tag{3.162}$$

and the full stiffness matrix linking the stress increments with the total strain increments can be obtained by substitution in (3.108) and (3.109). The hardening quantity H is given by

$$H = -\frac{\partial f}{\partial p'_o} \frac{\partial p'_o}{\partial \epsilon_p^p} \frac{\partial g}{\partial p'} = -(-p') \left(\frac{vp'_o}{\lambda - \kappa} \right) (2p' - p'_o) \tag{3.163}$$

and the full elastic-plastic stiffness relationship is given by

$$
\begin{pmatrix} \delta p' \\ \delta q \end{pmatrix} =
$$

$$
\left[\begin{pmatrix} K & 0 \\ 0 & 3G \end{pmatrix} - \frac{\begin{pmatrix} K^2 \left(2p' - p'_o\right)^2 & \frac{6GKq\left(2p'-p'_o\right)}{M^2} \\ \frac{6GKq\left(2p'-p'_o\right)}{M^2} & \frac{36G^2q^2}{M^4} \end{pmatrix}}{\left[K\left(2p' - p'_o\right)^2 + \frac{12Gq^2}{M^4} + \frac{vp'p'_o\left(2p'-p'_o\right)}{\lambda - \kappa} \right]} \right] \begin{pmatrix} \delta\epsilon_p \\ \delta\epsilon_q \end{pmatrix} \quad (3.164)
$$

where $K = vp'/\kappa$.

Whether the compliance form (3.162) or the stiffness form (3.164) is used it is evident that the controlling matrix is symmetric: this results from the assumption of associated flow in which the vectors of plastic strain increment are assumed to be normal to the yield locus at the current effective stress causing yield.

Study of (3.162) shows that the magnitude of the plastic strains is controlled largely by $\lambda - \kappa$. It will be the difference between these two soil parameters (rather than the absolute value of either of them) that will have to be varied in order to match available experimental data. Some qualitative statements about the nature of the stress:strain response can be made.

What happens as $\eta \to M$? The top line of the compliance matrix (3.164) shows that as the stress ratio approaches the value M so the plastic volumetric strains become smaller and smaller. Since the plastic hardening depends only on the plastic volumetric strain it can be deduced correspondingly that the change in p'_o in any stress increment has to tend to zero as the stress ratio approaches M. The bottom line of the compliance matrix shows that the shear compliance tends to infinity, or in other words the shear stiffness tends to zero. In fact, an asymptotic perfectly plastic condition is predicted in which distortional strains continue but with no further changes in size of yield locus, stresses or volumetric strains. Such an ultimate state has been termed a critical state (§2.6.1).

$$
\eta \to M: \quad \delta\epsilon^p_p \to 0; \quad \delta p'_o \to 0; \quad \frac{\delta\epsilon^p_q}{\delta q} \to \infty \quad (3.165)
$$

The value of the soil parameter M can therefore be related to the ultimate value of the angle of shearing resistance for the soil ϕ_c in triaxial compression:

$$
M = \frac{6\sin\phi_c}{3 - \sin\phi_c} \quad (3.166)
$$

The Cam clay model responds stably to yielding with stress ratio $\eta < M$ and under such conditions it does not matter whether the problem is driven by stress changes or by strain changes: it is often conceptually easier to think of the response to stress changes because the model has been described in terms of a yield locus in a stress plane. In a typical compression test the deviator stress rises steadily towards the ultimate value (low overconsolidation ratio in

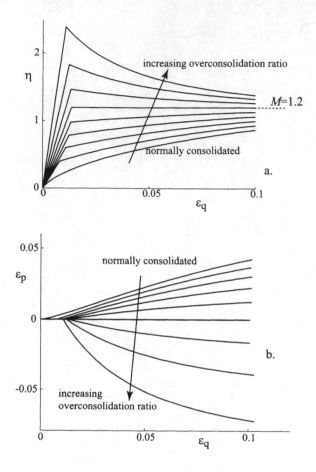

Figure 3.54: Cam clay: (a) stress:strain and (b) volumetric strain response in drained triaxial compression tests with constant mean stress ($\delta p' = 0$) ($\kappa = 0.05$, $G = 1500$ kPa, $\lambda = 0.25$, $M = 1.2$) (overconsolidation ratios p'_o/p'_i in range 1-5, $p'_o = 100$ kPa)

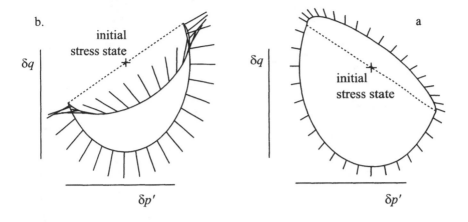

Figure 3.55: Cam clay: stress response envelopes (a) $\eta < M$, $p_i'/p_o' = 0.75$; (b) $\eta > M$, $p_i'/p_o' = 0.25$ ($\kappa = 0.1$, $\nu = 0.3$, $\lambda = 0.25$, $M = 1.2$) ($v_i = 2.5$, $p_o' = 200$ kPa)

Fig 3.54). A typical stress response envelope for this regime is shown in Fig 3.55a. The two elliptical segments are now tangential to each other for stress increments which imply neutral loading with the stress increment tangential to the yield locus. This is a consequence of the assumption of associated flow.

However, if the soil is yielding with stress ratio $\eta > M$ then study of (3.162) shows that the distortional compliance is negative and continuing shearing with $\delta \epsilon_q^p > 0$ implies $\delta \epsilon_p^p < 0$, $\delta p_o' < 0$ and $\delta q < 0$ which implies strain softening (high overconsolidation ratio in Fig 3.54). The stress ratio $\eta = M$ is still an ultimate asymptote but the soil now approaches this stress ratio from above rather than from below. Consideration of the equation for the yield locus (3.148) shows that yielding with stress ratios greater than M is only possible for values of p'/p_o' less than 1/2—overconsolidation ratios greater than 2. Such behaviour is characteristic of dense or heavily preloaded materials which are so tightly packed that they have to expand in order that the particles should be able to move relative to each other and allow the material to distort (§2.6).

This is a real phenomenon, but as noted in §3.4.1, it can lead to numerical problems because of the uncertainty: does a reduction in shear stress imply an elastic unloading or a continuing plastic strain softening? The stress response envelope (Fig 3.55b) illustrates this ambiguity—as for the Mohr-Coulomb models with strain softening (Figs 3.46b and 3.50b) the response envelope is folded over on itself. All strain increments are possible and each strain increment implies an unambiguous stress increment. However, certain stress increments—those which attempt to escape from the *current* yield locus—are not possible and the section of the stress plane lying outside the yield locus in the region for which $\eta > M$ is inaccessible (Fig 3.56). Stress changes which move inside the current yield locus can be associated with either purely elastic or with elastic plus plastic strains. In analysis of such situations the soil response has to be

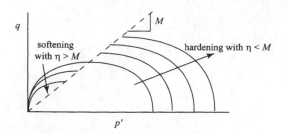

Figure 3.56: Cam clay: hardening of yield locus with $\eta < M$, softening of yield locus with $\eta > M$

driven by strain increments—which will make it quite clear whether unloading or plastic softening is implied—rather than stress increments precisely because many stress increments will in fact be physically either impossible or ambiguous. Numerically and physically such behaviour leads to the occurrence of localisation: as the material softens it becomes weaker and natural inhomogeneities lead to strain concentrations and formation of ruptures or shear bands through the material.

The Cam clay model has five material properties. There are two elastic properties κ and G or ν. The volumetric parameter κ is linked with swelling index C_s (3.142). There are two plastic properties M and λ which can be linked with angle of shearing resistance in triaxial compression ϕ_c (3.166) and compression index C_c (3.161) respectively.

The final soil parameter is a reference for volume in order that volumetric strains can be calculated. We have defined the equation of the *isotropic* normal compression line using a reference parameter N to indicate the specific volume for unit mean stress (Fig 3.52). However, results of predictions made using Cam clay are not usually very sensitive to plausible variations of N—so the reference volume can just as well be taken from one-dimensional compression data.

The isotropic normal compression line defines the values of specific volume when the stress state is always at the tip of the yield locus, $p' = p'_o$ and the soil is always yielding. More generally, for stress states inside the yield locus there is some implied elastic expansion from the normal compression line (Fig 3.52) and the specific volume is given by

$$v = N - \lambda \ln p'_o + \kappa \ln \left(\frac{p'_o}{p'} \right) \tag{3.167}$$

or, if the soil is yielding with stress ratio η

$$v = N - \lambda \ln p'_o + \kappa \ln \left(\frac{M^2 + \eta^2}{M^2} \right) \tag{3.168}$$

The value of N depends on the units used for measurement of stress. Users of Cam clay need to be vigilant. Here we will always take the value of N to correspond to a mean stress $p' = 1$ kPa.

Cam clay: effective stress path in undrained test

The effective stress path followed in an undrained test can be calculated in exactly the same way as for the Mohr-Coulomb models by requiring the elastic and plastic volumetric strain increments to be always equal and opposite. From (3.140) and (3.157) and the definitions of p'_o ((3.148) and (3.152)) and of stress ratio $\eta = q/p'$:

$$\kappa \delta p' + \frac{\lambda - \kappa}{M^2 + \eta^2} \left[\left(M^2 + \eta^2 \right) \delta p' + 2\eta p' \delta \eta \right] = 0 \qquad (3.169)$$

Integrating this expression, from an initial yielding stress state p'_i and η_i, and substituting

$$\Lambda = \frac{\lambda - \kappa}{\lambda} \qquad (3.170)$$

the effective stress path is found to be

$$\frac{p'_i}{p'} = \left(\frac{M^2 + \eta^2}{M^2 + \eta_i^2} \right)^\Lambda \qquad (3.171)$$

ending at a failure state with mean effective stress p'_f and stress ratio equal to M:

$$\frac{p'_i}{p'_f} = \left(\frac{2M^2}{M^2 + \eta_i^2} \right)^\Lambda \qquad (3.172)$$

For elastic stress changes the constant volume condition requires, for an isotropic elastic material, that there should be no change in mean effective stress. Effective stress paths for initially normally consolidated, lightly overconsolidated and heavily overconsolidated Cam clay are shown in Fig 3.57. For the isotropically normally consolidated soil ($\eta_i = 0$) the ratio of mean effective stresses at the start of the test (p'_i) and at failure (p'_f is 2^Λ. The undrained strength of the soil c_u is given by:

$$c_u = \frac{q_f}{2} = M\frac{p'_f}{2} = \frac{M}{2}\frac{p'_i}{2^\Lambda} \qquad (3.173)$$

and hence for isotropically normally consolidated soil the ratio of undrained strength to initial effective stress is a function only of soil constants M and Λ:

$$\frac{c_u}{p'_i} = \frac{M}{2^{1+\Lambda}} \qquad (3.174)$$

Cam clay: worked example

For the purposes of hand calculation using the Cam clay model it is usually easiest to consider problems as stress driven processes—though obviously this will break down unless special precautions are taken if yielding is occurring with $\eta > M$. The relevant equations are then the elastic and plastic compliance relationships: (3.146) and (3.162).

To use any elastic-plastic model we need to know about:

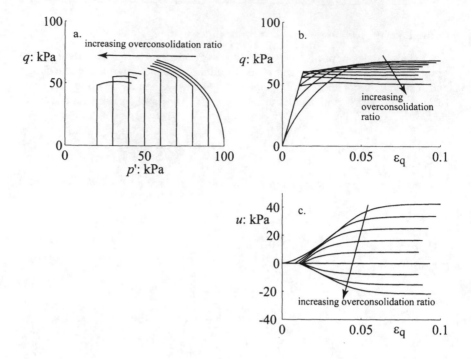

Figure 3.57: Cam clay: undrained triaxial compression tests conducted with constant total mean stress ($\delta p = 0$): (a) effective stress paths; (b) stress:strain response; (c) development of pore pressure ($\kappa = 0.05$, $G = 1500$ kPa, $\lambda = 0.25$, $M = 1.2$) (overconsolidation ratios p'_o/p'_i in range 1-5, $p'_o = 100$ kPa)

the *past*—because we must know how big is the current yield surface p'_o;

the *present*—because we need to know what are the current stresses (which must lie inside or on the yield surface) p'_i, q_i; and

the *future*—because we are trying to calculate the response of the model to some imposed perturbation.

A sample of Cam clay with properties $M = 1.2$, $\lambda = 0.3$, $\kappa = 0.06$, $N = 3.5$ (at $p' = 1$ kPa), $G = 2000$ kPa has been subjected to past stresses leaving it with a yield locus of size given by $p'_o = 100$ kPa. The sample is in pore pressure equilibrium under initial stresses $q_i = 0$, $p'_i = 75$ kPa and is then subjected to a conventional drained triaxial compression test.

1. What stress changes can be imposed before plastic strains start to occur?
2. What are the elastic strains at this stage?
3. What is the ratio of plastic strain increments (the mechanism of plastic deformation) immediately after yielding occurs?
4. What are the magnitudes of the strains for a further change of stresses $\delta q = 3$ kPa, $\delta p' = 1$ kPa?

1. The past is controlled by the size of the yield locus which is indicated by the initial value of p'_o. The present is indicated by the given initial effective stresses p'_i and q_i. The future is indicated by the specified test: in this case we are told that it is a conventional drained triaxial compression test. We need to use all three of these pieces of information in order to answer the first question.

From the specified drained stress path, the yield point is at:

$$p' = p'_i + x; \quad q = 3x$$

but also lies on the initial yield locus (3.149) with size $p'_o = 100$ kPa.

This equation can be solved either directly or by iteration to give a value of x—noting that only one of the two roots is plausible (the other root corresponds to a drained triaxial extension test at constant radial effective stress in which $q < 0$). This gives $x = 13$ kPa and hence $p'_y = 88$ kPa; $q_y = 39$ kPa; and the stress ratio at yield is $\eta_y = 0.443$.

2. The initial specific volume of the clay is given by (3.167):

$$v_i = N - \lambda \ln p'_o + \kappa \ln \left(\frac{p'_o}{p'_i} \right) = 2.136$$

The elastic strains can be calculated in a single increment using (3.146) and perhaps taking the average value of the initial and yield values of mean effective stress:

$$\delta \epsilon^e_p = \frac{\kappa \delta p'}{v p'} = \frac{0.06 \times 13}{2.136 \times \frac{1}{2}(75 + 88)} = 0.0045 \quad (0.45\%)$$

$$\delta \epsilon^e_q = \frac{\delta q}{3G} = \frac{39}{6000} = 0.0065 \quad (0.65\%)$$

From the definition of the volumetric and distortional strain increments, we can then deduce the corresponding axial and radial strain increments, $\delta \epsilon^e_a =$

0.8%; $\delta\epsilon_r^e = -0.175\%$. The new value of specific volume is 2.126, calculated from the definition of volumetric strain.

3. The ratio of *plastic* strain increments after yield is given by the flow rule (3.155):

$$\frac{\delta\epsilon_p^p}{\delta\epsilon_q^p} = \frac{M^2 - \eta^2}{2\eta} = \frac{1.2^2 - 0.443^2}{2 \times 0.443} = 1.404$$

and the volumetric strains have become more dominant. Note that the ratio of *total* strain increments depends on the direction of the stress increment because this will affect the elastic strains that are generated.

4. Once the actual details of a further stress change have been specified the strain increments can be calculated. The stress increments are: $\delta q = 3$ kPa, $\delta p' = 1$ kPa. The elastic strain increments are calculated as before:

$$\delta\epsilon_p^e = \frac{0.06 \times 1}{2.126 \times 88} = 0.0003; \quad \delta\epsilon_q^e = \frac{3}{6000} = 0.0005$$

The plastic strain increments are calculated using the plastic compliance matrix (3.162) using appropriate values of soil parameters, stresses and stress ratio:

$$
\begin{aligned}
\delta\epsilon_p^p &= \frac{\lambda - \kappa}{vp'\left(M^2 + \eta^2\right)}\left[\left(M^2 - \eta^2\right)\delta p' + 2\eta\delta q\right] \\
&= \frac{0.24}{2.126 \times 88 \times \left(1.2^2 + 0.443^2\right)}\left[\left(1.2^2 - 0.443^2\right) \times 1 + 2 \times 0.443 \times 3\right] \\
&= 0.0031
\end{aligned}
$$

$$\delta\epsilon_q^p = \frac{0.0031}{1.404} = 0.0022$$

The total strain increments are then: $\delta\epsilon_p = 0.34\%$; $\delta\epsilon_q = 0.27\%$. These can be converted to axial and radial increments: $\delta\epsilon_a = 0.38\%$; $\delta\epsilon_r = -0.02\%$. The resulting deformation is now very nearly one dimensional with almost no radial strain.

3.5   Modelling non-monotonic loading

In section §2.5.3 we described some of the kinematic aspects of stiffness of soils and we showed a schematic variation of stiffness with non-monotonic loading in Fig 3.17. How far have we progressed towards being able to reproduce this character of response?

Although our elastic-hardening plastic models are an evident improvement on the elastic-perfectly plastic models in that they provide for a steady decrease of tangent stiffness after yield occurs, rather than an immediate drop to zero (Fig 3.18), there is still the dramatic fall in stiffness as the stress path crosses the yield surface (Fig 3.58a) whereas real soils tend to show much more gradual stiffness changes. There is also a significant difference on unloading (Fig 3.58b). The elastic-plastic models described here predict that the yield surface will expand as the stress state pushes it outwards—and the more it expands the larger the elastic region that remains. In fact, our kinematic observation suggests that,

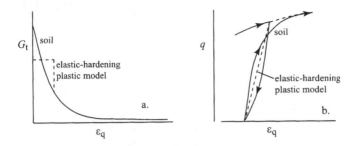

Figure 3.58: Comparison of typical capability of elastic-hardening plastic model with actual soil behaviour (a) variation of tangent stiffness with monotonic shearing; (b) unload-reload response

though the elastic region may indeed change in size as the stresses push it around, it is the change in position that is possibly more significant. Unloading paths develop plasticity in a way that the Cam clay model cannot describe.

Such response can be described using a kinematic hardening extension of the hardening plasticity models. Our models are essentially isotropic hardening models: the Cam clay yield locus retains its shape and orientation, and always passes through the origin of stress space whatever the stress path that interacts with it[6]; the yield locus of the Mohr-Coulomb model becomes a progressively more open cone.

A kinematic hardening extension of a Cam clay-like model is illustrated in Fig 3.59 (Al-Tabbaa and Muir Wood, 1989). The elastic region is now confined to an elastic 'bubble' which floats around in stress space with the current stress state. Plastic strains occur whenever the 'bubble' moves but the plastic stiffness is controlled by the separation, b, of the 'bubble' and some outer 'bounding' surface and falls as the 'bubble' approaches this 'bounding' surface. A translation rule is introduced to describe the way in which the 'bubble' decides how much to change in size and how much to change in position as the stress engages with it. With appropriate formulation this model can be made to behave identically to Cam clay when the soil is being loaded with the 'bubble' in contact with the bounding surface (which then looks rather like the Cam clay yield surface—but is not actually a yield surface because it does not control the onset of development of plastic strains) (Fig 3.59). There is thus a hierarchical development of the model, adding desirable features (smooth variation of stiffness, plasticity on stress reversal) to an already somewhat familiar model, Cam clay.

---

[6]Strictly, because the *centre* of the Cam clay yield locus moves as the soil hardens (or softens), Cam clay already contains a kinematic hardening element. However, the constraints imposed by the insistence that the yield locus should always pass through the origin of stress space and should have its centre on the mean effective stress p' axis make this model unable to reproduce the full kinematic character of response seen in experiments.

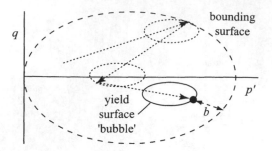

Figure 3.59: Kinematic hardening extension of Cam clay: 'bubble' bounds elastic region and moves with stress history

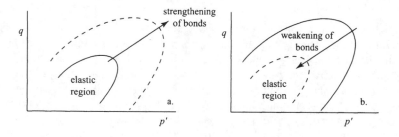

Figure 3.60: Effect of cementation or bonding as feature added to elastic-hardening plastic model

3.6   Modelling cementation and structure

We can add further effects in a similarly hierarchical way. Weak rocks have particles which are bonded together. However, we expect that with mechanical (or chemical) action these bonds can be damaged and the rock turned eventually into a soil. Natural soils often contain a certain amount of structure which may also manifest itself as a bonding between particles which can be destroyed with mechanical or chemical damage. Quick clay slides occur because a change of pore water chemistry upsets the particle bonds leaving a metastable structure which can easily be destroyed. When these quick clays collapse they flow like liquid—their structure has been entirely lost.

We can postulate that the bonded material might be described by an extended Cam clay type of model in which the yield surface has an increased size as a result of the bonding (given a rather general shape in Fig 3.60). With plastic straining (or chemical weathering) the yield surface gradually shrinks to the Cam clay-like surface, appropriate to the remoulded, structureless material. Such an approach is adopted by Nova *et al.* (2003) as an extension of a Cam clay-like model to describe effects of chemical weathering of rocks. A similar approach is adopted as an extension of the 'bubble' kinematic extension of Cam clay by Rouainia and Muir Wood (1999) and Callisto, *et al.* (2002) to simulate the behaviour of natural clays damaged only by plastic straining.

3.7 Modelling rate effects

If we perform a very slow triaxial test on a sample of clay with the drainage connections open then we can be reasonably confident that no pore pressure will build up in the soil and we will observe a fully drained response of the soil. If we repeat this test *extremely* fast then, even if the drainage connections are open, there will be no possibility for pore water to move into or out of the sample and we will observe a more or less fully undrained response of the soil. Although we seem to have discovered a rate effect—in that the response that we see depends on the rate at which we apply the load—this is in fact a system effect and not a material effect. If we were to test an infinitesimally small element of soil then the drainage path length would be essentially zero and drainage, even in the very rapid test, would occur instantly. We can explain this apparent rate effect entirely in terms of restricted pore water flow—which will obviously be more significant in an impermeable soil such as clay than in a relatively permeable soil such as a sand. However, what constitutes 'extremely' fast can only be determined in relation to the permeability and dimensions of the material system being tested.

There are other effects which can be ascribed to truly rate dependent elements of the constitutive response of soils. Clays left at constant effective stress creep and develop secondary consolidation strains. A helpful picture of the character of this response can be given using 'isotaches' (Fig 3.61) (Šuklje, 1957). For a one-dimensional configuration (such as the oedometer) these form a family of curves (in general) linking strain and effective stress with each isotache corresponding to a specific strain rate but with a somewhat logarithmic spacing. At constant stress, strain develops at a decreasing rate as the clay moves down across the family of isotaches, AB, and this creep can be assumed to have occurred over geological time for samples presently in the ground. If the sample is now placed in an oedometer and the total stress increased, the initial pore pressure will dissipate allowing the effective stress to increase and, as it does so, the strain rate to increase (BC). However, the strain rate will subsequently decrease again as the creep strains dominate over the effects of the deformation linked with the dissipation of residual pore pressures (CD). Description of an isotache model and examples of its application to the estimation of creep under an embankment on soft clay and long term secondary settlement of reclamation on soft clay can be found in Nash (2001) and Nash and Ryde (2001).

If we perform a compression test at a constant rate of strain then we will track down one of the isotaches (AB in Fig 3.62). If we suddenly increase or reduce the strain rate we will jump rapidly across to the isotache corresponding to this new rate (CDEF). The slope of the path followed in the stress:strain diagrams of Figs 3.61 and 3.62 is directly related to the stiffness. We might deduce from the changes in stiffness that are being seen that the clay is yielding—switching from high stiffness elastic to low stiffness plastic response—but actually this

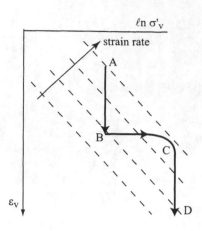

Figure 3.61: Family of isotaches for one-dimensional compression of clay

stiffness change is just the result of the interaction of the history of the soil with the viscoplastic isotache model.

If we collect data of stiffness change for a wider range of probing stress paths applied to a series of identical samples in a triaxial apparatus (for example) then we can discern a 'yield' locus for the clay with that particular initial history (Fig 3.63). Exploring this 'yielding' process with different strain rates will lead to a family of 'yield' loci, each corresponding to a different strain rate (Fig 3.63)—again we expect a somewhat logarithmic spacing. This implies a viscoplastic underpinning model for the soil response: there is a rate dependence of the irrecoverable deformation of the soil (the plasticity) that occurs when the stress state tries to go beyond the boundary of some region of stiff elastic response (the yield surface).

The effect of this viscoplastic interpretation of rate effects in clay on the stress response envelopes that were used to illustrate the history dependence of stiffness is probably something like that shown in Fig 3.64. With time, the small-strain stress response envelopes drift away a bit from the current stress state. Thus if creep equilibrium (the attainment of a tolerably low strain rate) is sought before probing to determine the response envelope begins, rather similar initial, very small strain, stiffness will be seen for all directions of stress probes. This is supported by the careful experimental observations of Clayton and Heymann (2001).

Sands have classically been regarded as completely free from viscous or rate effects. However, with increasingly accurate laboratory measurements—and instrumentation that is stable over long time periods—it has become evident that there are small but possibly significant rate effects in such materials too. There is evidence of the sort of effect shown in Fig 3.64—with sand stiffening when loads are left for a while so that the sand ages, even over the timescales of laboratory testing—but this effect is also familiar from set-up of driven piles in sand. Matsushita *et al.* (1999) show results from triaxial and plane strain

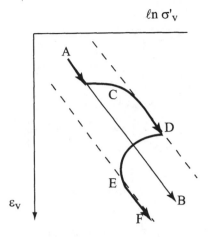

Figure 3.62: Effect of strain rate on shear stress:strain response of clay

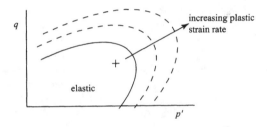

Figure 3.63: Strain rate dependence of 'yield' loci for clay

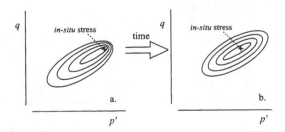

Figure 3.64: Effect of viscoplasticity on stress response envelopes for clay

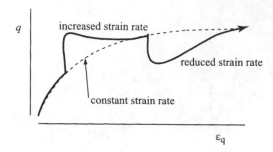

Figure 3.65: Effects of strain rate on shear stress:strain response of sand

tests on two sands which appear to show that: for testing at constant rate of strain the stress:strain response is rather independent of strain rate; if the strain rate is suddenly changed then the response overshoots the constant strain rate response but reverts to it with time (Fig 3.65); under constant stress creep occurs. The models that are being devised to reproduce these results are complex (DiBenedetto *et al.*, 2002) and introduce a viscosity which decays with time.

Clearly this is an area of active current research where data need to be gathered for many different soil types and under different test conditions. Physical explanation might well refer back to the force chains observed in particulate systems (Fig 2.6). These chains are immediately fragile when first formed and any small amount of particle movement—encouraged by ambient vibrations or by local crushing of particle contacts even at a microscopic scale—will allow adjustment of the chains leaving a more stable arrangement of particles. The challenge is evidently to build such physical thoughts into applicable constitutive models.

3.8   Design of programmes of laboratory tests

Parameters for soil models need to be obtained by calibration against laboratory tests—and perhaps *in-situ* tests and observations of performance of geotechnical systems. It should be clear that soils are nonlinear, history-dependent materials and that it is quite likely that any given soil model will only be able to give an approximate description of the actual mechanical behaviour of a particular soil. We may expect that the more complex the model that is adopted the more extensive will be the range of soil behaviour that it is able satisfactorily to reproduce. However, the models generally available for application in accessible numerical analysis programs will usually be at the simpler end of the modelling spectrum. Typically these would include the elastic-perfectly plastic Mohr-Coulomb model and the elastic-hardening plastic Cam clay model.

The subtlety of soil response should lead us to try to commission programmes of laboratory testing that follow stress paths which bear *some* resemblance to those that will be experienced by significant elements of soil around a geotechnical structure (Wood, 1984)—but, as we have seen in §2.3, the testing possibilities are in fact somewhat limited. In practice, most of the laboratory data with which we will be expected to calibrate our models will come from axisymmetric triaxial tests—very often (conventionally) conducted with constant cell pressure. However, if we can estimate the stress paths for typical elements in a prototype geotechnical system then we may have some idea of the sorts of initial stresses and stress changes that are likely to be relevant, even given the limited range of laboratory testing configurations that are available.

Just as for application of constitutive models to estimate response of soil elements—and also for performance of numerical analysis of the response of geotechnical systems—we have to think about the *past*, the *present* and the *future*. So far as the present is concerned we will usually be able to make a reasonably good estimate of the vertical total stress at any point in the ground, simply from the weight of overburden. To estimate the vertical effective stress we will then need some additional information about the pore water regime. Then, in order to estimate the horizontal stress—and hence estimate the *in-situ* shear stresses—we need to have some idea about the past history of the soil: how has it got to its present position? For soils which have had at least a somewhat one-dimensional history we can use empirical expressions for the earth pressure coefficient at rest, K_o, to estimate the horizontal stress provided we have some idea about the history of overconsolidation (see Muir Wood, 1990 for a more detailed discussion). For normally consolidated soils the value of K_{onc} is linked with angle of shearing resistance ϕ':

$$K_{onc} \approx 1 - \sin \phi' \qquad (3.175)$$

For overconsolidated soils, with overconsolidation ratio $n = \sigma'_{vmax}/\sigma'_v$, we can obtain an initial estimate using the expression

$$K_o \approx K_{onc}\sqrt{n} \qquad (3.176)$$

The value of K_o builds up with increasing overconsolidation: for heavily overconsolidated soils the value of K_o can approach the passive pressure coefficient.

So far as the future is concerned we will consider four examples. In each of them we are concerned to make order-of-magnitude estimates of many quantities which we will need to confirm through more detailed testing—but of course we are concerned to ensure that that detailed testing is as relevant as possible. There are no precise answers to these examples: the important thing is to think through the stress changes that are likely and make choices for testing that can be logically defended.

Example 1: A strip footing of width 2 m is to be founded at a depth of 0.5 m in a sandy soil. The water table is at a depth which will not influence the response of the footing.

Let us consider typical elements A and B beneath the footing and to the side of the footing, at a depth equal to half the width of the footing, and hence 1.5 m below the original ground level, as shown in Fig 3.66a. Of course the influence of the footing will extend to greater depths but much of the significant action will occur near the surface and any failure mechanism would be expected to extend to a depth no more than 1.5-2 times the width of the footing. We have

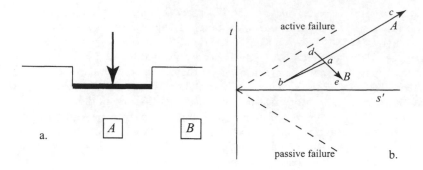

Figure 3.66: (a) Shallow footing on sand; (b) stress paths for elements A (abc) and B (ade)

to estimate a typical unit weight for the sand: we guess about 20 kN/m^3 so that the vertical stress at A and B is about $\sigma'_v \approx 30$ kPa. For a dense sand the angle of shearing resistance might be about 40° so that, from (3.175), $K_o \approx 0.35$ and $\sigma'_h \approx 10.5$ kPa. For this plane strain problem we can convert these stresses to a plane strain mean stress $s' = (\sigma'_v + \sigma'_h)/2 \approx 20$ kPa and plane strain shear stress $t = (\sigma'_v - \sigma'_h)/2 \approx 10$ kPa. We notice directly that the initial stresses at this typical element are quite low—much lower than the stresses which might routinely be used for laboratory testing.

Excavation removes a vertical stress of about 10 kPa. At A the horizontal stress will also reduce very slightly—we could estimate a new value from (3.176). The horizontal stress at B will reduce by roughly the same amount while the vertical stress remains unchanged. The initial stage of the stress paths at A and B is shown, exaggerratedly, in Fig 3.66b: paths ab and ad respectively.

Loading of the footing increases the vertical stress at A, and also increases the horizontal stress by some unknown amount. Evidently there will be less lateral restraint than there would be for one-dimensional oedometric loading so the stress path (bc in Fig 3.66b) will be somewhat steeper than the K_o path. Element B will experience similar changes in horizontal stress with no change in vertical stress (de in Fig 3.66b).

In designing our laboratory testing programme we accept that we probably cannot demand plane strain tests. We should choose triaxial tests with initial stresses corresponding to those estimated—perhaps $p' = 20$ kPa, $q = 20$ kPa. To model element A we might impose a compression stress path with slope $\delta q/\delta p' \approx 1.5 - 2$ (somewhat steeper than K_o) and for element B an extension stress path with constant cell pressure $\delta q/\delta p' = -1.5$. We might expect to make some allowance for the larger strengths expected in plane strain than in triaxial conditions in interpreting the results of these tests for design. If we wish to model the kinematic evolution of the incremental stiffness of the soil (§2.5.3), then we should probably include little excursions in the opposite direction before we set off on these stress paths because the stiffness always increases after significant changes in strain path direction.

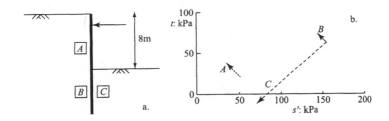

Figure 3.67: (a) Flexible retaining wall; (b) stress paths for elements A, B, C

Example 2: A numerical analysis is to be performed in order to estimate the deformations that might develop in the ground around a flexible retaining wall propped near the ground surface (Fig 3.67a) which is required to stabilise an excavation of depth 8 m in a free draining dense sand. The water table is at great depth. Typical elements A, B and C are shown.

Element A is at mid-height of the wall. Assume a unit weight $\gamma = 18$ kN/m^3, and angle of shearing resistance $\phi' - 35°$. Then vertical stress $\sigma'_v = 72$ kPa, $K_o = 1 - \sin \phi' = 0.43$, horizontal stress $\sigma'_h = 30$ kPa, $s' = 51$ kPa, $t = 21$ kPa. There will be little change in vertical stress, but reduction in horizontal stress as excavation proceeds: $\Delta t / \Delta s' \approx -1$, $\Delta t > 0$.

Element C is below the excavated soil, at a depth of 12 m, say. The initial vertical stress is $\sigma'_v = 216$ kPa, horizontal stress $\sigma'_h = 92$ kPa, $s' = 154$ kPa, $t = 62$ kPa. If the wall does not move then the major effect of the excavation is to reduce the vertical stress: $\Delta t / \Delta s' \approx +1$, $\Delta t < 0$. However, in fact the horizontal stress will fall somewhat so that $\Delta t / \Delta s' < 1$.

Element B behind the toe of the wall has an initial stress similar to element C. The vertical stress does not change much with excavation but horizontal stress falls more or less in step with the horizontal stress for element C: $\Delta t / \Delta s' \approx -1$, $\Delta t > 0$. Hence the stress paths shown in Fig 3.67b.

Triaxial tests with cell pressures between 30 kPa and 90 kPa might be reasonable. It would be a good idea to start with initial stress states matching the *in-situ* stress ratio. Conventional compression and extension with constant cell pressure might be acceptable but really it would be better to perform special tests in which the vertical stress is held constant while the horizontal stress is reduced to mimic elements A and B. Elements nearer the surface will of course have lower stress levels.

Example 3: An excavation is to be made in a clay slope to provide a building platform as shown in Fig 3.68a. It is anticipated that the retaining structure may form part of the eventual building. The interaction of soil and structure will influence the support forces and other structural resultants. Construction may be rapid and essentially undrained but eventually drainage will occur for long term effects. Numerical modelling is proposed and a programme of laboratory tests is required in order to obtain data which can be used to calibrate appropriate constitutive models. Estimate total and effective stress paths for typical elements A, B and C located approximately as shown.

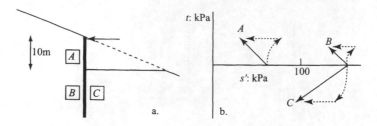

Figure 3.68: (a) Schematic diagram of excavation in clay slope; (b) stress paths for elements A, B, C (full lines: total stress paths; dotted lines: effective stress paths)

Soil-structure interaction is the central topic of Chapter 8: it is driven by stiffness or deformation properties rather than by strength and the testing needs to concentrate on these. We have seen in section §2.5.3 how sensitive stiffness can be to the detail of recent stress paths.

First ignore the slope. Element A is 5 m deep (say), $\gamma = 20$ kN/m^3; $\sigma_v = 100$ kPa, $u \approx 40$ kPa(?), $\sigma_v' \approx 60$ kPa, $K_o = 1 - 1.5$(?). Obviously we need to have a bit more knowledge of how the slope was formed, and of the current hydrological regime, in order to obtain a better indication of the *in-situ* stresses. We expect initial undrained plane strain response, with subsequent drainage as pore pressure equilibrium is established, probably with a modified flow regime. Excavation implies reduction in horizontal stress, with more or less constant vertical stress so that t increases and s decreases and $\Delta t/\Delta s \approx -1$. For such a path we expect the drained strength to be lower than the undrained strength.

Element B is 15 m deep (say), $\sigma_v = 300$ kPa, $u \approx 150$ kPa?, $\sigma_v' \approx 150$ kPa, $K_o = 1 - 1.5$? There will be some reduction in horizontal stress, with more or less constant vertical stress: t increases, s decreases and $\Delta t/\Delta s \approx -1$.

Element C has similar initial stresses to B. There will be reduction in vertical stress, some reduction in horizontal stress and a resulting passive/extension path.

So, we might propose a programme of at least consolidated undrained tests with pore pressure measurement. We should include compression and extension tests over an initial stress range 60-150 kPa. We should perform tests with a total stress path with reducing mean stress—this is a key deduction from consideration of the stress paths for elements A, B, C. Perform tests in which consolidation (drainage) is allowed at various stages towards the appropriate total stress path—these will provide data which can be used to calibrate a constitutive model which can be used to describe the long term response. (We can assume rough equivalence of s and p, t and $q/2$ in the first instance.)

One of the main problems will be the directions of principal stresses. Even in the slope before excavation they will not be vertical and horizontal. There will be the problem of initial anisotropy of samples if they are taken vertically (§2.5.4, Fig 2.52). There will be the usual problem of interpreting plane strain response from axially symmetric tests. We will need to think about the significance of

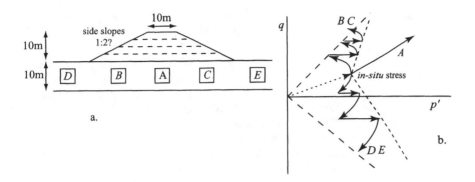

Figure 3.69: (a) Embankment constructed in stages; (b) stress paths for elements A, B, C, D, E

divergence of laboratory and field stress paths in the deviatoric π-plane (Figs 2.20b, 2.71).

Example 4: A long embankment with eventual height of 10 m and crest width 10 m is to be constructed at a site where the soil consists of 10 m of soft alluvial silty clays underlain by sands and gravels. It is intended to build this embankment in stages, allowing time for consolidation and strengthening of the soft soils between each stage of embankment loading.

- Identify typical soil elements within the soft soils which will influence the performance and design of the embankment and discuss, with appropriate sketches of stress paths, the changes in effective stress that are likely to occur at these elements.

- To what extent will tests in a conventional triaxial apparatus be useful for calibration of constitutive models for this application?

- Design a programme of laboratory tests that could be used to establish soil properties for this application.

- What other properties of the soil would you wish to explore in some detail before completing the design of the staged construction of this embankment?

The embankment is sketched in Fig 3.69a and typical elements A, B, C, D, E are shown. Stress paths are indicated in Fig 3.69b. For all elements there will be episodes of undrained or partially drained loading followed by consolidation. Support for the suggested paths can be found in numerical analysis of similar geotechnical systems (Almeida *et al.*, 1986; Muir Wood, 1990).

At element A the dominant effect is continuing confined compression—which may be quite close to one-dimensional compression as noted for the footing in *Example 1*. The eventual vertical stress increase is $10 \times 18 = 180$ kPa, estimating the unit weight of the fill $\gamma_{fill} = 18$ kN/m^3. The change in horizontal effective stress $\Delta\sigma'_h \approx K_o \Delta\sigma'_v \approx 0.7 \times 180 = 126$ kPa.

At elements D and E the dominant effect will be increase of horizontal stress by something less than 126 kPa with no change in vertical stress. This results in an extension or passive type of stress path.

Elements B and C are more difficult. At these locations, under the sides of the embankment, rotation of principal axes will certainly be important but we cannot easily study this in the standard laboratory except, to some extent, using simple shear testing which may not be available. Numerical studies (Almeida *et al.*, 1986) have shown that the undrained shearing stages for these elements under the sides of the embankment are much more damaging than those at A—they come closer to failure. The effect of consolidation after these undrained loadings will affect the subsequent undrained strength and this will control how high the next stage of the embankment can be constructed. The effective stresses reached at these elements after reconsolidation will certainly see less lateral constraint than one-dimensional compression—perhaps draining back to an effective stress path with zero horizontal stress change.

So far as our programme of triaxial tests is concerned, we should take samples from various depths—for example, 3, 6, 9 m—and reestablish *in-situ* stresses with an estimate of the *in-situ* value of K_o. Then subject these samples to one-dimensional compression for element A; constant axial stress extension with undrained episodes for elements D, E; multistage undrained tests almost to failure followed by reconsolidation to a constant total horizontal stress path for elements B and C.

We will certainly need information about *in-situ* permeability since this will control the rate at which consolidation occurs and hence the rate at which additional embankment layers can be added.

3.9 Selection of soil parameters: calibration of models

As a simple exercise in parameter selection, we will show how soil parameters might be selected to match the response observed in a single drained triaxial compression test on normally consolidated Weald clay (Fig 3.70).

The most commonly required parameter selection is certainly the most subjective: the choice of parameters for an elastic-perfectly plastic Mohr-Coulomb model. We have seen that this model can only decribe a constant linear elastic response up to yield/failure—and then the tangent stiffness falls to zero. During the (isotropic) elastic phase volume changes only occur if the imposed stress path includes change in mean effective stress; once plastic failure occurs volume change occurs at a continuing steady rate—either compressive or expansive—indefinitely. We have to decide in choosing the soil parameters whether we are attempting to match the overall response moderately or certain aspects of the response in detail. One possible fitting is shown in Fig 3.70 (EPP). The elastic properties are chosen to give a good match on average: the initial stiffness is underestimated. The plastic properties give a slight underestimate of the strength, and evidently indicate volumetric compression continuing much longer than is actually observed. There are obviously many perfectly defensible alternative

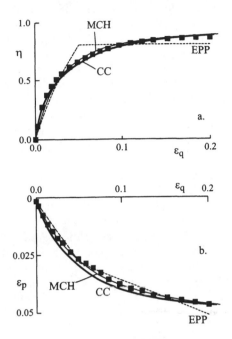

Figure 3.70: Conventional drained triaxial compression test on Weald clay (data from Bishop and Henkel, 1962) and fitting of constitutive models (initial specific volume $v_i = 1.64$, confining pressure $\sigma'_r = 207$ kPa (EPP: elastic-perfectly plastic Mohr-Coulomb model; MCH: extended (hardening) Mohr-Coulomb model; CC: Cam clay)

sets of parameters for this model. For the simulation shown: $G = 1500$ kPa, $K = 2800$ kPa, $M = 0.8$ and $M^* = -0.15$. The value of Poisson's ratio is $\nu = 0.273$.

It turns out that with a hardening plastic model—whether the extended Mohr-Coulomb model or Cam clay—it is quite possible, by trial and error, to obtain really quite a good match to both the stress:strain and the volumetric response of the soil in this test. In this particular case the Mohr-Coulomb model with its hyperbolic distortional hardening law is perhaps slightly better than Cam clay with its logarithmic volumetric hardening law—but with further perseverance in trial and error selection of parameters better fits might be obtained.

The values of soil parameters used for the Mohr-Coulomb model (MCH in Fig 3.70) are: $G = 3800$ kPa, $K = 6000$ kPa, $M = 0.91$, $a = 0.015$ and $\eta_p = 0.95$. The value of Poisson's ratio implied by the elastic properties is $\nu = 0.238$. The elastic stiffnesses are higher because—as seen in the worked example using the extended Mohr-Coulomb model (§3.4.1)—plastic strains occur right from the start of the test. Deformations that are being entirely described by the elastic properties in the elastic-perfectly plastic model are now being described as a combination of elastic and plastic effects.

For the Cam clay model (CC in Fig 3.70) the parameters giving a similar quality of fit are: elastic properties $G = 3500$ kPa and $\kappa = 0.015$—implying an initial bulk modulus $K = vp'/\kappa = 22632$ kPa and Poisson's ratio $\nu = 0.426$; and plastic properties $\lambda = 0.055$ and $M = 0.9$. The value of the intercept N on the normal compression line is $N = 1.933$. Whereas the values of shear modulus G are similar for these two hardening plastic models, Cam clay predicts a lot of plastic volumetric strain at the start of the test—the plastic strain increment, normal to the elliptical yield locus, indicates only plastic volumetric strain to start with—and the elastic properties indicate near incompressibility ($\nu \to 0.5$) in order to ensure that there is negligible additional elastic volumetric strain. The values of M are similar.

We conclude that we cannot determine the optimum selection of model by fitting data for a single test. We have seen that in the simple form presented here, the Mohr-Coulomb models will not predict significant volumetric strain for stress paths which load the soil at more or less constant stress ratio. If we believe that such paths are going to be important in the behaviour of our geotechnical system then we need to ensure that we have data from special triaxial tests with which to calibrate our model. The more sets of data that we attempt to fit simultaneously the less likely it is that we will be able to achieve a fit as close as that shown in Fig 3.70. Often, faced with data from tests of varying reliability we may wish to weight differently the several sets of data and perhaps attempt some algorithmic best overall fit to give greater objectivity to our parameter selection (see, for example, Muir Wood *et al.*, 1993)—visual fitting may introduce some unconscious bias.

We deduce the importance of trying to ensure that the paths followed in our laboratory tests bear some resemblance to the range of significant paths that will be followed in our geotechnical system (§3.8). The stiffness characteristics of soils are so sensitive to the detail of history and stress path (§2.5.3) that using a constitutive model to extrapolate from inappropriate limited laboratory testing may not lead to reliable estimates of response of geotechnical systems—especially under working loads, far from failure, where the detail of rather small strains will be crucial.

4

Numerical modelling

4.1 Introduction

There are many books which deal in great detail with the application of numerical methods—usually finite element methods—to engineering problems in general or to geotechnical problems in particular (eg Zienkiewicz and Taylor, 2000; Cook, Malkus and Plesha, 1989; Cook, 1995; Livesley, 1983; Smith and Griffiths, 1988; Britto and Gunn, 1987; Potts and Zdravković, 1999). It is intended here to provide merely a brief introduction to numerical modelling: enough for the reader to be able to understand some of the language of numerical modelling, some of the issues that need to be confronted when setting about numerical modelling of a geotechnical problem, and some of the pitfalls that may confront the numerical modeller.

Chapter 3 has presented in some detail some of the constitutive models that might be used to describe the mechanical response of soils. It is clear from the discussion of key aspects of soil behaviour in section §2.5 that elastic models are unlikely to be especially satisfactory except in limited applications. If the material can be deemed to be linearly elastic then many of the details of numerical analysis become rather straightforward—and, in particular, there are many existing solutions for distributions of stresses and displacements in elastic systems that can be readily adapted (see, for example, Poulos and Davis, 1974). For more nonlinear and history dependent (elastic-plastic) materials numerical analysis is almost certain to be required except for the most trivial of applications.

We start by deriving the governing equations for mechanical and flow problems in one dimension. This apparently trivial beginning allows us to illustrate the development of a number of aspects of the finite element approximation which can be readily extended to two and three dimensions. The governing equations are also presented for the two-dimensional problem: parallels with the one-dimensional equations will be drawn. The finite difference approximation to differential equations will be described briefly—this will often be needed for solution of problems involving time, such as dynamic loading or transient flow.

4.2   Field problems

Our concern is with the solution of field problems for which we are able to
write down governing partial differential equations which describe the way in
which quantities of interest (field variables) must vary within a particular region and must satisfy boundary conditions at the edges of that region. We
will concentrate particularly on problems of stress analysis where the quantities
of interest are stresses and displacements but we might also be concerned with
other geotechnical field problems such as analysis of seepage and flow (where the
field variable is the pressure head) and the coupled flow and mechanical response
that governs the consolidation process (where the field variables now combine
pore pressures with stresses and displacements). More generally we might be
concerned with flow of heat or migration of pollutants, so that field variables
would include temperature or pollutant concentration respectively. And it can
be expected in the end that all of these effects might interact and require analytical coupling: there will be obvious mechanical consequences of the changes
in dimension that accompany temperature changes; changes in the chemical
constitution of the fluid in the pores of a soil may well influence its mechanical
characteristics. However, provided we can assemble a set of physically reasonable equations which describe the various interactions and flows then we are
well on the way towards setting up a numerical analysis of the problem.

4.2.1   One-dimensional problem

We start by developing the equations which govern the behaviour of a one-
dimensional problem and can then generalise these equations to a fully three-
dimensional problem. Consider the element shown in Fig 4.1a. Equilibrium
tells us that the gradient of total stress must satisfy the equation:

$$\frac{\partial \sigma_z}{\partial z} - \gamma_z = 0 \tag{4.1}$$

where γ_z is the body force acting in the negative z direction—in this case the
unit weight of the soil. Throughout this chapter we will regard *tensile* stresses
and strains as positive. This is not the usual soil mechanics convention but
makes the development of the mathematics more straightforward.[1]

In general the stresses in the soil will be associated with displacements and
the definition of strain allows us to write down a compatibility equation (Fig
4.1b):

$$\frac{\partial u_z}{\partial z} = \epsilon_z \tag{4.2}$$

where u_z is the displacement in the z direction and *tensile* strains are regarded
as positive.

---

[1]When preparing a chapter on the present topic one becomes all too aware of the distress-
ingly finite nature of the Greek and Roman alphabets. In an attempt to reduce confusion
somewhat (but at the expense of inconsistency with other chapters) the symbol ϱ will be used
for pore pressure so that u always represents displacement.

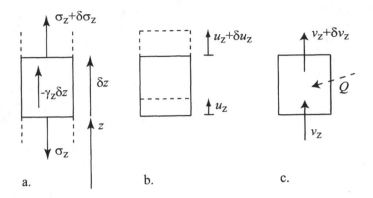

Figure 4.1: (a) Equilibrium of one-dimensional element; (b) displacements of one-dimensional element; (c) flow through one-dimensional element

We understand that changes in strains in soils arise because of changes in *effective* stresses so we cannot progress without introducing the principle of effective stress:

$$\sigma_z = \sigma'_z - \varrho \tag{4.3}$$

linking total stress σ_z, effective stress σ'_z and pore pressure ϱ. Tensile stresses are regarded as positive but pore pressure is, as usual, positive for pressures above ambient pressure.

Then we have a constitutive equation which links changes in effective stress and development of strain which, for this simple one-dimensional problem, introduces a one-dimensional constrained stiffness E_{oed} (which will, in general, not be a soil constant but will vary with volumetric compression of the soil):

$$\frac{\partial \sigma'_z}{\partial \epsilon_z} = E_{oed} \tag{4.4}$$

Now in general we may have some flow of pore fluid occurring through our soil element. We can write down one equation describing the conservation of volume changes of the element linked with this flow (Fig 4.1c):

$$-\frac{\partial v_z}{\partial z} + Q = \frac{\partial \epsilon_z}{\partial t} + \frac{n}{K_f}\frac{\partial \varrho}{\partial t} \tag{4.5}$$

where v_z is the velocity of flow in the positive z direction, Q is the flow per unit volume *into* the element (the source), n is porosity and K_f is the bulk modulus of the fluid. The porosity n indicates the proportion of the volume of the element that is taken up with the fluid. (Note again that tensile strains are regarded as positive.) This equation recognises that changes in pore pressure will lead to changes in the volume of the fluid stored in a soil element. For steady flow the right hand side of (4.5) is zero. For undrained conditions the left hand side is zero (no flow into or out of the soil element) and a small volume change is associated with the small but non-zero compressibility of the pore fluid.

We assume that the flow of water through the element is governed by the permeability of the soil k and by the gradient of some potential, the total head H, through Darcy's law:

$$v_z = -k\frac{\partial H}{\partial z} = -\frac{k}{\gamma_f}\frac{\partial \varrho_w}{\partial z} \tag{4.6}$$

where Bernoulli's equation distinguishes the total head H (and corresponding fluid pressure ϱ_w) that drives flow from the pressure head ϱ/γ_f and the elevation head z:

$$H = \frac{\varrho_w}{\gamma_f} = \frac{\varrho}{\gamma_f} + z \tag{4.7}$$

where γ_f is the unit weight of the pore fluid.

Combination of these equations produces two simultaneous partial differential equations in the two field variables, displacement u_z (or, in fact, strain ϵ_z since the displacement only enters through its spatial gradient) and pore pressure ϱ, which we would need to solve satisfying imposed boundary conditions:

$$-E_{oed}\frac{\partial^2 u_z}{\partial z^2} + \frac{\partial \varrho}{\partial z} + \gamma_z = 0 \tag{4.8}$$

$$\frac{\partial^2 u_z}{\partial t \partial z} - \frac{k}{\gamma_f}\frac{\partial^2 \varrho}{\partial z^2} + \frac{n}{K_f}\frac{\partial \varrho}{\partial t} - Q = 0 \tag{4.9}$$

We have thus combined statements of equilibrium (4.1), strain compatibility (4.2) and conservation of volume (4.5), with constitutive laws governing the stress-strain response of the soil (4.4) and the flow characteristics of the soil (4.6) in order to provide sufficient equations to be able, in principle, to deduce the values of our field variables. We might in general wish to know the variation of stress with position and stress is in principle an additional field variable. However, it is clear from (4.2) and (4.4) that once we know the displacement, and more particularly the gradient of displacement, we can calculate the stresses without further ado.

Terzaghi's equation of one-dimensional consolidation can be deduced from (4.9). The total stress is assumed constant so that changes in pore pressure and effective stress are equal (recall the sign convention in (4.3) and

$$\frac{\partial^2 u_z}{\partial t \partial z} + \frac{n}{K_f}\frac{\partial \varrho}{\partial t} = \frac{\partial \epsilon_z}{\partial t} + \frac{n}{K_f}\frac{\partial \varrho}{\partial t} = \frac{1}{E_{oed}}\frac{\partial \sigma'_z}{\partial t} + \frac{n}{K_f}\frac{\partial \varrho}{\partial t} = \left(\frac{1}{E_{oed}} + \frac{n}{K_f}\right)\frac{\partial \varrho}{\partial t} \tag{4.10}$$

and thence, from (4.9) with $Q = 0$, if we assume that the pore fluid is incompressible so that $K_f = \infty$,

$$\frac{\partial \varrho}{\partial t} = \frac{kE_{oed}}{\gamma_f}\frac{\partial^2 \varrho}{\partial z^2} = c_v\frac{\partial^2 \varrho}{\partial z^2} \tag{4.11}$$

where $c_v = kE_o/\gamma_f$ is the coefficient of consolidation.

For a drained equilibrium analysis in which flow is of no concern, and hence H is constant (4.7), the gradient of pore pressure is given by

$$\frac{\partial \varrho}{\partial z} = -\gamma_f \tag{4.12}$$

and the equilibrium equation (4.8) becomes

$$-E_{oed}\frac{\partial^2 u_z}{\partial z^2} + (\gamma_z - \gamma_f) = 0 \tag{4.13}$$

or

$$-\frac{\partial \sigma'_z}{\partial z} + (\gamma_z - \gamma_f) = 0 \tag{4.14}$$

and the difference of unit weights, $(\gamma_z - \gamma_f)$, would be described as the buoyant unit weight of the soil.

4.2.2 Two-dimensional problem

Many geotechnical systems can be seen as two-dimensional problems which can be analysed in plane strain. There are obviously additional degrees of freedom by comparison with the one-dimensional problem. However, subsequent extension to three dimensions merely increases the number of degrees of freedom without particularly influencing the structure of the governing equations.

We assume that the problem is defined within cartesian axes $(x\ y)$, where the y axis will typically be vertical but this will not be a necessary restriction. We start by defining a vector of stresses, $\boldsymbol{\sigma} = (\sigma_{xx}\ \sigma_{yy}\ \tau_{xy})^T$ and a corresponding vector of strains $\boldsymbol{\epsilon} = (\epsilon_{xx}\ \epsilon_{yy}\ \gamma_{xy})^T$. There will in general be a vector of body forces per unit volume $\boldsymbol{F} = (F_x\ F_y)^T$. If the body forces come purely from the unit weight of the soil then we would expect $\boldsymbol{F} = \gamma\hat{\boldsymbol{g}}$ where γ is the total unit weight of the soil and $\hat{\boldsymbol{g}}$ is a unit vector in the direction of gravitational acceleration. Typically, with vertical y axis, $\hat{\boldsymbol{g}} = (0\ -1)^T$. The vector differential $\boldsymbol{\nabla}$:

$$\boldsymbol{\nabla} = \begin{pmatrix} \partial/\partial x \\ \partial/\partial y \end{pmatrix} \tag{4.15}$$

and the differential matrix $\boldsymbol{\partial}$:

$$\boldsymbol{\partial} = \begin{pmatrix} \partial/\partial x & 0 \\ 0 & \partial/\partial y \\ \partial/\partial y & \partial/\partial x \end{pmatrix} \tag{4.16}$$

will be useful.

Equilibrium then requires that (compare (4.1)):

$$\boldsymbol{\partial}^T \boldsymbol{\sigma} + \boldsymbol{F} = 0 \tag{4.17}$$

Kinematic compatibility (the definition of strain) implies that (compare (4.2)):

$$\boldsymbol{\epsilon} = \boldsymbol{\partial} \boldsymbol{u} \tag{4.18}$$

where $\boldsymbol{u} = (u_x\ u_y)^T$ is a vector displacement.

The definition of effective stress (compare (4.3)) becomes:

$$\boldsymbol{\sigma} = \boldsymbol{\sigma}' - \boldsymbol{\mu}\varrho \tag{4.19}$$

introducing the vector $\boldsymbol{\mu} = (1\ 1\ 0)^T$.

With a constitutive link between stresses and strains, $\sigma' = D\epsilon$ our equilibrium equation (4.17) becomes:

$$\partial^T D \partial u - \partial^T \mu \varrho + F = 0 \qquad (4.20)$$

Flow through a soil is controlled by the permeability of the soil, which in general can be described by a permeability matrix k linking flow velocities $v = (v_x \; v_y)^T$ with the gradient of total head H. In general

$$k = \begin{pmatrix} k_{xx} & k_{xy} \\ k_{yx} & k_{yy} \end{pmatrix} \qquad (4.21)$$

to allow for anisotropy of permeability. Then

$$v = -k\nabla H \qquad (4.22)$$

where the total head H is given from Bernoulli's equation:

$$H = \frac{\varrho}{\gamma_f} - r^T \hat{g} \qquad (4.23)$$

where ϱ is the pore pressure, γ_f is the unit weight of the pore fluid, and $r = (x \; y)^T$ is the position vector.

Continuity requires a volume balance (strictly a mass balance but we will neglect density changes in the flowing fluid). There will in general be sources or sinks giving a nett flow Q per unit volume into a soil element. The soil element will change in volume because it undergoes changes in effective stress and the pore fluid may itself change in volume as the pore pressure changes. Combining these effects, and invoking (4.22 and 4.23) (compare (4.5)):

$$-\nabla^T v + Q = \nabla^T k \nabla H + Q = \frac{\partial \epsilon_p}{\partial t} + \frac{n}{K_f} \frac{\partial \varrho}{\partial t} \qquad (4.24)$$

We can link the volumetric strain ϵ_p with the general two-dimensional strain ϵ:

$$\epsilon_p = \mu^T \epsilon \qquad (4.25)$$

Hence (compare (4.9))

$$\mu^T \partial \frac{\partial u}{\partial t} - \frac{1}{\gamma_f} \nabla^T k \nabla \varrho + \frac{n}{K_f} \frac{\partial \varrho}{\partial t} - Q = 0 \qquad (4.26)$$

The presence of displacement u in (4.26) and the presence of pore pressure ϱ in (4.20) lead to coupling between the flow and mechanical effects in the soil.

Various special cases can be extracted from these general equations. Terzaghi's consolidation equation is obtained if we eliminate sources and sinks, $Q = 0$, assume the pore fluid to be incompressible and specify that the total stress is held constant. The volumetric strain in the element arises because the effective stress changes directly with the pore pressure:

$$\epsilon_p = \mu^T D^{-1} \mu \varrho \qquad (4.27)$$

and

$$\frac{1}{\gamma_f}\boldsymbol{\nabla}^T\boldsymbol{k}\boldsymbol{\nabla}\varrho = \boldsymbol{\mu}^T\boldsymbol{D}^{-1}\boldsymbol{\mu}\frac{\partial\varrho}{\partial t} \qquad (4.28)$$

With incompressible fluid $K_f = \infty$ and steady state conditions, so that all time differentials are zero,

$$\frac{1}{\gamma_f}\boldsymbol{\nabla}^T\boldsymbol{k}\boldsymbol{\nabla}\varrho + Q = 0 \qquad (4.29)$$

which is Laplace's equation, as expected.

4.3   One-dimensional finite elements

We take a continuum approach to geotechnical systems and assume that our field variables vary continuously throughout our region of interest. We know that in certain circumstances the governing equations (4.8) and (4.9) can be solved analytically. The situations where this will be possible will be much more frequent for one-dimensional problems than for fully three-dimensional problems but even for one-dimensional problems analytical solutions may become tricky if the stiffness properties (encapsulated in E_{oed}) are nonlinear and/or history dependent. However, our concern here is to consider approximate solutions where, instead of discovering the values of our field variables at *every* point within our continuum, we aim to find the values at a finite number of points only. We will concentrate on equilibrium problems where our aim is to discover a field of displacements and stresses.

Let us divide the one-dimensional problem up into a series of elements of typical length ℓ connected at their nodes (Fig 4.2). The displacements at the bottom and top of a typical element are u_{z1} and u_{z2} and we assume that we have some description of the variation of displacement within the element using so-called interpolation or shape functions N_1 and N_2 such that, within the element

$$u_z = N_1 u_{z1} + N_2 u_{z2} \quad \text{or} \quad u = \begin{pmatrix} N_1 & N_2 \end{pmatrix}\begin{pmatrix} u_{z1} \\ u_{z2} \end{pmatrix} \quad \text{or} \quad u = \boldsymbol{Nd} \quad (4.30)$$

An obvious simple form for these shape functions, in terms of a *local* coordinate z for a given element (Fig 4.2b), might be

$$N_1 = \frac{\ell - z}{\ell}; \quad N_2 = \frac{z}{\ell} \qquad (4.31)$$

These describe a linear variation of displacement within the element and have the evidently desirable characteristic that $N_1 = 0$ for $z = \ell$ and $N_2 = 0$ for $z = 0$. The strain at any point within the element is then given by:

$$\epsilon_z = \frac{\partial N_1}{\partial z}u_{z1} + \frac{\partial N_2}{\partial z}u_{z2} \quad \text{or} \quad \epsilon_z = \frac{\partial \boldsymbol{N}}{\partial z}\boldsymbol{d} \qquad (4.32)$$

Accompanying the nodal displacements there will be nodal forces F_1 and F_2 at the ends of the elements transferring stresses from one element to the

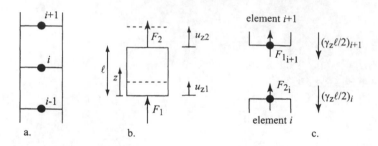

Figure 4.2: (a) One-dimensional elements connected at nodes; (b) forces and displacements at nodes; (c) nodal equilibrium of connected elements

next—assuming unit cross sectional area for the one-dimensional elements. We suppose that all external loads, including body forces, are applied at the nodes in some appropriately distributed form—for this simple one-dimensional element we can divide the body force $\gamma_z \ell$ equally between the two ends of the element. Equilibrium is applied at the node as the force leaving one element becomes the force entering the next element (Fig 4.2c). Equilibrium at the connection between element i and element $i + 1$ then tells us that

$$(F_2)_i + (F_1)_{i+1} - \left(\frac{\gamma_z \ell}{2}\right)_i - \left(\frac{\gamma_z \ell}{2}\right)_{i+1} = 0 \qquad (4.33)$$

In general, the lengths and unit weights of successive elements might be different.

Information can only be passed between elements at the nodes. We take the displacements of the nodes as the independent variables and therefore have to look for ways in which we can calculate the nodal forces as dependent variables. We are searching for a link between the nodal forces and the nodal displacements in the form of a stiffness matrix K:

$$\left(\begin{array}{c} F_1 \\ F_2 \end{array}\right) = \left(\begin{array}{cc} k_{11} & k_{12} \\ k_{21} & k_{22} \end{array}\right) \left(\begin{array}{c} u_{z1} \\ u_{z2} \end{array}\right) \quad \text{or} \quad F = Kd \qquad (4.34)$$

For this simple one-dimensional element it is not difficult to draw the simplest link between nodal forces and nodal displacements, through the stiffness properties of the material in the element, and deduce

$$\left(\begin{array}{c} F_1 \\ F_2 \end{array}\right) = \frac{E_{oed}A}{\ell} \left(\begin{array}{cc} 1 & -1 \\ -1 & 1 \end{array}\right) \left(\begin{array}{c} u_{z1} \\ u_{z2} \end{array}\right) \quad \text{or} \quad F = Kd \qquad (4.35)$$

maintaining a careful sign convention that forces and displacements are positive in the positive z direction. The cross-sectional area A is included in (4.35) even though we are considering a unit section, in order to remind ourselves of the necessary dimensional consistency of the expression.

For more elaborate elements for analysis of two- and three-dimensional problems it is not easy to deduce the stiffness matrix by this direct route. A more

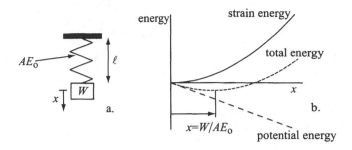

Figure 4.3: (a) Spring extended by weight W; (b) components of energy for extending spring

general procedure is obtained by thinking about strain energy in the element. Imagine a weight W being applied to an unstressed spring of length ℓ and stiffness AE_o (Fig 4.3a). As the weight displaces slowly through a distance x it loses potential energy $-Wx$. As the spring extends it stores strain energy $\frac{1}{2}AE_ox^2$. The total energy of the system is then

$$V = \frac{1}{2}AE_ox^2 - Wx \qquad (4.36)$$

and this has a minimum (Fig 4.3b) at

$$x = \frac{W}{AE_o} \qquad (4.37)$$

which is the expected stable extension of the spring. (An identical result can be obtained using a virtual work approach (Livesley, 1983).)

This is an example of a general principle of stationary potential energy which states that: *Among all admissible configurations of a conservative system, those that satisfy the equations of equilibrium make the potential energy stationary with respect to small admissible variations of displacement* (Cook *et al.*, 1989). The configurations of interest to us are the values of nodal displacements. The principle as stated applies to an entire system: we will make the assumption that the same principle can also be applied to individual elements within the system.

As for the simple spring in Fig 4.3, we have two components of potential energy: the strain energy in the element and the work done by the forces acting on the element. The strain at any point within the element is given by (4.32), the stress is then

$$\sigma_z = E_{oed}\frac{\partial N}{\partial z}d \qquad (4.38)$$

and the strain energy in the element of length ℓ and cross sectional area A is

$$V_E = \int_{vol} \frac{1}{2}\left(\frac{\partial N}{\partial z}d\right)^T E_{oed}\left(\frac{\partial N}{\partial z}d\right)\,d(\text{vol}) =$$

$$\frac{1}{2}A\boldsymbol{d}^T \left[\int_0^\ell \left(\frac{\partial \boldsymbol{N}}{\partial z} \right)^T E_{oed} \left(\frac{\partial \boldsymbol{N}}{\partial z} \right) \mathrm{d}z \right] \boldsymbol{d} \qquad (4.39)$$

The potential energy of the nodal forces is

$$V_P = -\boldsymbol{d}^T \boldsymbol{F} \qquad (4.40)$$

Seeking a minimum of the total potential energy with respect to the unknown nodal displacements we find

$$A\delta\boldsymbol{d}^T \left\{ \left[\int_0^\ell \left(\frac{\partial \boldsymbol{N}}{\partial z} \right)^T E_{oed} \left(\frac{\partial \boldsymbol{N}}{\partial z} \right) \mathrm{d}z \right] \boldsymbol{d} - \boldsymbol{F} \right\} = 0 \qquad (4.41)$$

which must be satisfied by all possible displacements $\delta\boldsymbol{d}$. Hence we have a set of equations

$$\boldsymbol{F} = A \left[\int_0^\ell \left(\frac{\partial \boldsymbol{N}}{\partial z} \right)^T E_{oed} \left(\frac{\partial \boldsymbol{N}}{\partial z} \right) \mathrm{d}z \right] \boldsymbol{d} \qquad (4.42)$$

This is an expression of the virtual work principle which states that the total work of internal and external forces must vanish for any *admissible* infinitesimal displacement from an *equilibrium* configuration.

From our definition of $\boldsymbol{N}$ we know that

$$\frac{\partial \boldsymbol{N}}{\partial z} = (\ -1/\ell \quad 1/\ell \ ) \qquad (4.43)$$

so that

$$A \left[\int_0^\ell \left(\frac{\partial \boldsymbol{N}}{\partial z} \right)^T E_{oed} \left(\frac{\partial \boldsymbol{N}}{\partial z} \right) \mathrm{d}z \right] =$$

$$= E_{oed} A\ell \left(\begin{array}{c} -1/\ell \\ 1/\ell \end{array} \right) (\ -1/\ell \quad 1/\ell \ ) = \frac{E_{oed} A}{\ell} \left(\begin{array}{cc} 1 & -1 \\ -1 & 1 \end{array} \right) \qquad (4.44)$$

which exactly matches the stiffness matrix $\boldsymbol{K}$ defined in (4.35).

4.4   Two-dimensional finite elements

This route to deduction of the form of the stiffness matrix from energy considerations can be extended to two and three dimensional systems. The detailed demonstration will not be shown here but can be found in standard texts on the finite element method (see §4.1). We have to note that the expression in (4.44) from which the stiffness matrix is generated is an integration over the volume of the element of the product of three terms. The first and third terms $(\partial \boldsymbol{N}/\partial z)$ represent, in general, the matrix of expressions which convert nodal displacements to strains within the element, where $\boldsymbol{N}$ are now more general shape functions which describe the variation of displacement within the element in terms of the values of displacement at the nodes. The term E_{oed} becomes a more general stiffness matrix for the material.

For example, in a two-dimensional problem we might have a field of displacements u_x and u_y in the x and y directions respectively. The strains are then

$$\epsilon_x = \frac{\partial u_x}{\partial x}; \quad \epsilon_y = \frac{\partial u_y}{\partial y}; \quad \gamma_{xy} = \frac{\partial u_x}{\partial y} + \frac{\partial u_y}{\partial x} \tag{4.45}$$

or

$$\epsilon = \partial u \tag{4.46}$$

Within the element the displacement field u is linked to the nodal displacements through as many shape functions as there are nodes:

$$\left(\begin{array}{c} u_x \\ u_y \end{array} \right) = \left(\begin{array}{ccccc} N_1 & 0 & N_2 & 0 & \cdots \\ 0 & N_1 & 0 & N_2 & \cdots \end{array} \right) \left(\begin{array}{c} u_{x1} \\ u_{y1} \\ u_{x2} \\ u_{y2} \\ \cdots \end{array} \right) \quad \text{or} \quad u = Nd \tag{4.47}$$

Then

$$\epsilon = \partial Nd \quad \text{or} \quad \epsilon = Bd \quad \text{where} \quad B = \partial N \tag{4.48}$$

With a stiffness matrix D linking the changes in stresses that result from changes in strains (developed from one of the candidate models described in Chapter 3, for example), the stiffness matrix for the two-dimensional finite element becomes:

$$K = \int_V B^T DB \mathrm{d}V \tag{4.49}$$

where V is the volume of the element.

For problems involving flow, as in (4.47), we write both displacement and pore pressure in terms of nodal values of these quantities invoking shape functions for displacement N and for pore pressure N_ϱ which will in general be different:

$$u = Nd \quad \text{and} \quad \varrho = N_\varrho \varrho_w \tag{4.50}$$

where ϱ_w is the vector of nodal values of pore pressure. Ultimately, through argument similar to that just adopted (see Smith and Griffiths, 1988 or Potts and Zdravković, 1999), we can convert the governing equations ((4.26) and (4.20)) into equations involving integrals of soil properties over the elements:

$$Kd - T\varrho_w + F = 0 \tag{4.51}$$

and

$$T^T \frac{\partial d}{\partial t} - S \frac{\partial \varrho_w}{\partial t} - R\varrho_w - Q = 0 \tag{4.52}$$

where K was given in (4.49) and

$$T = \int_V B^T \mu N_\varrho \mathrm{d}V \tag{4.53}$$

$$S = \int N_\varrho^T \frac{n}{K_f} N_\varrho \mathrm{d}V \tag{4.54}$$

$$R = \frac{1}{\gamma_f} \int\limits_V (\nabla N_\varrho)^T k (\nabla N_\varrho) dV \tag{4.55}$$

The effects of flow have thus been presented in a way that is compatible with the effects of stress change and the problem is defined in terms of a number of standard integrals.

Analysis of flow—even steady flow—contains the possible complexity of the need to model a free fluid surface within the soil separating saturated from unsaturated or dry soil. The location of this boundary to the flow is initially unknown—and in unsteady flow this will be a moving unknown boundary. Potts and Zdravković (1999) suggest that this is an area where robust algorithms do not yet exist and further research is required. The user of a finite element code is warned to take care in setting up the analysis and interpreting the results.

The choice of shape function is an important consideration. It will very often be hidden from the user of a finite element program, or there may be very little choice available. However, the type of element and associated shape function will have a major influence on the accuracy with which continuous strain fields can be reproduced and hence on the accuracy of numerical results which are obtained.

4.4.1 Example: Constant strain triangle

A typical triangular element is shown in Fig 4.4. This element has three nodes located at its vertices and can be imagined to be attached to adjacent elements only at these nodes. The displacement field within the element is given by:

$$u_x = \alpha_1 + \alpha_2 x + \alpha_3 y; \quad u_y = \alpha_4 + \alpha_5 x + \alpha_6 y \tag{4.56}$$

so that the strain field within the element is:

$$\epsilon_x = \frac{\partial u_x}{\partial x} = \alpha_2; \quad \epsilon_y = \frac{\partial u_y}{\partial y} = \alpha_6; \quad \gamma_{xy} = \frac{\partial u_y}{\partial x} + \frac{\partial u_x}{\partial y} = \alpha_5 + \alpha_3 \tag{4.57}$$

and components α_1 and α_4 merely generate rigid body displacements. The displacement varies linearly so that the strains are constant within the element. The link between the coefficients $\alpha_1, \cdots, \alpha_6$ and the nodal displacements is somewhat tedious to derive (see, for example, Cook et al., 1989) but in the end the strain:displacement relationship can be written in the form $\epsilon = Bd$:

$$\begin{pmatrix} \epsilon_x \\ \epsilon_y \\ \gamma_{xy} \end{pmatrix} = \frac{1}{2A} \begin{pmatrix} y_{23} & 0 & y_{31} & 0 & y_{12} & 0 \\ 0 & x_{32} & 0 & x_{13} & 0 & x_{21} \\ x_{32} & y_{23} & x_{13} & y_{31} & x_{21} & y_{12} \end{pmatrix} \begin{pmatrix} u_{x1} \\ u_{y1} \\ u_{x2} \\ u_{y2} \\ u_{x3} \\ u_{y3} \end{pmatrix} \tag{4.58}$$

where $x_{ij} = x_i - x_j$; $y_{ij} = y_i - y_j$; x_i, y_i $(i = 1, 2, 3)$ are nodal coordinates, numbered sequentially anticlockwise round the element as shown in Fig 4.4, and $2A$ is twice the area of the element, so that $2A = x_{21} y_{31} - x_{31} y_{21}$ (or any other

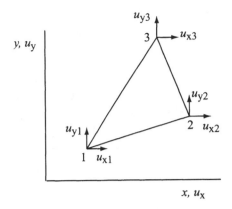

Figure 4.4: Constant strain triangle element

expression obtained by appropriate permutation). (If the nodes are numbered sequentially in the clockwise direction round the element then the area A will be negative and the sign in (4.58) has to be changed accordingly.) Since there is no variation of strain within the element, the integral in expression (4.49) for the element stiffness can be written exactly

$$K = B^T DBtA \qquad (4.59)$$

where t is the element thickness.

It will be evident that the constant strain triangle is not going to be well suited to analysis of problems which contain significant gradients of strain: it is not particularly good for problems which involve bending.

4.4.2 Example: Linear strain triangle

If we want to be able to describe more elaborate variations of strain within elements then we will usually need to link the elements together at additional side nodes in addition to the vertices. For example, the linear strain triangle (Fig 4.5) has additional nodes at the mid-points of each side. This element can sustain a full quadratic displacement field

$$
\begin{align}
u_x &= \alpha_1 + \alpha_2 x + \alpha_3 y + \alpha_4 x^2 + \alpha_5 xy + \alpha_6 y^2 \qquad (4.60)\\
u_y &= \alpha_7 + \alpha_8 x + \alpha_9 y + \alpha_{10} x^2 + \alpha_{11} xy + \alpha_{12} y^2 \qquad (4.61)
\end{align}
$$

and a corresponding strain field

$$
\begin{align}
\epsilon_x &= \frac{\partial u_x}{\partial x} = \alpha_2 + 2\alpha_4 x + \alpha_5 y \qquad (4.62)\\[6pt]
\epsilon_y &= \frac{\partial u_y}{\partial y} = \alpha_9 + \alpha_{11} x + 2\alpha_{12} y \qquad (4.63)\\[6pt]
\gamma_{xy} &= \frac{\partial u_y}{\partial x} + \frac{\partial u_x}{\partial y} = (\alpha_8 + \alpha_3) + (\alpha_5 + 2\alpha_{10}) x + (2\alpha_6 + \alpha_{11}) y \qquad (4.64)
\end{align}
$$

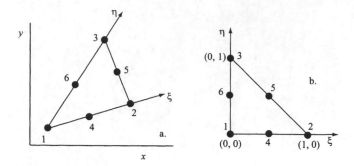

Figure 4.5: (a) Linear strain triangle; (b) isoparametric coordinates

Because all the nodes of the linear strain triangle have separate degrees of freedom and are free to move independently, the sides of the triangle do not necessarily remain straight as the element deforms. However, connectivity between elements ensures that no gaps open along the curving boundary between adjacent elements.

We can map the general triangular element of Fig 4.5a onto a general 'isoparametric' element defined in terms of 'natural' coordinates (ξ, η), with non-orthogonal axes as shown. The expressions that link global coordinates and natural coordinates, interpolating between the coordinates of the nodes (x_i, y_i), introduce shape functions which are the same as those used to interpolate the displacements within the element from the nodal displacements (u_{xi}, u_{yi}).

$$
\begin{array}{llll}
x & = & \sum N_i x_i; & y & = & \sum N_i y_i \\
u_x & = & \sum N_i u_{xi}; & u_y & = & \sum N_i u_{yi}
\end{array}
\tag{4.65}
$$

where the shape functions N_i are

$$
\begin{array}{rcl}
N_1 & = & (1 - \xi - \eta)(1 - 2\xi - 2\eta) \\
N_2 & = & \xi(2\xi - 1) \\
N_3 & = & \eta(2\eta - 1) \\
N_4 & = & 4\xi(1 - \xi - \eta) \\
N_5 & = & 4\xi\eta \\
N_6 & = & 4\eta(1 - \xi - \eta)
\end{array}
\tag{4.66}
$$

Note that $N_i = 1$ at node i and $N_i = 0$ at every other node—this is a general property of shape functions.

(For the three node constant strain triangle of the previous section:

$$
N_1 = 1 - \xi - \eta; \quad N_2 = \xi; \quad N_3 = \eta
\tag{4.67}
$$

omitting nodes 4, 5, 6 and their associated shape functions, and leaving a purely linear interpolation of displacements.)

In order to calculate the general element stiffness matrix we need, on the way, to calculate the strain-displacement matrix $\boldsymbol{B} = \partial \boldsymbol{N}$ (4.48). However, the displacements are calculated from (4.65) and (4.66) as functions of (ξ, η) and not of (x, y). We have to start by differentiating displacements as a function of (ξ, η) and then use the chain rule. For example,

$$\begin{pmatrix} \partial u_x / \partial \xi \\ \partial u_x / \partial \eta \end{pmatrix} = \begin{pmatrix} \partial x / \partial \xi & \partial y / \partial \xi \\ \partial x / \partial \eta & \partial y / \partial \eta \end{pmatrix} \begin{pmatrix} \partial u_x / \partial x \\ \partial u_x / \partial y \end{pmatrix} \tag{4.68}$$

where the 2×2 matrix is the coordinate linking Jacobian matrix $\boldsymbol{J}$:

$$\boldsymbol{J} = \begin{pmatrix} \partial x / \partial \xi & \partial y / \partial \xi \\ \partial x / \partial \eta & \partial y / \partial \eta \end{pmatrix} \tag{4.69}$$

with

$$\frac{\partial x}{\partial \xi} = \sum \frac{\partial N_i}{\partial \xi} x_i \quad \text{etc} \tag{4.70}$$

Equation (4.68) can be solved to give the necessary derivatives from which the strains can be calculated

$$\begin{pmatrix} \partial u_x / \partial x \\ \partial u_x / \partial y \end{pmatrix} = \boldsymbol{J}^{-1} \begin{pmatrix} \partial u_x / \partial \xi \\ \partial u_x / \partial \eta \end{pmatrix} \tag{4.71}$$

The element stiffness matrix is obtained by integrating over the volume of the element. This integration is again performed in terms of natural coordinates

$$\boldsymbol{K} = \int_0^1 \int_0^1 \boldsymbol{B}^T \boldsymbol{D} \boldsymbol{B} t \, |\boldsymbol{J}| \, \mathrm{d}\xi \mathrm{d}\eta \tag{4.72}$$

where t is the element thickness and $|\boldsymbol{J}|$ is the determinant of the Jacobian matrix which provides a scale factor between areas: $\mathrm{d}x\mathrm{d}y = |\boldsymbol{J}|\mathrm{d}\xi\mathrm{d}\eta$.

4.4.3 Quadrilateral elements

We can quickly introduce the shape functions for some of the quadrilateral elements that are used. The procedures for implementing them in the construction of stiffness matrices are exactly the same as that just described.

The mapping of global coordinates (x, y) onto natural coordinates (ξ, η) is illustrated in Fig 4.6a. For a four noded quadrilateral the shape functions are

$$\begin{array}{rcl} N_1 & = & \frac{1}{4}(1 - \xi)(1 - \eta) \\ N_2 & = & \frac{1}{4}(1 + \xi)(1 - \eta) \\ N_3 & = & \frac{1}{4}(1 + \xi)(1 + \eta) \\ N_4 & = & \frac{1}{4}(1 - \xi)(1 + \eta) \end{array} \tag{4.73}$$

These interpolation functions lead to a linear variation of ϵ_x with η and a linear variation of ϵ_y with ξ. The shear strain γ_{xy} has linear variations with both ξ and η.

Because the sides of this four noded element always deform as straight lines it cannot describe the strain field associated with bending—which would require

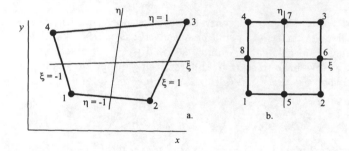

Figure 4.6: (a) Four noded quadrilateral element; (b) eight noded quadrilateral element

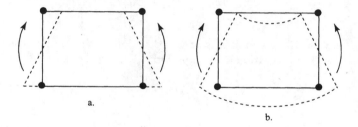

Figure 4.7: (a) 'Bending' of four noded quadrilateral; (b) desired shape of rectangular element subjected to pure bending

curvature of the sides. If these elements are subjected to pure moment loading (Fig 4.7) then, although the linear variation of direct strain with position is correctly described, no right angles in the element are preserved and shear stresses must be generated on all surfaces. The stiffness of such elements tends to be too high.

Greater freedom can be obtained by adding additional nodes (Fig 4.6b), allowing the sides of the element to take up general quadratic shapes in conformity with neighbouring elements. The interpolation functions are now

$$
\begin{aligned}
N_1 &= -(1-\xi)(1-\eta)(1+\xi+\eta)/4 \\
N_2 &= -(1+\xi)(1-\eta)(1-\xi+\eta)/4 \\
N_3 &= -(1+\xi)(1+\eta)(1-\xi-\eta)/4 \\
N_4 &= -(1-\xi)(1+\eta)(1+\xi-\eta)/4 \\
N_5 &= \left(1-\xi^2\right)\left(1-\eta\right)/2 \\
N_6 &= \left(1+\xi\right)\left(1-\eta^2\right)/2 \\
N_7 &= \left(1-\xi^2\right)\left(1+\eta\right)/2 \\
N_8 &= \left(1-\xi\right)\left(1+\eta^2\right)/2
\end{aligned}
\tag{4.74}
$$

Each of the three strains now contains *some* quadratic variation—but there is no variation of ϵ_x with ξ^2, for example. In its rectangular form this element can exactly represent bending states.

4.4.4 Comparison of elements

In order to illustrate the relative advantage of using higher order elements—those with more degrees of freedom and greater ability to match spatially varying strain fields—it is convenient to analyse a problem for which the exact analytical result is known (Livesley, 1983). The deep cantilever of depth d and length L in Fig 4.8a carries a transverse load W at its tip. The tip deflection δ is:

$$
\delta = \frac{WL^3}{3EI}\left[1 + \left(2 + \frac{5}{2}\nu\right)\frac{d^2}{4L^2}\right]
\tag{4.75}
$$

where I, E and ν are second moment of area, Young's modulus and Poisson's ratio respectively (Timoshenko and Goodier, 1970).

The calculated fraction of this exact tip deflection is shown in Fig 4.8b as a function of the number of free nodes in the numerical analysis for different types of element—including comparison of triangular elements laid out on a square grid or a rectangular grid. The results illustrate clearly that any numerical approximation of this type will be too stiff—it is not able to deform as freely as the continuum that it is trying to represent. The higher order elements which permit internal variation of strain converge rapidly towards the correct result: in fact using eight noded squares or rectangles a very small number of elements is required. The six noded triangles are clearly superior to the constant strain, three noded, triangles and to the four noded squares.

Evidently more complex elements require more computing time for any element calculation. However, in general the greatly reduced number of elements that can be used outweighs this apparent computational disadvantage and leads to a lower overall computing time for a given accuracy of result.

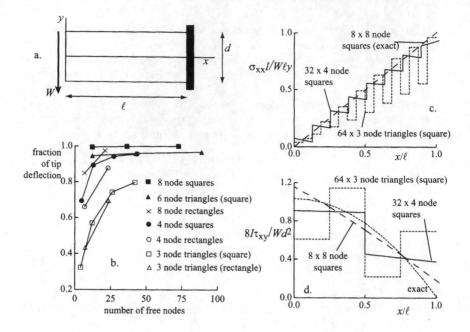

Figure 4.8: (a) Rectangular cantilever subjected to tip loading; (b) comparison of tip deflections predicted using different element types; (c) comparison of distributions of σ_{xx} at $y = 0.375d$ for different element types; (d) comparison of distribution of shear stress τ_{xy} on vertical plane for different element types (after Livesley, 1983)

It is often tempting in assessing results of numerical analysis to look only at the output quantity of direct interest—in this case perhaps just the tip deflection. It is, however, always instructive to probe more deeply. The difference between the different elements can be seen more strikingly when internal stresses within the deep beam are investigated. Fig 4.8c shows the variation of the longitudinal stress σ_{xx} along the beam at a position $y = 0.375d$ for three distributions of elements, all using the same nodal positions. Only the eight noded square is able to produce a displacement field which matches the result $\sigma_{xx} = Wxy/I$: the numerical calculation for these elements is indistinguishable from the exact result. It proves even harder to match the shear stresses shown in Fig 4.8d for a transverse plane near the centre of the cantilever. The shear stresses vary with y^2, but even the eight noded square element has none of the y^3 terms in the displacement field that could provide this desired variation. Nevertheless these elements are still very much superior to the other elements used in the comparison.

4.5 Integration—Gauss points

The element stiffness matrix is obtained in (4.49) or (4.72) by integration of a possibly quite elaborate function over the volume of the element. While this may be possible for simple models and simple geometries, in general it may be computationally exhausting. For a nonlinear material—such as soil—the components of the material stiffness matrix D will vary from point to point. Gauss quadrature provides an efficient route to numerical integration. The exact integral is replaced by the sum of a number of weighted terms

$$I = \int_{-1}^{+1} \phi \mathrm{d}\xi \quad \text{becomes} \quad I \approx \sum_{1}^{n} w_i \phi_i \qquad (4.76)$$

The function ϕ is evaluated at n carefully chosen points and each value of the function is weighted by a corresponding factor w_i. If $\phi = \phi(\xi)$ is a polynomial, then use of n sampling points gives an exact result for polynomials of degree not greater than $2n-1$. The locations of one, two and three sampling points and the corresponding weightings for this one-dimensional integration are shown in Fig 4.9. Evidently sampling at the centre is sufficient—and exact—if the function is linear. For non-polynomial functions the accuracy will improve as the number of sampling points increases.

The locations of the Gauss points can be identified in just the same way in two dimensions (Fig 4.10). The metamorphosis (4.76) is now

$$I = \int_{-1}^{+1} \int_{-1}^{+1} \phi \mathrm{d}\xi \mathrm{d}\eta \quad \text{becomes} \quad I \approx \sum_{i=1}^{n} \sum_{j=1}^{m} w_{ij} \phi\left(\xi_i, \eta_j\right) \qquad (4.77)$$

Usually $n = m$ and the same numbers of sampling points are used in each direction. For a single central Gauss point $n = m = 1$, $\xi = 0$, $\eta = 0$, $w = 4$ and $I \approx 4\phi_1$. For four point and nine point quadrature the summation rules are:

$$I \approx \phi_1 + \phi_2 + \phi_3 + \phi_4 \quad \text{(four)} \qquad (4.78)$$

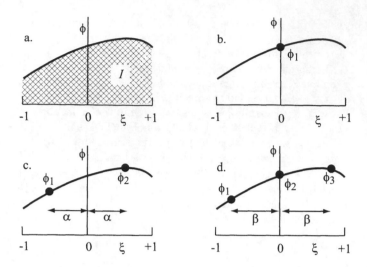

Figure 4.9: Gauss integration points: (a) $I = \int_{-1}^{+1} \phi d\xi$; (b) single point integration $I \approx 2\phi_1$; (c) two point integration $I \approx \phi_1 + \phi_2$, $\alpha = 1/\sqrt{3}$; (d) three point integration $I \approx \frac{5}{9}\phi_1 + \frac{8}{9}\phi_2 + \frac{5}{9}\phi_3$, $\beta = \sqrt{3/5}$

and

$$I \approx \frac{25}{81}(\phi_1 + \phi_2 + \phi_3 + \phi_4) + \frac{40}{81}(\phi_5 + \phi_6 + \phi_7 + \phi_8) + \frac{64}{81}\phi_9 \quad \text{(nine)} \quad (4.79)$$

4.5.1   Reduced integration

Given a particular form of shape function describing the internal variation of displacement within an element, we will obtain an accurate calculation of the stiffness matrix of the element if we use a Gaussian integration rule which is compatible with the polynomial degree of the shape function. Thus with the eight noded quadrilateral element (Fig 4.6b, (4.74)) the shape functions imply polynomial interpolation of degree 3 and the integral of (4.72) implies a polynomial of degree 4 which requires a 3×3 Gauss point formula (Fig 4.10b) for exact evaluation of the stiffness matrix. (In fact, once this element deforms and the sides are no longer straight, then the Jacobian J varies across the element and there are polynomial terms in the denominator of J^{-1} which enter the terms of B through (4.71). Numerical integration can then never be exact.)

Because a finite element model only permits a finite number of degrees of freedom it will be usually be stiffer than the continuum reality that it is trying to describe (see Fig 4.8). This excessive stiffness is usually worsened when additional Gauss points are used because these resist higher order deformation

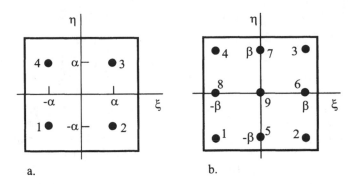

Figure 4.10: Gauss integration points in two dimensions: (a) four point, $\alpha = 1/\sqrt{3}$; (b) nine point, $\beta = \sqrt{3/5}$

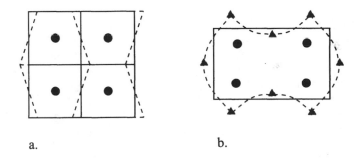

Figure 4.11: (a) Instability mode for four node square elements with single Gauss (b) 'hourglass' instability mode for eight node rectangular element with four Gauss points (after Cook, 1995)

modes which are released in lower order integration rules. Paradoxically, increased accuracy in the computation of the element stiffness matrix K can lead to reduced accuracy in the outcome of the finite element analysis (as well as involving more cumbersome calculation). 'Reduced integration' then implies using a lower order Gaussian integration rule than would apparently be associated with the interpolation function (shape function) of the element. It is in fact common practice to use a 2×2 Gauss point formula to compute K for four and eight node quadrilateral elements.

The disadvantage of using lower order integration rules is that there is the possibility of the introduction of spurious element deformation modes which are able to occur without any change in strain energy—'zero energy modes'—because the nodal displacements are somehow dissociated from the displacements of the Gauss points, and hence there is no stiffness associated with these mechanisms. Classic examples are shown in Fig 4.11 for 4-noded elements integrated with a 1-point Gauss rule, and for 8-noded elements integrated with a 4-point Gauss rule. Cook *et al.* (1989) note that the instability mode shown in Fig 4.11b is not communicable in a mesh of such elements and may therefore not be of concern. However, it can generate problems at an interface between materials of widely differing stiffness. Sudden jumps in material properties can often lead to numerical problems anyway.

Strains and hence stresses calculated from the displacement field within an element are often most accurate at Gauss points: Gauss point values of strains and stresses will typically be presented as the output of a numerical, finite element analysis. To calculate the stresses at nodes or other points in an element it is then necessary to extrapolate from the known values at Gauss points. This extrapolation can be achieved using a polynomial of degree appropriate to the number of Gauss points that is available and, if reduced integration has been used, may not necessarily contain as much nonlinearity as is contained in the shape functions N_i which were used to build up the element stiffness matrix in the first place.

4.6   Nodal forces and external loads

Finite elements can only convey information to each other, and to the boundaries of the problem being analysed, at their nodes. It has already been mentioned (§4.3) that body forces have to be divided among the nodes bounding an element. Any external loading then also has to be converted into equivalent nodal quantities even if it is conceived—in the design of the problem—as a distributed load. As in the generation of other aspects of finite element theory, considerations of *work* control the conversion of distributed loadings to nodal quantities.

For an element having only two nodes on each boundary we can at most describe a linear variation of transverse displacement u_y and a linear variation of loading q (Fig 4.12)[2]. We require to establish the nodal loads F_A and F_B to give the correct work:

$$\int_0^\ell u_y q \mathrm{d}x =$$

$$\int_0^\ell \left[\left(1 - \frac{x}{\ell}\right) u_{yA} + \frac{x}{\ell} u_{yB} \right] \left[\left(1 - \frac{x}{\ell}\right) q_A + \frac{x}{\ell} q_B \right] \mathrm{d}x =$$

$$F_A u_{yA} + F_b u_{yB} \quad (4.80)$$

where the linear interpolation functions are evidently those that we previously used for the one-dimensional proto-element (4.31). Hence:

$$\begin{pmatrix} F_A \\ F_B \end{pmatrix} = \frac{\ell}{6} \begin{pmatrix} 2 & 1 \\ 1 & 2 \end{pmatrix} \begin{pmatrix} q_A \\ q_B \end{pmatrix} \quad (4.81)$$

---

[2]For convenience we associate the x direction with the boundary linking the nodes and apply the loads and the corresponding displacements in the orthogonal y direction. However, the result is general so that the same result would also apply for tangential as opposed to orthogonal boundary loading.

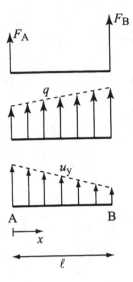

Figure 4.12: Nodal loads for element with two boundary nodes

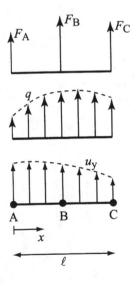

Figure 4.13: Nodal loads for element with three boundary nodes

For an element with three nodes along its boundary we can exactly reproduce quadratic variation of displacement and of loading (Fig 4.13) and the resulting nodal load equivalence is:

$$\begin{pmatrix} F_A \\ F_B \\ F_C \end{pmatrix} = \frac{\ell}{30} \begin{pmatrix} 4 & 2 & -1 \\ 2 & 16 & 2 \\ -1 & 2 & 4 \end{pmatrix} \begin{pmatrix} q_A \\ q_B \\ q_C \end{pmatrix} \tag{4.82}$$

In many finite element programs the conversion of distributed loading to equivalent nodal quantities is achieved automatically. A corresponding inverse process is required in order to convert nodal forces calculated by a finite element program, and presented as output, into equivalent distributed boundary loading.

4.7   Dynamic analysis

For dynamic analyses we are concerned with variations of acceleration within the soil accelerations lead to equivalent forces through application of Newton's laws. We need to be able to write equations of motion in terms of the time derivatives of the nodal displacements and we have to discover some way of assigning the mass of the element to the individual nodal degrees of freedom.

Our statement of minimum potential energy, or zero virtual work, that was used to deduce the general form of the stiffness matrix for any chosen finite element (4.49) can also be used to generate the mass matrix. The virtual work from the acceleration forces subjected to a virtual displacement field δu is

$$\int_V (\delta u)^T \rho \ddot{u} \, dV$$

and, with the usual link between internal displacements u and nodal displacements d and their derivatives

$$u = Nd; \qquad \dot{u} = N\dot{d} \qquad \ddot{u} = N\ddot{d}$$

we obtain integrals of the form

$$\delta d^T \left[\int_V N^T \rho N \, dV \right] \ddot{d}$$

and the mass matrix to be used in analysis is

$$m = \int_V N^T \rho N \, dV \tag{4.83}$$

It can be shown that an exactly similar form of integral is required to produce the damping matrix:

$$c = \int_V N^T \eta N \, dV \tag{4.84}$$

where η is the material viscosity, linking stresses and rates of displacement.

The element equations of motion then take the form

$$\boldsymbol{K}\boldsymbol{d} + \boldsymbol{c}\dot{\boldsymbol{d}} + \boldsymbol{m}\ddot{\boldsymbol{d}} = \boldsymbol{F} \tag{4.85}$$

where $\boldsymbol{F}$ is the applied force, now potentially varying with time.

The mass matrix computed using (4.83) is called a 'consistent' mass matrix precisely because it is determined rigorously from considerations of energy. However, it is sometimes computationally expedient—though usually (but not always) somewhat less accurate—to work with a purely diagonal 'lumped' mass matrix in which the mass of the element $\boldsymbol{m}$ is simply assigned to the individual nodes in a very discontinuous way. For a three noded homogeneous triangular element the consistent mass matrix is

$$\boldsymbol{m} = \frac{m}{12}\begin{pmatrix} 2 & 0 & 1 & 0 & 1 & 0 \\ 0 & 2 & 0 & 1 & 0 & 1 \\ 1 & 0 & 2 & 0 & 1 & 0 \\ 0 & 1 & 0 & 2 & 0 & 1 \\ 1 & 0 & 1 & 0 & 2 & 0 \\ 0 & 1 & 0 & 1 & 0 & 2 \end{pmatrix} \tag{4.86}$$

whereas the lumped mass matrix is

$$\boldsymbol{m} = \frac{m}{3}\begin{pmatrix} 1 & 0 & 0 & 0 & 0 & 0 \\ 0 & 1 & 0 & 0 & 0 & 0 \\ 0 & 0 & 1 & 0 & 0 & 0 \\ 0 & 0 & 0 & 1 & 0 & 0 \\ 0 & 0 & 0 & 0 & 1 & 0 \\ 0 & 0 & 0 & 0 & 0 & 1 \end{pmatrix} \tag{4.87}$$

which is obviously somewhat different.

Other procedures have been proposed for producing more suitable diagonal mass matrices but there does not appear to be any one algorithm which guarantees accurate results. Zienkiewicz and Taylor (2000) suggest that in some ways lumping mass is equivalent to increasing the material viscosity which thus leads to somewhat smoother (more damped), if less accurate, solutions.

Damping is often included in dynamic analyses as a numerical device. It is known that there are dissipative effects present—we expect these to be primarily associated with real material hysteretic nonlinearities arising from irrecoverable plastic deformations and frictional dissipation within the material. However, the computational costs of performing full dynamic analyses using advanced constitutive models of the type described, or hinted at, in Chapter 3 may be such that engineers prefer to use more commonly available (more extensively verified) simpler models—such as elastic-perfectly plastic models—and then add in some extra damping to allow for dissipation of energy in the elastic region.

A classical way of doing this is through the use of Rayleigh damping, assuming (arbitrarily) that the damping matrix is a linear combination of the mass and stiffness matrices:

$$\boldsymbol{c} = \alpha\boldsymbol{m} + \beta\boldsymbol{K} \tag{4.88}$$

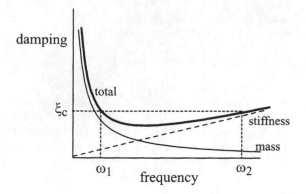

Figure 4.14: Rayleigh damping

The resulting effective damping ratio is

$$\xi = \frac{\alpha}{2\omega} + \frac{\beta\omega}{2} \qquad (4.89)$$

and this varies with frequency ω in a way that damping effects produced by material plasticity would not be expected to vary. However, if the range of frequencies of importance in a particular dynamic analysis can be estimated then the values of α and β can be chosen to give tolerably constant damping over this frequency range (Fig 4.14). With damping ratio $\xi = \xi_c$ at $\omega = \omega_1$ and at $\omega = \omega_2$

$$\frac{\alpha}{2\omega_1} + \frac{\beta\omega_1}{2} = \xi_c = \frac{\alpha}{2\omega_2} + \frac{\beta\omega_2}{2} \qquad (4.90)$$

and hence

$$\alpha = \frac{2\xi_c\omega_1\omega_2}{\omega_1 + \omega_2} \qquad \beta = \frac{2\xi_c}{\omega_1 + \omega_2} \qquad (4.91)$$

Evidently the mass damping operates primarily at low frequencies and the stiffness damping operates more at higher frequencies.

4.8   Finite differences

Numerical modelling is required as a vehicle for the solution of the field equations that govern geotechnical problems. Finite element schemes provide a powerful and much adopted treatment of the spatial discretisation of a problem. Finite difference schemes provide an alternative route to the conversion of continuum field equations into relationships between discrete numerical values—a link with finite element discretisation will be noted for the spatial domain. However, we have just encountered equations which introduce the time domain for transient or dynamic problems. Problems are usually spatially finite—or at least can be treated as spatially finite—and the spatial boundary conditions consist of prescribed loadings or displacements. In the time domain our concern is usually to march into the infinite future from some initial condition and, while there may be some asymptotic condition to which we expect to tend (for example, an eventual state of zero excess pore pressure once consolidation is complete, or an eventual renewed state of rest), we expect the numerical modelling to tell us how fast and by what route we will get there.

We imagine that we are calculating the values of a function ϕ at intervals separated by finite time steps Δt (though the stepping could be in space rather than time). Using Taylor's series expansion we can write down expressions for the values of the function one step ahead of, ϕ_{n+1}, and one step behind, ϕ_{n-1}, the current value, ϕ_n, in terms of the *current* values of the function and its derivatives:

$$\phi_{n+1} = \phi_n + \Delta t \frac{\partial \phi}{\partial t} + \frac{\Delta t^2}{2!} \frac{\partial^2 \phi}{\partial t^2} + \frac{\Delta t^3}{3!} \frac{\partial^3 \phi}{\partial t^3} + \dots \tag{4.92}$$

$$\phi_{n-1} = \phi_n - \Delta t \frac{\partial \phi}{\partial t} + \frac{\Delta t^2}{2!} \frac{\partial^2 \phi}{\partial t^2} - \frac{\Delta t^3}{3!} \frac{\partial^3 \phi}{\partial t^3} + \dots \tag{4.93}$$

From these we can deduce so called 'central difference' approximate expressions for the first and second derivatives

$$\frac{\partial \phi}{\partial t} \approx \frac{\phi_{n+1} - \phi_{n-1}}{2 \Delta t} \tag{4.94}$$

$$\frac{\partial^2 \phi}{\partial t^2} \approx \frac{\phi_{n+1} - 2\phi_n + \phi_{n-1}}{\Delta t^2} \tag{4.95}$$

where we have ignored terms in the Taylor series involving Δt^3 and higher powers. These expressions thus have second order accuracy: if the time step is halved then the error is reduced roughly by a factor of 4. A graphical interpretation (Fig 4.15) confirms our expectation that (4.94) will be more accurate than a 'forward difference' approximation of first derivative

$$\frac{\partial \phi}{\partial t} \approx \frac{\phi_{n+1} - \phi_n}{\Delta t} \tag{4.96}$$

or a 'backward difference' form

$$\frac{\partial \phi}{\partial t} \approx \frac{\phi_n - \phi_{n-1}}{\Delta t} \tag{4.97}$$

though it may sometimes be necessary to make use of these forms. However, the central difference approximation of the second derivative (4.95) is just the difference between these forward and backward approximations of first derivative.

Spatially, we would naturally estimate strains in a simple triangular element (Fig 4.16) using

$$\begin{aligned} \epsilon_x &= \frac{\partial u_x}{\partial x} \approx \frac{u_{x_2} - u_{x_1}}{x_2 - x_1} \\ \epsilon_y &= \frac{\partial u_y}{\partial y} \approx \frac{u_{y_3} - u_{y_1}}{y_3 - y_1} \\ \gamma_{xy} &= \frac{\partial u_x}{\partial y} + \frac{\partial u_x}{\partial y} \approx \frac{u_{y_2} - u_{y_1}}{x_2 - x_1} + \frac{u_{x_3} - u_{x_1}}{x_3 - x_1} \end{aligned} \tag{4.98}$$

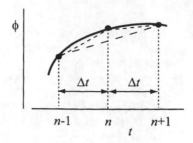

Figure 4.15: Finite difference aproximations to slope of a function

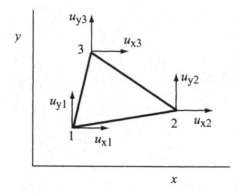

Figure 4.16: Strains in triangular element from nodal displacements

and these are evidently first order finite difference approximations to the displacement derivatives. They are exactly equivalent to the approximations implicit in the constant strain triangle finite element (4.58).

Typically, in dynamic problems, we will have some governing equation of the form

$$m\ddot{\phi} + c\dot{\phi} + \lambda\phi = F \tag{4.99}$$

where m, c and λ are mass, damping and stiffness, F is a driving force and ϕ will usually be some displacement variable. In this equation, these symbols can equally represent systems of equations for a finite element description of the problem (see, for example, (4.85)).

Combining (4.94) and (4.95) and (4.99) we can obtain an expression for the future value of the variable. ϕ_{n+1} in terms of present and past information:

$$\left(\frac{1}{\Delta t^2}m + \frac{1}{2\Delta t}c\right)\phi_{n+1} = F_n - \lambda\phi_n + \frac{1}{\Delta t^2}m\left(2\phi_n - \phi_{n-1}\right) + \frac{1}{2\Delta t}c\phi_{n-1} \tag{4.100}$$

This 'explicit' equation shows us how we can predict the future using only present and past information. The solution requires starting information ϕ_o and

$\dot{\phi}_o$ at $t = 0$ which will usually be available. This explicit prediction equation is only conditionally stable (see, for example, Cook *et al.*, 1989) and the time step to be used for numerical integration must be less than a critical value:

$$\Delta t \leq \Delta t_{crit} = \frac{2}{\omega_{max}} = \frac{T_{min}}{\pi} \qquad (4.101)$$

where ω_{max} is the maximum natural frequency (T_{min} is the minimum period) of the system. For a simple single degree of freedom equation of motion,

$$\omega_{max}^2 = \frac{\lambda}{m} \qquad (4.102)$$

but higher oscillation modes are possible for a system of connected elements.

For a system of spatially discretised finite elements the need for a critical time step can be interpreted in a slightly different way. If an element has typical length ℓ then in-plane vibration involves displacement waves travelling across the element at a speed which will be equal to the speed of sound in the material—given in §3.2.3 as the one-dimensional compression wave velocity $v_p = \sqrt{(K + 4G/3)/\rho}$. The time step must be sufficiently small that information cannot travel across the element within a single time step:

$$\Delta t < \frac{\ell}{v_p} \qquad (4.103)$$

and we expect that, whatever the mode of vibration, the element will form an integral number of half wavelengths in the direction of travel so that the natural frequency is

$$\omega_{max} \propto \frac{v_p}{\ell} \qquad (4.104)$$

It is evident that the choice of time step will be influenced both by the material properties (which control the compression wave velocity) and by the size of elements. Small elements of stiff material will necessitate small time steps. In fact, convergent numerical results will often require the use of time steps considerably smaller than the theoretical limit for numerical stability (Itasca, 2000).

On the other hand the element size must be sufficiently small that the dynamic motion of the system can be adequately reproduced. A vibrating soil layer (Fig 4.17) will need at least ten elements per wavelength to give adequate detail (Kuhlemeyer and Lysmer, 1973).

Implicit integration schemes are more numerically stable than the explicit scheme so far described. Implicit schemes introduce future values into the predictive formula. For example, the Newmark methods use a relationship

$$\phi_{n+1} = \phi_n + \Delta t \dot{\phi}_n + \frac{\Delta t^2}{2} \left[(1 - 2\beta) \ddot{\phi}_n + 2\beta \ddot{\phi}_{n+1} \right] \qquad (4.105)$$

$$\dot{\phi}_{n+1} = \dot{\phi}_n + \Delta t \left[(1 - \gamma) \ddot{\phi}_n + \gamma \ddot{\phi}_{n+1} \right] \qquad (4.106)$$

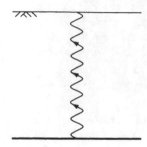

Figure 4.17: Shear wave propagating up through soil layer

where β and γ are numbers that can be selected by the user. Substitution into the equation of motion (4.99) for time t_{n+1} gives

$$\left(\frac{m}{\beta \Delta t^2} + \frac{c\gamma}{\beta \Delta t} + \lambda\right) \phi_{n+1} =$$
$$F_{n+1} + \frac{m}{\beta \Delta t^2} \phi_n + \frac{c\gamma}{\beta \Delta t} \phi_n + \frac{m}{\beta \Delta t} \dot{\phi}_n - c\dot{\phi}_n + \frac{c\gamma}{\beta} \dot{\phi}_n +$$
$$\frac{m}{2\beta} (1 - 2\beta) \ddot{\phi}_n - c\Delta t (1 - \gamma) \ddot{\phi}_n + \frac{c\gamma \Delta t}{2\beta} (1 - 2\beta) \ddot{\phi}_n \quad (4.107)$$

and we are able to compute the new value of ϕ_{n+1} in terms of the information available at time t_n (the external force F_{n+1} is assumed to be given by some known time history).

For nonlinear problems λ and c may well depend on the state of the system at time t_{n+1}—which is not known in advance—and some iterative technique will be required which may delay convergence and increase the computational cost. However, this procedure is unconditionally stable for $2\beta \geq \gamma \geq 1/2$. A popular choice is $\gamma = 1/2$ and $\beta = 1/4$ which leads to the 'constant average acceleration' method. For $\gamma > 1/2$ and $\beta = (\gamma + 1/2)^2 /4$ the method provides some artificial algorithmic damping but the accuracy of the solution is reduced (Cook et al., 1989). Of course, as noted by Cook (1995), just because the procedure is numerically stable for any value of the time step, this does not mean that the results are guaranteed to be accurate if large time steps are used and an argument like that associated with Fig 4.17 suggests that the time steps should be small enough to pick up the detail of the motion at the highest frequency of the system that is believed to be important.

The one-dimensional consolidation equation has already been encountered in various forms (§4.2.1):

$$c_v \frac{\partial^2 \varrho}{\partial z^2} = \frac{\partial \varrho}{\partial t} \quad (4.108)$$

We can use the central difference approximation for the first and second derivatives ((4.94) and (4.95)) to write this as

$$\varrho_{i,t+\Delta t} \approx \varrho_{i,t-\Delta t} + \frac{2c_v \Delta t}{\Delta z^2} (\varrho_{i-1,t} - 2\varrho_{i,t} + \varrho_{i+1,t}) \quad (4.109)$$

where the subscript $_i$ indicates spatial location. This equation can again be solved explicitly to give prediction of future pore pressure variation at each node at time $t + \Delta t$ in terms of information that is available now at times t and $t - \Delta t$. However, there are constraints on the choice of time step and of spatial grid dimension if numerical stability of the prediction into the future is to be guaranteed:

$$\frac{c_v \Delta t}{\Delta z^2} \leq \frac{1}{2} \qquad (4.110)$$

and small time steps are required to guarantee stability.

A similar approach can be used to write the wave equation in finite difference form

$$\frac{\partial^2 \phi}{\partial x^2} = \frac{1}{v_p^2} \frac{\partial^2 \phi}{\partial t^2} \qquad (4.111)$$

for compression waves, where $v_p = \sqrt{(K + 4G/3)/\rho}$ is the compression wave speed. The explicit finite difference form of this equation becomes

$$\phi_{i,t+\Delta t} \approx 2\phi_{i,t} - \phi_{i,t-\Delta t} + \frac{v_p^2 \Delta t^2}{\Delta x^2} \left(\phi_{i-1,t} - 2\phi_{i,t} + \phi_{i+1,t} \right) \qquad (4.112)$$

Again there are limits on stability of the explicit solution

$$\frac{v_p \Delta t}{\Delta x} < 1 \qquad (4.113)$$

and again the size of time step depends on a material property—the wave speed v_p—and a dimension of the discretisation grid. The finer the grid or the stiffer the material (the higher the wave speed) the smaller the time step. The wave velocity v_p for compression waves is the speed of sound through the material and this constraint on time step is again essentially requiring the time step to be smaller than the time required for the compression wave to travel across the element.

It has been noted that the simple finite difference expressions for derivatives in terms of nodal values imply that strains are uniform between nodes and hence there is an equivalence to the constant strain shape functions assumed in finite element analysis. The ability of finite difference grids to represent rapidly varying quantities will be similar to that of meshes of constant strain finite elements and very fine meshes will often be required in regions (spatial or temporal) of high gradients of these quantities. Use of finite difference equations in explicit form has the advantage that the mathematics and hence programming are very simple and this simplicity and consequent efficiency of computer storage may often the disadvantage of needing very large numbers of elements and time steps.

4.9 Solution schemes

For linear problems, such as the application of working loads to systems of elastic materials, the finite element solution of the problem of determination of the

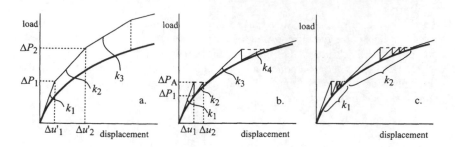

Figure 4.18: Finite element solution schemes: (a) tangent stiffness; (b) Newton-Raphson; (c) modified Newton-Raphson

field of displacements (and thence strains and stresses) is relatively straight-forward. It matters not whether we write the governing equations in terms of stresses or increments of stresses. The resulting displacements will be directly proportional to any applied loading (assuming that the resulting displacements do not change the geometry of the problem significantly). However, we know that soils are extremely nonlinear materials with mechanical response which is history dependent. We can expect that the displacement of a geotechnical system, reached at the end of a loading process, will depend on the detail of that process, and not just on the final loads that are imposed. The analysis of such nonlinear systems poses challenges which are not so different from those linked with attempts to use fully explicit schemes to predict the future in time domain analyses of dynamic problems. We can illustrate some of the possible solution schemes with reference to a single load:displacement relationship (Fig 4.18).

We assume that there is a correct, true link between load and displacement which we are trying to recover. Inevitably we have to discretise the loading in some way into finite (as opposed to infinitesimal) steps. The simplest tangent stiffness algorithm would be satisfied with a prediction based on the current tangent (incremental) stiffness of the material. Thus, given an initial tangent stiffness k_1, the application of a load ΔP_1 would be predicted to produce a displacement $\Delta u'_1 = \Delta P_1/k_1$ (Fig 4.18a). At this new displacement we are able to calculate the tangent stiffness to be k_2 so that the effect of applying a second load increment ΔP_2 is to produce an additional displacement $\Delta u'_2 = \Delta P_2/k_2$. Evidently the predicted load:displacement response drifts away from the correct relationship.

In order to obtain results which are accurate to within some specified tolerance it is necessary to use very small loading steps. The accuracy of the method can be particularly poor if the material changes from elastic to plastic during a single increment—but the calculation assumes a single (high) initial stiffness for the entire increment.

The nonlinearity necessitates an iterative approach. The Newton-Raphson scheme (Fig 4.18b) produces a first prediction of displacement Δu_1 resulting from load increment ΔP_A using the initial stiffness k_1 in the same way as the tangent stiffness method. However, the constitutive model is used to calculate

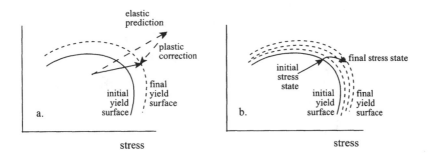

Figure 4.19: (a) Return strategy for initial prediction of inadmissible stress increment; (b) substepping strategy for elasto-plastic increment

the true load ΔP_1 corresponding to this displacement and the difference between this load ΔP_1 and ΔP_A becomes a load imbalance which is now applied using a tangent stiffness k_2 calculated at the displacement Δu_1. This leads to a new incremental displacement Δu_2 and a new load imbalance. The procedure is repeated until the load and displacement satisfy the constitutive model within the specified tolerance.

This Newton-Raphson method can be expensive computationally because the tangent stiffness matrix should be recalculated after each iteration. In the modified Newton-Raphson scheme (Fig 4.18c) the stiffness used for each iterative computation of displacement within an overall load increment is the same—in some cases the initial elastic stiffness is used throughout. Evidently, this is likely to increase the number of iterations required to reach a satisfactory solution but the saving in not recalculating and inverting stiffness matrices may make this computationally desirable.

There is an inevitable paradox in these iterative methods that the load imbalance has to be calculated from a finite displacement increment applied to a model which is constructed (as in Chapter 3) for infinitesimal increments. The model has in fact to be integrated along an implied strain path. Potts and Zdravković (1999) describe substepping procedures for achieving this—again with a specified convergence tolerance being imposed to limit the size of the substeps within a displacement increment. This substepping scheme must also be iterative in order to make some adequate accommodation for the variation in stiffness occurring over a substep. These iterations have to be performed for each Gauss point in each element of the problem because of course in general the displacement gradients and hence the strains and stresses will vary within each element.

The number of substeps can be reduced if instead an 'implicit' return strategy is used in which plastic strains are calculated from the conditions at the (initially unknown) end of the increment. Typically an elastic prediction of the stress changes is used to start the iteration and an algorithm is then used to return the stress state to the (hardened or softened) yield surface (Fig 4.19), if the predicted stress state is found to have violated the yield condition. Once

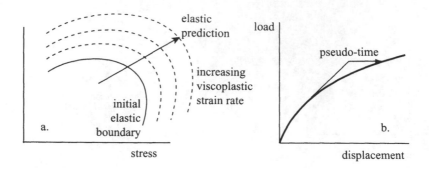

Figure 4.20: Viscoplastic strategy for interpretation of inadmissible stress states

again the computational advantage of rapid convergence and reduced number of iterations has to be set against the computational disadvantage of a more complex driving algorithm. Potts and Zdravković discuss the particular desirability of integrating along a path which is admissible throughout its length (Fig 4.19b) as opposed to a path which is certainly inadmissible (Fig 4.19a).

When using commercial software packages, engineers may not have much opportunity to choose the solution scheme—still less the return strategy—which is to be adopted. However, they *should* be able to vary the size of the load steps that they use to drive the problem in order to convince themselves that these steps are sufficienlty small to give a convergent and presumably accurate result.

Numerical solution of geotechnical problems is driven by time—either explicitly, as in transient seepage or dynamic problems, or notionally as a sequence of construction stages in pseudo-static problems. In either case the purpose of the numerical subterfuges is to accommodate the finite size of time steps (or the notional arrow of time, as loading stages) in the same way that the finite element spatial discretisation divides a continuous field problem into a finite number of chunks. Two other techniques have been used to achieve the same result by the application of (pseudo?) physical principles.

It has already been noted that the solution difficulties arise because the geotechnical materials are nonlinear and our assertion in this book is that this nonlinearity is primarily the result of plasticity. In the elastic-hardening plastic models (§3.4), the consistency condition (3.103) specifies that the only permissible stress states are on or within the current yield surface. The finite element solution algorithm has to find some way of ensuring admissible states of the soil *during* finite loading increments.

Elastic-viscoplastic models provide a different constraint on response. There is still a yield surface which forms a boundary to the region of elastically attainable states (Fig 4.20) but there is then a family of progressively larger surfaces (perhaps all of the same shape) each corresponding to a higher plastic strain rate. With a yield function $f(\sigma, \chi)$ and plastic potential $g(\sigma)$ we have

$$\delta\epsilon^p = \mu\frac{\partial g}{\partial\sigma} \qquad (4.114)$$

$$\mu = \gamma\langle\phi(f)\rangle \qquad (4.115)$$

where $\langle\ \rangle$ implies that

$$\langle\phi(f)\rangle = 0 \quad \text{if} \quad f \le 0 \qquad (4.116)$$

$$\langle\phi(f)\rangle = \phi(f) \quad \text{if} \quad f > 0 \qquad (4.117)$$

γ is some viscosity parameter and ϕ is a monotonic function of f.

Of course, as seen in §3.7, some soils have mechanical properties which can legitimately be described as viscoplastic. For any soil, however, we can use a notional numerical viscoplasticity as a means of coping with finite loading steps. The initial elastic stress increment prediction is no longer seen as leading to an inadmissible stress state (Fig 4.19) but rather to a high plastic strain rate stress state (Fig 4.20a). Using notional time to march through the viscoplastic straining, the plastic strains produce hardening of all the yield surfaces until eventually the calculated stress states imply viscoplastic strain rates which are deemed to be below an acceptable tolerance (though it is implicit that, for a constant load which has caused yielding, truly zero strain rates can only be obtained at infinite time) (Fig 4.20b).

A second quasi-physical algorithm, dynamic relaxation, also presents itself. If forces on a mass are not in equilibrium then Newton's laws of motion tell us that the mass will accelerate and move. If we take a simple lumped mass approach to the distribution of mass at nodes within our finite element discretisation, we can calculate out-of-balance nodal forces at any stage and again march through time using the deformations of the mesh to generate new internal stresses (from the constitutive law—and we have seen in Chapter 3 that moving from strain increments to stress increments will usually be a well defined process even for strain softening materials), and hence new out-of-balance nodal forces, and hence new accelerations. The masses provide distributed inertia but usually some additional numerical or constitutive damping is needed to speed up the attainment of equilibrium (when the accelerations become tolerably small). Such a procedure could of course also be used for real dynamic analyses (Fig 4.21, 4.22) using real material damping rather than notional numerical damping properties.

The computer program FLAC (Fast Lagrangian Analysis of Continua) (Itasca, 2000) adopts this solution strategy. Although it is programmed as a finite difference code the spatial discretisation is handled in essentially the same way as for constant strain finite element triangles and we can deduce that reliable results will require a mesh containing a large number of small elements. The advantage, for nonlinear problems, is that the computational processes involved in each time step are extremely simple.

Evidently for both this dynamic relaxation method and for the viscoplastic solution it is rather necessary to choose the time step for calculation carefully. For dynamic relaxation we are limited by the time it takes for information to travel across the smallest/stiffest element (recall section 4.8). In both cases a pseudo-physical analogy has been used to cope with the finite size of loading

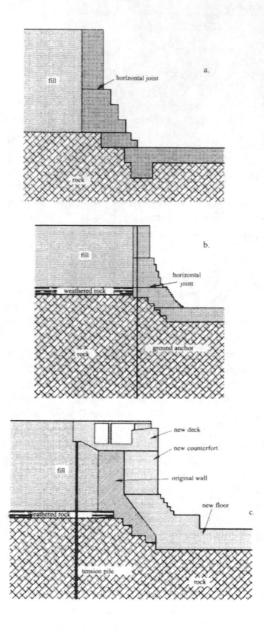

Figure 4.21: (a) Original dock wall; (b) dock wall strengthened with ground anchors; (c) dock wall strengthened with counterforts and tension piles (Mair and Muir Wood, 2002)

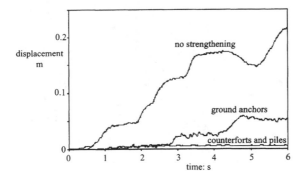

Figure 4.22: Dynamic response of dock wall under simulated earthquake loading (Mair and Muir Wood, 2002)

steps. The results obtained will be no more reliable than the detail of the implied stress:strain response within each step. Potts and Zdravković (1999) come down firmly in favour of the modified Newton-Raphson solution algorithm and this is probably the numerical approach that has received the most attention precisely because it is a purely numerical algorithm. It is not obvious that either the viscoplasticity or the dynamic relaxation algorithm could not be developed to the same degree of constitutive rigour. One of the difficulties associated with these approaches is that prior to equilibrium the solution method may require the calculation of the gradient of the plastic potential in illegal or erroneous parts of stress space and therefore the stress path during the increment, as it heads to a convergent state, may be spurious. The error associated with this will evidently depend on the local curvature of the plastic potential.

An example of the computational efficiency of different solution procedures is given in Fig 4.23, taken from Potts and Zdravković (1999). The settlement of the edge of an excavation is shown as a function of number of increments. The modified Newton-Raphson algorithm is found to be fast and accurate—a result confirmed for other classes of geotechnical system. The tangent stiffness algorithm is usually quite slow to converge to the correct result.

4.10 Conduct of numerical modelling

Having embarked upon the numerical analysis of a geotechnical problem, how can an engineer be confident that he or she has obtained the right answer to the problem? What does it mean to seek the 'right' answer to the problem? There seem to be three aspects to consider.

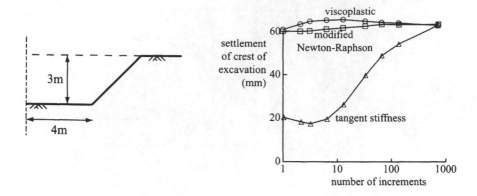

Figure 4.23: Settlement of crest of excavation: comparison of different finite element solution methods (after Potts and Zdravković, 1999)

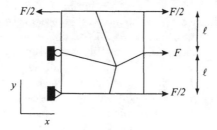

Figure 4.24: Patch test for group of four-node elements of unit thickness: boundary loading implies uniform stress state $\sigma_x = F/l$, $\sigma_y = \tau_{xy} = 0$

4.10.1   Verification: Is the program doing what it claims to be doing?

Evidently the numerical analysis program itself should be correctly coded and implemented. If it is a widely used commercially available or publicly accessible program then there is a good chance that the openness of access will with time lead to discovery and correction of coding errors. There may be checks that the engineer should undertake in order to satisfy himself or herself—or his or her client—that this is indeed the case. This is a verification exercise.

A patch test provides some sort of check on the coding of the elements (Fig 4.24). A small group of elements with at least one irregularly placed internal node is subjected to the least amount of boundary constraint that is necessary, and subjected to boundary loads which are compatible with a calculable elementary uniform state of stress and strain. Thus the loading in Fig 4.24 is compatible with a uniform normal stress σ_x in the x direction and zero shear stress τ_{xy} and direct stress σ_y. That stress distribution should be independent of

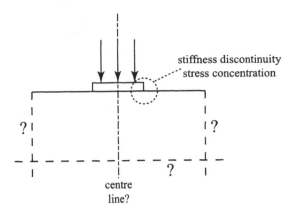

Figure 4.25: Location of boundaries for finite element analysis of footing

the soil model. For an elastic soil the strains in the y direction and the x direction should also be related by Poisson's ratio—and this could be checked. At all Gauss points in the patch the stress and strain components should correspond to the precalculated homogeneous values.

It should be possible to test individual elements under appropriate boundary conditions in order to check that constitutive models have been correctly implemented. In general, results for any constitutive model will either be calculable theoretically or, as published, will have been obtained by numerical algorithms which (it is hoped) are different from those implemented in the numerical analysis program thus ensuring a certain independence in the verification process.

Some boundary value problems are capable of closed form solution and attempts can be made to model these using finite elements. Many elastic analyses are available (for example in the wonderful compendium compiled by Poulos and Davis, 1974). Some of these—for example, stress and displacement distributions within elastic layers or half-spaces (§7.2)—will impose the challenge of deciding where to put the boundaries of the problem (Fig 4.25) in order to minimise their influence. The bottom boundary may represent a real geological stiffness discontinuity, but the location of lateral boundaries may often be somewhat arbitrary. A finite problem will usually be stiffer than its infinite counterpart. (Alternatively, it may be possible to use so-called 'infinite' elements which deliberately set out to reproduce boundaries at infinite distance (see, for example, Zienkiewicz and Taylor, 2000).)

There are some problems in plasticity which are capable of exact analysis—for example, the expansion or contraction of a cylindrical cavity (pressuremeter or tunnel, §8.8)—though again the location and effect of the boundaries needs to be considered carefully. However, as the constitutive model becomes more elaborate the likelihood that any closed form analysis of any useful boundary value problem will exist diminishes.

If the program claims to be able to model interfaces which allow concentrated relative displacement between adjacent blocks of deforming material, then it should be possible to set up simple analyses to demonstrate that the mode of mechanical operation of the interfaces that is described in the program manual can indeed be reproduced. Exactly the same can be said about structural elements—such as beams, bars, shells, cables.

Problems of transient flow—consolidation—or steady state flow—seepage— can also be checked against independent analyses for certain simple situations in order to reassure the user that algorithms for these typical geotechnical applications have been correctly implemented.

There may be published benchmark results for relatively simple problems that have been obtained with other programs which can be used as at least a first check. However, some reported studies (Schweiger, 2003) suggest that, even when the same benchmark problem is analysed by different people using the same numerical analysis program (PLAXIS), and the same constitutive model with the same soil parameters (so that the subjectivity in matching soil model to laboratory test data is supposedly eliminated), then the same results are not necessarily obtained (Fig 4.26) probably because of differences in the assumptions governing the detailed numerical description of the past history and the future perturbations. Even for such a 'simple' excavation problem there are nevertheless decisions to be made in locating the right hand and bottom boundaries and in the precise way in which the excavation is modelled. This sort of result should obviously make users cautious and aware that numerical modelling of complex problems is not to be undertaken lightly.

4.10.2 Are we getting the answers that we think we are getting?

The computer has reached the end of a finite element analysis without producing any error messages; post-processing of the output has produced colourful plots of stresses and displacements. We heave a sigh of relief and move on to the next project. Unfortunately, it is rarely as simple as this: users of numerical analysis programs need to do rather more than produce one successful analysis.

First there are the details of the numerical model. We have already mentioned boundaries (Fig 4.25). The bottom boundary of a model will often be defined by some known geological stratum with high stiffness and continuity which can be considered to provide a rigid base to the problem. Lateral boundaries are less easily fixed by natural features of the problem under investigation and it will be necessary to repeat the analysis with different widths of the numerical model to see how far away the boundary needs to be placed for its influence to become negligible. Evidently, the cost of numerical analysis will increase with the size of the problem being studied—the nearer the boundaries can be located the better. We can clearly economise by taking advantage of symmetries: the centre line of a foundation may form an obvious plane of reflection so that only half the problem needs to be analysed (Fig 4.25); or, for a circular structure symmetrically loaded, the analysis can treat one segment.

There are boundary conditions to consider. A plane of symmetry can sustain no shear stress and must be modelled as a smooth boundary with constraint on normal but not tangential motion. The bottom and distant side boundaries

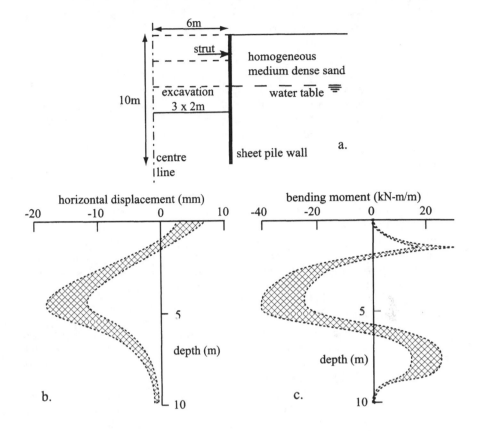

Figure 4.26: Benchmark comparisons of results of numerical analysis of strutted sheet pile wall retaining dense sand using PLAXIS: (a) outline of problem analysed; (b) approximate range of predictions of horizontal displacement of wall; (c) approximate range of predictions of bending moment in wall (data provided by Helmut Schweiger, 2003)

might often be described as fully rough—a check on the influence of releasing the tangential constraint might be appropriate.

Then there are the details of the mesh to be chosen. There may be limited choice of element type. Probably, higher order elements, if available, are to be preferred to simple elements, especially if high strain gradients are anticipated. More, smaller elements need to be placed where gradients are expected to be highest, and at regions of stress concentration. Many physical problems involve severe discontinuities: a footing has finite width (Fig 4.25), a retaining wall has finite height and at the ends of these structural loading elements there will be sudden changes in boundary loads and element stiffnesses. Sophisticated programs will permit adaptive mesh refinement during an analysis so that the mesh is progressively automatically refined in the areas where the highest strain gradients develop. Alternatively a user might inspect the output obtained with a simple mesh and repeat the analysis with a new mesh with strategically placed extra elements. It has to be noted that discontinuities are often associated with infinities in analytical results and these are always going to prove disturbing to numerical analysis. By the same token, analytical economy can be obtained by using large elements in regions where strain gradients are expected to be low—for example, towards the lateral boundary in Fig 4.25.

The shape of elements is also important to ensure that the stiffness equations are well conditioned: long thin elements tend to be unsatisfactory.

Data checking forms a vital part of the process—most programs will permit plotting of the mesh and the boundary conditions and constraints in ways which will readily reveal any obvious errors. A simple numerical error in one piece of input data may produce some very visible distortion of the mesh.

We have seen how significant the discretisation of the loading process can be (Fig 4.23). It is beholden on the user to ensure that the loading has been broken down into sufficiently small steps that further subdivision of steps produces no further improvement in the result.

Convergence will often be sought in a key output—for example, the load:-displacement response of a footing or horizontal displacement of the top of a retaining structure. Confirmation of reliability of the response requires closer inspection of the output. For example, it is valuable to plot contours or profiles of displacements, strains, and stresses across the model—both horizontally and vertically—to seek departures from expected monotonic or at least smooth variations. If the analysis has been performed with an assumed centreline symmetry, then contours of any variable should be orthogonal to this centreline. It is also valuable to plot paths for active elements through the course of the analysis—either strain paths or stress paths—again to demonstrate that these are in accord with expectation (and with the underlying modelling assumptions discussed in the next section).

4.10.3   Validation: Are we getting the answers that we need?

The third issue in interpreting the results of numerical modelling is much more fundamental and relates to the modelling itself. There are several stages that we should go through in planning a campaign of numerical modelling.

First we need to understand the problem. What effects are likely to be important? We need to have identified the key controlling influences—soil properties or features of the construction process—in order to ensure that these are included in the modelling. In performing numerical analysis there is always the danger that the complexity of the analysis—and more especially the apparent finality of the colourful output—will obscure the physical understanding of the system response. Before we embark on the analysis we should have some idea of what we expect to happen: we should be able to perform a back-of-the-envelope calculation to give us some idea of the magnitude of the effects that we expect to see. If the numerical analysis produces results which are initially counterintuitional then we should nevertheless be able to produce new back-of-the-envelope calculations to support what we have actually seen. It is unlikely that results for which we have no physical explanation are correct.

We must not be seduced by the precision with which a computer is able to report numerical results into thinking that the results are in fact imbued with this level of accuracy. We know that there are approximations in the finite element method. There are approximations in the assumptions about the constitutive response of the soil and the detailed description of the numerical model and its boundary conditions.

The constitutive model may well be the weakest link in performing numerical modelling. All programs used for analysis of geotechnical problems will include elastic models—perhaps permitting anisotropic elasticity. Any program that is seriously intended for geotechnical application should certainly permit the use of the elastic-perfectly plastic Mohr-Coulomb model. Whether such a model will permit the rather necessary luxury of nonassociated flow (dilation $\neq$ friction) is less certain though rather crucial—the numerical advantage of symmetrical stiffness matrices may have taken precedence over physical plausibility. Many programs will permit the use of Cam clay models. That will probably be the end of standard constitutive modelling provision. Availability of kinematic hardening models—or even Mohr-Coulomb models with pre-failure plastic nonlinearity—is likely to be restricted.

As a result, the possibilities of accurate and reliable matching of experimental data with the constitutive models actually available are limited. We may well have to make the type of approximation illustrated in Fig 4.27 (§3.3.4)—and as we will see in Fig 4.29 even an elastic-perfectly plastic model can generate a smooth transition from initial elastic to ultimate plastic behaviour of a system (see also §7.5.1). However, the resulting analysis can hardly be expected to give a six figure precision for the way in which the real soil would behave.

Even if we are able to make use of a constitutive model which provides a reasonable fit to data obtained from a range of laboratory stress paths we must recognise that almost every element in our numerical model will be undergoing stress changes which are quite different from those that we are able to apply in the laboratory—we are certainly extrapolating towards the unknown region. We need to beware that the constitutive model does not contain any hidden secrets leading to unexpected modes of response, which were not intended by the developer of the model, when it is used in the analysis for stress paths

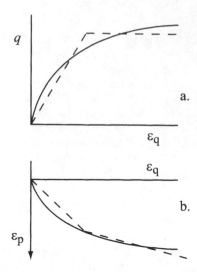

Figure 4.27: Elastic-perfectly plastic approximation to actual stress:strain response of soil

for which its behaviour has never been calibrated or investigated. It may be worthwhile to take a few key elements and study the predicted stress:strain response in detail to check that it does make sense in the spirit in which the constitutive model was created.

In geotechnical analyses we know that initial conditions are as important as constitutive properties in computing response under working loads (the value of ultimate, failure, loads will be less affected). We rarely have great detail of the *in-situ* stress conditions. We may have to attempt to model the full geological processes by which the soil has reached its present condition—but we certainly need to recognise the associated uncertainty.

In discussing physical modelling in Chapter 5 we will see how an efficient programme of physical modelling can be designed if we have a clear idea about those parameters of a problem and those material properties which are likely to control the behaviour. We hope that we can generate dimensionless groups which will characterise the system response and permit specific models to be interpreted for generic application. The same ideas should guide our numerical modelling too. Numerical parametric studies should be much more rapid than physical parametric studies but in terms of presentation of results the route to distillation should be the same.

With nonlinear history dependent materials such as soils it is to be expected that the final response that is calculated will depend on the route by which it is obtained. All aspects of the modelling need to be considered: the past, and how the soil has reached its present state, and the detail of the future changes that are expected. Uniqueness of ultimate response without attention to intermediate detail cannot be guaranteed.

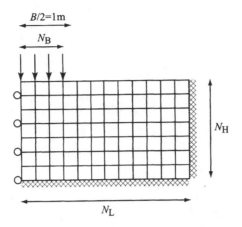

Figure 4.28: FLAC model of footing on surface of layer of elastic-perfectly plastic Mohr-Coulomb soil

We have made very little mention of three-dimensional analyses. Everything that has been said about two-dimensional elements for analysis of plane strain geotechnical problems can be extended to three dimensions. Three-dimensional models are tedious to prepare and time consuming to intepret. They should certainly be preceded by some carefully chosen two-dimensional models. For three-dimensional numerical models even more than for two-dimensional models there is the danger that the apparent beauty of the result—*any* result—will distract attention from *all* the assumptions on which it is dependent. The need to be able to produce a simple qualitative physical explanation for the predicted behaviour remains vital.

In the end, it is the engineer who uses the software who is responsible for the results. The aim of this book is to encourage engineers to take an interest in—because they will probably have to take ownership of—the whole process of numerical modelling and not just the one result—footing settlement, structural displacement, tunnel lining movement—which he or she thinks is necessary for a design application.

4.10.4   Exercise in numerical modelling: FLAC analysis of footing on Mohr-Coulomb soil

As an exercise in model validation and interpretation—and a demonstration of some of the features that may influence the results of numerical modelling—a simple student activity has been devised using the finite difference program FLAC (Itasca, 2000). This is intended to be illustrative of some of the issues that might be encountered in numerical modeling and not intended to provide a definitive analysis of the footing.

The modelling of the load:settlement response of a strip footing on the surface of a layer of elastic-perfectly plastic soil has been kept deliberately simple.

Table 4.1: Analysis of footing on Mohr-Coulomb soil: soil properties

bulk modulus	K	200 MPa	
shear modulus	G	100 MPa	
density	ρ	1.8 Mg/m^3	
friction	ϕ	30°	also 20°, 40°
dilation	ψ	0°	also 10°, 20°, 30°
earth pressure coefficient	K_o	1	also 0.33, 3

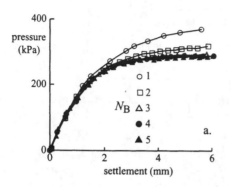
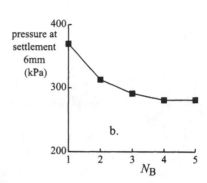

Figure 4.29: Effect of number of elements under footing N_B on footing response($N_L/N_B=$ 3, $N_H/N_B = 2$)

The material properties used for the analysis are shown in Table 4.1. Notwithstanding what was written earlier (§4.4.1, §4.8, §4.10.2), the soil is discretised as a series of square elements (Fig 4.28). The numbers of elements under the footing (N_B), across the model (N_L), and over the depth of the model (N_H) can be varied. The load is applied to the nodes bounding the N_B elements so that the half footing width (which is kept fixed at 1 m) corresponds to $N_B + 1/2$ elements. The left boundary is a smooth plane of symmetry. The other boundaries are perfectly rough.

The footing is loaded by pushing it into the soil at a slow constant velocity—the pseudo-dynamic nature of FLAC has been described above. The chosen velocity is a compromise between calculation time and accuracy of results.

First, the effect of spatial discretisation with constant model proportions is explored by varying N_B while keeping N_L/N_B and N_H/N_B constant at 3 and 2 respectively. Typical results for the load:settlement response are shown in Fig 4.29. It is noted that with $N_B > 4$ little further improvement is obtained.

Next, keeping N_B and N_H constant (at 4 and 8 respectively), N_L is varied to explore the influence of the proximity of the right hand boundary. Typical results are shown in Fig 4.30. In the limit, when $N_L = N_B$, there is no possibility of soil displacement to the side under the footing and the very stiff response discovered is that of an elastic material under one-dimensional oedo-

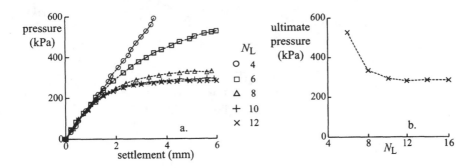

Figure 4.30: Effect of model width on footing response: variation of N_L with $N_B = 4$ and $N_H = 8$

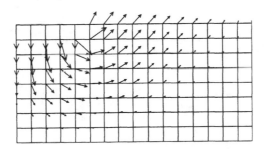

Figure 4.31: Displacement vectors around footing ($N_B = 4$, $N_L = 16$, $N_H = 8$)

metric loading. As the boundary moves further away its influence reduces until eventually the model is able to accommodate the displacement pattern for the failure mechanism for this frictional soil (Fig 4.31). For $N_L/N_B > 3$ no further significant change in the bearing capacity is obtained.

Keeping N_B and N_L constant (at 4 and 12 respectively), the depth of the model N_H is changed (Fig 4.32). There are two effects to observe. The initial stiffness of the footing (P/ρ) is dependent on the ratio of soil layer thickness to footing width B. Results can be extracted from Poulos and Davis (1974) to show that

$$P/\rho = E/\omega \tag{4.118}$$

where ω is a function of H/B and Poisson's ratio ν (Table 4.2). The results of the FLAC analysis can be used to confirm that the calculated initial footing stiffnesses correspond reasonably well with the theoretical values as H/B is changed (Fig 4.32b). The ultimate load also varies with H/B (when $N_H < N_B$ the layer being compressed is very thin and the ultimate load would be extremely high—and this has not been reached in the example shown in Fig 4.32 for $N_H = 2$) but reaches a limiting value once the soil layer is thick enough to include the full plastic mechanism (Figs 4.31, 4.32).

Table 4.2: Values of ω (equation (4.118)) for calculation of settlement of footing width B on elastic layer of thickness H with Poisson's ratio ν

	ν		
H/B	0.005	0.3	0.45
1	0.7900	0.6684	0.4170
2	1.1959	1.0685	0.7618
3	1.5015	1.3523	0.9940

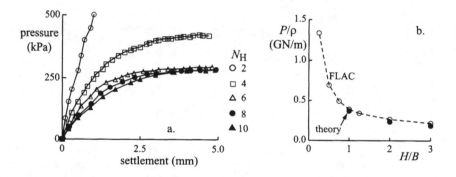

Figure 4.32: (a) Effect of model depth on load:settlement response ($N_B = 4$, $N_L = 12$) (b) comparison of calculated and theoretical initial footing stiffness as function of H/B

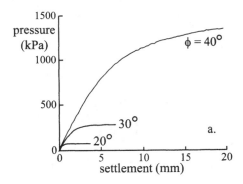
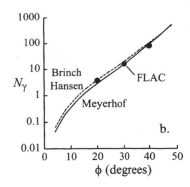

Figure 4.33: (a) Effect of ϕ on load:settlement response; (b) comparison of bearing capacity factors deduced from numerical analysis with expressions proposed by Meyerhof (1963) and Hansen (1968) (analyses performed with $N_B = 4$, $N_L = 16$, $N_H = 8$)

Having explored issues of discretisation and boundary location, this simple model can be used to compare predicted and calculated bearing capacities for this frictional soil. In the absence of cohesion and surcharge we expect the ultimate footing load to be given by

$$\frac{P}{B} = \frac{1}{2}\gamma B N_\gamma \qquad (4.119)$$

where N_γ is a function of angle of shearing resistance ϕ. The deduced values of N_γ are compared in Fig 4.33 with the values predicted using formulae proposed by Meyerhof (1963):

$$N_\gamma = \left(\frac{1+\sin\phi}{1-\sin\phi}\mathrm{e}^{\pi\tan\phi} - 1\right)\tan 1.4\phi \qquad (4.120)$$

and by Hansen (1968):

$$N_\gamma = 1.8\left(\frac{1+\sin\phi}{1-\sin\phi}\mathrm{e}^{\pi\tan\phi} - 1\right)\tan\phi \qquad (4.121)$$

The two expressions give rather similar results and the match with the values calculated using FLAC is satisfactory. Increasing the angle of shearing resistance increases the distance from the initial stress state to failure so that the elastic region is greatly increased and the displacement required to reach the ultimate load is also increased.

The soil model used in these analyses is an elastic-perfectly plastic Mohr-Coulomb model (section 3.3.4). The analyses presented so far have been performed with a zero angle of dilation $\psi = 0$ implying that plastic deformation occurs at constant volume. The effect of varying ψ on the load:displacement relationship is shown in Fig 4.34. The bearing capacity increases slightly as

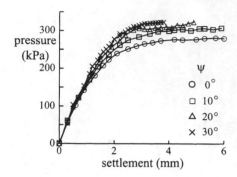

Figure 4.34: Effect of angle of dilation ψ on load:settlement response

ψ increases: this is not the place to dwell on the implications of the value of angle of dilation on bearing capacity of frictional soil. The calculation becomes somewhat wobbly when $\psi = \phi$: it was noted in section 3.3.4 that when angles of friction and dilation are equal there is no plastic energy dissipation in the material—which is physically unreasonable and numerically disconcerting particularly for a pseudo-dynamic calculation procedure which will be upset by deformations which are able to occur without any energy change.

The emphasis, in understanding elastic-plastic constitutive models, on the need to specify the past (the resulting initial size of the yield surface), the present (the initial stress state in relation to the yield surface), and the future (the nature of the loading to which the geotechnical system is to be subjected) applies to a numerical model just as much as to a single soil element. The analyses described so far have assumed that before the footing is loaded the ratio K_o of horizontal (effective) stress to vertical (effective) stress in the dry frictional soil is given by

$$\frac{\sigma'_h}{\sigma'_v} = K_o = 1 \tag{4.122}$$

Evidently the earth pressure coefficient at rest K_o can in fact take any value between the limits of the active pressure ratio K_a and the passive pressure ratio K_p:

$$K_a = \frac{1}{K_p} = \frac{1 - \sin\phi'}{1 + \sin\phi'} \tag{4.123}$$

which, with $\phi = 30°$, take the values of 0.33 and 3 respectively.

The effect of varying K_o between these limits is shown in Fig 4.35a. The effect is not great, but on close inspection it can be seen that as K_o falls the stiffness at intermediate loads also falls. The value of K_o controls the location of the initial stress state relative to the Mohr-Coulomb failure loci (Fig 4.35b). Underneath the footing—these are the soil elements that particularly control the load:settlement response—the vertical stress increases rapidly as the footing is loaded, with much less change in horizontal stress. If K_o is initially low, so that the soil is close to active failure (A in Fig 4.35b), then the onset of failure

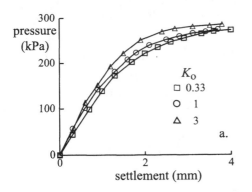
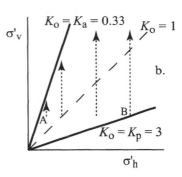

Figure 4.35: Effect of K_o on load:settlement response

occurs early as the footing load is increased, the purely elastic region finishes earlier and the overall footing stiffness falls. As K_o increases (towards B in Fig 4.35b), the attainment of local failure is delayed and the intermediate stiffness increases. As expected, varying K_o has negligible effect on the ultimate footing load: this is the result of the development of a plastic mechanism throughout the soil and the stresses associated with failure are essentially independent of the starting point.

4.11 Closure

Potts (2003) poses the question: 'Is numerical modelling just an advanced toy for academics and the privileged few, or is it in a position to provide a genuine tool for routine geotechnical analysis?' He observes that, while many geotechnical engineers have had some involvement with numerical modelling, few have been sufficiently engaged in the detail of the modelling to appreciate the complexities and subtleties that its use implies. In debating a motion that numerical modelling is well placed to play a central role in much geotechnical design he observes that numerical analysis can:

- do everything that conventional analysis can do—all the theoretical modelling strategies that are described in Chapter 7 can be reproduced in numerical modelling;

- accommodate realistic soil behaviour—in principle any of the models described or alluded to in Chapter 3 can be invoked in numerical modelling;

- account for complex soil stratigraphy;

- describe mechanisms of system response for comparison with physical modelling (Chapters 5 and 6);

- describe soil-structure and structure-soil-structure interaction as will be illustrated in Chapter 8; and

- accommodate three-dimensional geometries—whereas many of the simpler theoretical models find this difficult.

On the other hand he also observes that there are difficulties with numerical modelling:

- because there is no standard numerical strategy for implementation of nonlinear models;

- because some constitutive models seem to be unable to give reasonable predictions—and we have noted some obvious deficiencies in Chapter 3; and

- because, even for apparently simple problems, the results of numerical modelling can be very dependent on the decisions made by the user.

He further suggests that useful numerical modelling requires skilled operators who:

- have a detailed understanding of soil mechanics and the underpinning theory for the numerical algorithms;

- understand the limitations of constitutive models; and

- are familiar with the software that is being used for the numerical modelling.

If the message seems gloomy, it is more that the warning is important because numerical modelling, much more than physical modelling—which requires equipment and laboratories and technical support—or theoretical modelling—which requires conscious adaptation and simplification to fit analytical capabilities—is apparently available at the touch of a keyboard on *any* computer in *any* design office. The potential for disaster through misuse is certainly great.

We have attempted here to give some clues to the problems that may be associated with numerical modelling in the hope that the users of tomorrow will be more alert to the challenges to which Potts refers.

5

Physical modelling

5.1 Introduction

Physical modelling is performed in order to study particular aspects of the behaviour of prototypes. Full-scale testing is in a way an example of physical modelling where all features of the prototype being studied are reproduced at full scale. However, most physical models will be constructed at much smaller scales than the prototype precisely because it is desired to obtain information about expected patterns of response more rapidly and with closer control over model details than would be possible with full-scale testing. This usually implies that parametric studies should be performed in which key parameters of models are varied in order to discover their effect. This itself implies that many model tests will be required and in addition it is often desirable to repeat individual tests in order to gain greater confidence in the results that are obtained.

If the model is not constructed at full scale then we need to have some idea about the way in which we should extrapolate the observations that we make at model scale to the prototype scale. If the material behaviour is entirely linear and homogeneous for the loads that we apply in the model and expect in the prototype then it may be a simple matter to scale up the model observations and the details of the model may not be particularly important but, as will be shown, this still depends on the details of the underlying theoretical model which informs our physical modelling. Dimensional analysis is particularly important.

However, if the material behaviour is nonlinear, or if the geotechnical structure to be studied contains several materials which interact with each other, then the development of the underlying theoretical model will become more difficult. It then becomes even more vital to consider and understand the nature of the expected behaviour so that the details of the model can be correctly established and the rules to be applied for extrapolation of observations are clear. In short we need to understand the scaling laws.

5.2   Dimensional analysis

Dimensional analysis is a method for deducing elements of the form of a theoretical relationship from consideration of the variables and parameters that make up that relationship. The underlying premise is simply that any phenomenon can be described by a dimensionally consistent equation linking the controlling variables. Dimensional analysis of a problem then leads to a reduction in the number of variables that must be studied in order to understand the problem. The key is to seek to create dimensionally homogeneous equations whose form does not depend on the units of measurement. Governing equations cannot just be plucked from the air: they must come from an underlying insight into the phenomenon that is being modelled. If dimensional homogeneity appears elusive then this is probably an indication that some key variables or parameters have been omitted.

The theory of dimensional analysis is encapsulated in Buckingham's theorem: *If an equation is dimensionally homogeneous, it can be reduced to a relationship among a complete set of dimensionless products.* Once the application of this theorem has been understood it may appear to be intuitive but it can in fact be supported by rigorous mathematical proof (Langhaar, 1951). A further general conclusion can be drawn: *A set of dimensionless products of given variables is complete if each product in the set is independent of the others and every other dimensionless product of the variables is a product of powers of dimensionless products in the set.*

Dimensional analysis does not reveal the form of the relationships between the dimensionless products but correct use of the dimensionless products makes parametric studies more efficient by revealing which variables are truly independent and also forms the basis for extrapolating from one scale of observation to another.

There are various different ways in which dimensions of variables can be defined but the most commonly used fundamental system reduces everything to combinations of length [L], mass [M], time [T]. Where thermal or electrical effects are important then it is necessary also to add in temperature and charge respectively but those additions will not concern us here. For many geotechnical problems we are concerned with forces and stresses rather than masses and the dimension of time only comes in through the conversion of mass to force. Butterfield (1999) shows that application of classical theories of dimensional analysis in this situation can produce misleading results unless the alternative grouping for *force* $[MLT^{-2} = F]$ is used as a member of the fundamental system.

5.2.1   Slope in cohesive soil

Take as an example the factor of safety of a slope formed in purely cohesive soil. The variables that need to be considered are the factor of safety F, which is already dimensionless, and is expected to be a function of the geometry of the problem characterised by the height H of the slope (dimensions of length, L) and angle θ of the slope (dimensionless), together with the physical properties of the soil: its undrained cohesive strength c_u (dimensions of stress = force/area,

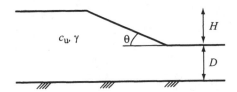

Figure 5.1: Slope in cohesive soil

FL^{-2}) and unit weight γ (dimensions of force/volume, FL^{-3}) (Fig 5.1). We can immediately write:

$$F = f\left(H, \theta, c_u, \gamma\right) \tag{5.1}$$

or in terms of dimensions

$$[1] = f\left([L], [1], [FL^{-2}], [FL^{-3}]\right) \tag{5.2}$$

A quick inspection of the dimensions of these variables indicates that the only dimensionless group that can be formed is: $c_u/\gamma H$. One can then conclude that the governing equation for this problem is

$$F = f\left(\frac{c_u}{\gamma H}, \theta\right) \tag{5.3}$$

and the number of variables that needs to be considered has been reduced by two. The factor of safety F is only of interest in association with the shear strength: it indicates the degree to which the shear strength can be reduced while still just ensuring slope stability ($F = 1$). So the result is in accord with our geotechnical experience: charts (Fig 5.2) presented by Taylor (1948), for example, show stability number $c_u/\gamma H$ as a function of slope angle θ and also of depth D to a strong layer below the slope (Fig 5.1)—but this is merely introducing a second dimensionless group D/H which characterises another aspect of the geometry of the slope: $c_u/F\gamma H = f(\theta, D/H)$.

A practical consequence from the point of view of physical modelling is that to maintain the same margin of safety in a model and prototype not only the geometry (the slope angle θ) but also the dimensionless group $c_u/\gamma H$ should be kept constant. If the slope height is reduced, as it usually will be in a small scale model, then, if the unit weight of the soil remains unchanged, as at first sight it must, the strength of the soil must be reduced in the same proportion.

In fact, the unit weight γ that we have introduced as a basic variable in our assessment of slope stability is not actually a fundamental quantity: it is calculated as a product of density ρ and gravitational acceleration g. Whereas the density of a material is a direct function of the packing of the particles for a soil or is a basic property for a metal, the gravitational acceleration can change from one celestial body to another (on the moon the acceleration due to lunar gravity is about 20% of the acceleration due to gravity at the surface of the earth) and can be artificially controlled in a geotechnical centrifuge, as

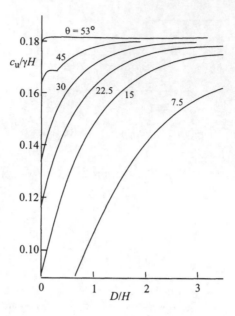

Figure 5.2: Stability chart for slope in cohesive soil (after Taylor, 1948)

described in more detail in Chapter 6. More fundamentally, therefore, we should write

$$\frac{c_u}{F \rho g H} = f\left(\theta, D/H\right) \tag{5.4}$$

and our assessment of the conditions necessary for a small physical model of a slope to maintain the same margin of safety as a prototype slope now concludes that, if we increase g as we reduce H then the soil properties, strength c_u and density ρ, can be kept unchanged.

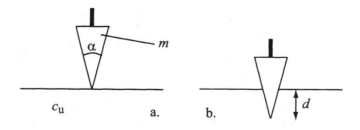

Figure 5.3: Fall-cone test in cohesive soil

5.2.2 Fall-cone

The fall-cone used as a quick measure of undrained strength in the laboratory can also be conveniently subjected to dimensional analysis (Fig 5.3). The penetration d of the cone, released from contact with the surface of a sample of soil and allowed to fall and penetrate the soil under its own weight will depend on the mass m of the cone, the gravitational acceleration g, the undrained strength c_u of the soil and the geometry of the cone, expressed by the tip angle α. (We neglect the buoyancy effect of the different densities of the soil and falling cone on the penetration process.) We conclude that

$$\frac{c_u d^2}{mg} = f(\alpha) = k_\alpha \qquad (5.5)$$

so that for a given cone angle α the dimensionless group $c_u d^2/mg$ should be a constant k_α. The fall-cone is of course used to determine the liquid limit of cohesive soils as the water content for which a cone with tip angle $\alpha = 30°$ and mass $m = 80$ g penetrates a distance $d = 20$ mm when allowed to fall under its own weight (§1.8). This analysis demonstrates the much more powerful use of the fall-cone as a strength measuring device and demonstrates too that the liquid limit test is itself a strength measuring test. It is found that for angle $\alpha = 30°$, $k_{30} = 0.85$ (Wood, 1985) so that at the liquid limit all cohesive soils have undrained strength $c_u = 1.7$ kPa. For a fall cone with angle $\alpha = 60°$, the cone factor $k_{60} = 0.29$. In Scandinavian countries the liquid limit is defined as the water content at which a 60 g, 60° cone penetrates 10 mm from rest. Knowing k_{60} we can deduce that the associated undrained strength is again $c_u = 1.7$ kPa. The fall-cone definition of the liquid limit is seen to be in reality a strength index—and thus helpfully relevant to geotechnical engineering practice.

5.2.3 Consolidation

In the two examples that we have just discussed, slope stability and penetration of the fall-cone, we do not know the exact form of the theoretical relationship. Dimensional analysis helps us to understand how we might efficiently explore it experimentally. There are other situations where we have a clear idea of the theoretical model which controls the phenomena in which we are interested. Consolidation is one such example.

Terzaghi's theory of one-dimensional consolidation tells us that temporal and spatial variations of pore pressure u are linked through a partial differential equation which introduces a coefficient of consolidation c_v (§4.2.1, §4.8):

$$\frac{\partial u}{\partial t} = c_v \frac{\partial^2 u}{\partial z^2} \qquad (5.6)$$

This equation is certainly dimensionally consistent because the coefficient of consolidation c_v has dimensions $L^2 T^{-1}$. We can expect to be able to characterise the solution in terms of pore pressure u as a proportion of some reference stress (initial pore pressure u_i or applied total stress change $\Delta\sigma$—the solution of the consolidation equation requires some particular boundary conditions for its completion) as a function of position z and time t. We must suppose that there is some typical length H that characterises the geometry of a series of similar consolidation problems. Then

$$u = f(u_i, z, t, H, c_v) \tag{5.7}$$

and consideration of the dimensions of these variables allows us to write more efficiently

$$\frac{u}{u_i} = f\left(\frac{z}{H}, \frac{c_v t}{H^2}\right) \tag{5.8}$$

which recovers the dimensionless time variable

$$T = \frac{c_v t}{H^2} \tag{5.9}$$

If we were also to write

$$U = \frac{u}{u_i} \tag{5.10}$$

$$Z = \frac{z}{H} \tag{5.11}$$

then (5.8) becomes $U = f(Z, T)$. Theory tells us, consistent with this, that the consolidation equation itself can then be written

$$\frac{\partial U}{\partial T} = \frac{\partial^2 U}{\partial Z^2} \tag{5.12}$$

and solutions of this equation will be entirely general and capable of application, by decoding the definitions of U, T and Z, to specific physical problems. We only need one set of observations for a known soil to be able to compute an expected response for *any* soil.

5.2.4   Fluid drag

In the examples of slope stability and of cone penetration, study of the groups of variables which controlled the problem indicated that there was really only one dimensionless group of interest. In other situations there will be many more variables and there is some element of choice in selection of the important dimensionless groups which should desirably be chosen in order to isolate particular effects. An example from fluid mechanics may help to illustrate this.

Let us suppose that we are concerned to study the drag exerted on a body by a fluid. We deduce that the drag force F is likely to depend on some or all of these variables: a typical dimension (assuming that we are considering geometrically similar bodies); the velocity of flow; the density, viscosity and elastic properties of the fluid. The gravitational field within which the flow is taking place may influence the drag effects, so we should include the acceleration due to gravity. There may be situations is which we have measured some pressure drop across the body and are interested in the correlation of this pressure drop with the drag force. If the object breaks the surface of the fluid then the surface tension will also be important. The variables and their dimensions are shown in Table 5.1.

Table 5.1: Fluid drag: summary of variables

symbol	variable	dimensions
F	force	ML/T^2
ℓ	dimension	L
V	velocity	L/T
ρ	density	M/L^3
μ	viscosity	M/LT
K	bulk modulus	M/LT^2
g	gravity	L/T^2
Δp	pressure change	M/LT^2
T	surface tension	M/T^2

Table 5.2: Fluid drag: different types of fluid force

types of force	
external force	F
fluid pressure force	$\Delta p \ell^2$
inertial force	$\rho v^2 \ell^2$
viscous force	$\mu v \ell$
gravitational force	$\rho g \ell^3$
elastic force	$K \ell^2$
surface tension force	$T\ell$

Since each of these variables can be expressed in terms of the three dimensions length, mass and time we can reduce the number of variables from nine to six dimensionless groups. It is ultimately a force that is of concern to us, so one logical route for generating these dimensionless groups is to think of ways in which force-like quantities can be generated. Bernoulli's equation tells us that inertial pressure is $\frac{1}{2}\rho v^2$ and hence inertial force is $\frac{1}{2}\rho v^2 \ell^2$. A viscous shear stress is proportional to a velocity gradient: force is stress × area. A gravitational force comes directly from the weight of an element of fluid. If the fluid is changing in volume then the link between stresses and deformations will be controlled by the elastic stiffness of the fluid. Surface tension gives a force proportional to the length of broken fluid surface. Expressions for the typical forces originating from these different sources are shown in Table 5.2.

A series of dimensionless groups can be obtained by comparing different types of forces. These are summarised in Table 5.3. It should not be surprising that five of the ratios (or quantities directly proportional to them) are familiar groupings of parameters used so regularly in fluid mechanics that they are given special names. Thus Reynolds Number is a ratio of inertial and viscous forces. Froude Number is a ratio of inertial and gravitational forces. Mach Number is a ratio of inertial and elastic forces. Euler Number is a ratio of inertial and pressure forces. Weber Number is a ratio of inertial and surface tension forces.

Table 5.3: Fluid drag: dimensionless groups

group		name
inertial/viscous	$\frac{\rho v \ell}{\mu}$	Reynolds
inertial/gravitational	$\frac{v^2}{g\ell}$	Froude^2
inertial/elastic	$\frac{\rho v^2}{K}$	Mach^2
inertial/pressure	$\frac{\rho v^2}{\Delta p}$	Euler^2
inertial/(surface tension)	$\frac{\rho v^2 \ell}{T}$	Weber
(external force)/inertial	$\frac{F}{\rho v^2 \ell^2}$	drag coefficient

Any other ratio of forces—for example, a drag coefficient—could be obtained as a ratio of two of these Numbers.

The choice of these particular ratios deliberately tries to isolate effects that in certain circumstances can be expected to be negligible. Mach Number is really about comparing flow velocity with the speed of sound in the fluid and obviously this ratio will often be extremely small and effects of fluid compressibility of no interest if we are concerned with low speed drag on an object under water.

This argument can also be presented in the opposite way. The identification of dimensionless groups provides a structure for the experimental investigation of the general problem of, in this instance, fluid drag. The dimensionless groups tell us, as before for the slope or the fall-cone, that certain combinations of variables are likely to influence the result in an interrelated way and that we can achieve an adequate coverage of the study of the problem provided we study appropriate ranges of the values of the dimensionless *groups* but we do not need to cover the full range of *each* individual variable. As a result of such a comprehensive experimental study we may well conclude that the drag, for example, is insensitive to Mach number over a certain range of values. In designing our physical models we would ideally want to ensure that we maintain in the models the same values of all dimensionless groups as in the prototype. This may be extremely restrictive on the design of exact small scale models. A discovery of insensitivity to one group tells us that we do not necessarily need to strive too carefully to retain the same value of this group in model and prototype.

5.2.5   Settlement of a footing

Returning to geotechnical applications, let us now explore the application of dimensional analysis to the modelling of the behaviour of a footing (Fig 5.4). First we can consider the settlement of a rigid circular footing on an elastic soil. We expect that the settlement ρ will be a function of the footing load P, the footing radius a, and the elastic properties of the soil, bulk modulus K and shear modulus G (or Young's modulus E and Poisson's ratio ν):

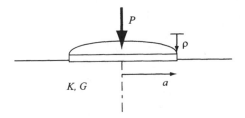

Figure 5.4: Settlement of circular footing on elastic soil

$$\rho = f(P, a, K, G) \tag{5.13}$$

Consideration of the dimensions of these quantities shows that we can expect a dimensionless relationship of the form:

$$\frac{\rho}{a} = f\left(\frac{P}{Ga^2}, \frac{G}{K}\right) \tag{5.14}$$

where the ratio G/K is directly related to Poisson's ratio. In fact, our prior knowledge of the behaviour of linear elastic materials leads us to expect that the settlement will be directly proportional to the load so that the form of relationship must actually be:

$$\frac{\rho}{a} = \frac{P}{Ga^2} f\left(\frac{G}{K}\right) \tag{5.15}$$

or

$$\frac{\rho Ga}{P} = f(\nu) \tag{5.16}$$

and, for a given material (constant value of ν) we would only need to perform one footing test at any scale to determine the particular value of $f(\nu)$.

In fact, this is a problem that is capable of exact elastic solution from integration of Boussinesq equations giving (§8.2.4):

$$\frac{\rho Ga}{P} = \frac{1 - \nu}{4} \tag{5.17}$$

Dimensional analysis cannot determine the constants and the details of the functions but it can set us on the right track.

For a rectangular footing of width B and length L (Fig 5.5) an exactly similar argument gives

$$\frac{\rho}{B} = \frac{P}{GLB} f\left(\nu, \frac{L}{B}\right) \tag{5.18}$$

Here there is no exact result but Poulos and Davis (1974) quote results of an approximate solution in the form (§8.2.4):

$$\frac{\rho G\sqrt{LB}}{P} = \frac{1 - \nu}{\beta_z} \tag{5.19}$$

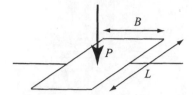

Figure 5.5: Settlement of rectangular footing on elastic soil

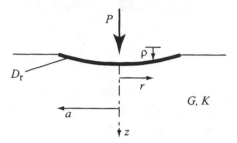

Figure 5.6: Flexible circular footing on elastic soil

where β_z is a function of both ν and L/B. This is directly consistent with our generalised dimensional deduction (5.18).

We can extend this analysis a little further by introducing the possibility of footing flexibility. This will be considered in greater detail in the chapter on soil-structure interaction (§8.4). Here we need to note that the settlement is no longer a single quantity but will vary with location on the footing. Let us consider a circular footing or raft for simplicity (Fig 5.6): the settlement will be a function of radius r. The footing is flexible so the structural material will be deforming and the behaviour will in some way depend on the elastic properties of the footing. We could introduce these in the form of Young's modulus E_r and Poisson's ratio ν_r (or shear modulus G_r and bulk modulus K_r) but, if we restrict ourselves to a class of footings which deform only by bending, we expect the stiffness properties of the footing of thickness h to enter our analysis through the flexural rigidity D_r

$$D_r = \frac{E_r h^3}{12 \left(1 - \nu_r^2\right)} \tag{5.20}$$

This immediately solves part of our problem since we no longer need to consider footing thickness, footing Young's modulus and footing Poisson's ratio independently. We can now write that we expect the settlement ρ at some radius r to be a function of the total load P on the footing, the elastic properties of the soil, G and K, the radius of the footing a, and the composite property, flexural rigidity of the footing, D_r:

$$\rho = f\left(r, a, P, G, K, D_r\right) \tag{5.21}$$

which from considerations of dimensional analysis becomes:

$$\frac{\rho}{a} = f\left(\frac{P}{Ga^2}, \frac{r}{a}, \frac{G}{K}, \frac{Ga^3}{D_r}\right) \tag{5.22}$$

and once again from our prior knowledge of behaviour of elastic systems we can combine two of the dimensionless groups to give (compare §8.4):

$$\frac{\rho Ga}{P} = f\left(\frac{r}{a}, \frac{G}{K}, \frac{Ga^3}{D_r}\right) \tag{5.23}$$

We are using a certain amount of background knowledge to underpin our description of the problem and guide our modelling. This now tells us that we can obtain similar settlements at corresponding points in model and prototype footings for appropriately scaled loads if the Poisson's ratio of the soil is the same in both and if the flexural rigidity of the footing is scaled appropriately with the soil stiffness and the radius of the footing.

A natural extension of this line of thought can be used to deduce rational expressions for the stress distribution as a function of the variables. We can argue from symmetry for a circular footing that the stresses will not depend on circumferential position. Hence depth z and radius r should be sufficient coordinates (Fig 5.6). The elastic change of a general stress component σ will be given by

$$\sigma = f(r, z, a, P, G, K, D_r) \tag{5.24}$$

which, in principle, becomes

$$\frac{\sigma a^2}{P} = f\left(\frac{r}{a}, \frac{z}{a}, \frac{P}{Ga^2}, \frac{G}{K}, \frac{Ga^3}{D_r}\right) \tag{5.25}$$

However, we can again introduce our prior understanding of elastic problems to indicate that, since we expect the elastic stress changes to be directly proportional to the applied load (and that guided our choice of the dimensionless group $\sigma a^2/P$), there can be no further dependence on the applied load and therefore no dependence on the dimensionless group P/Ga^2. The result is

$$\frac{\sigma a^2}{P} = f\left(\frac{r}{a}, \frac{z}{a}, \frac{G}{K}, \frac{Ga^3}{D_r}\right) \tag{5.26}$$

and at geometrically similar locations in model and prototype (same values of r/a and z/a) the normalised stress will be the same provided the Poisson's ratio of the soil is retained together with the *relative* stiffness of the footing—but the *absolute* value of soil stiffness is not important.

5.2.6 Bearing capacity of a footing

At the other extreme we can seek appropriate forms of relationship for the ultimate load on a footing (Fig 5.7). We can now suggest, from our physical understanding of the problem, that the deformation properties of the soil (and the footing) will not be important and that the ultimate load P_u will depend

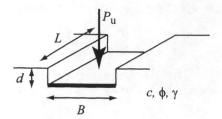

Figure 5.7: Bearing capacity of rectangular footing on cohesive-frictional soil

on the geometry and dimensions of the footing—for example, breadth B and length L for a rectangular footing—and the depth of embedment of foundation d, together with the strength properties of the soil—for example, cohesive strength c and frictional strength ϕ. We can imagine also that the unit weight γ of the soil will be relevant since it must be harder to push an object into a heavier medium.

$$P_u = f\left(L, B, d, c, \phi, \gamma\right) \qquad (5.27)$$

Rearranging into dimensionless groups this becomes

$$\frac{P_u}{\gamma B^2 L} = f\left(\frac{L}{B}, \frac{d}{B}, \frac{c}{\gamma B}, \phi\right) \qquad (5.28)$$

and for similarity in model and prototype we should not only maintain geometrical similarity—L/B, d/B—but also scale cohesive strength c with γB. As for the slope, this implies reduction of strength by the geometrical scale factor if we are testing at one gravity but implies that the strength can be maintained at its prototype value if the gravitational acceleration (and hence the unit weight) is increased by the same factor by which the linear dimensions are reduced.

We are more familiar with the bearing capacity expression

$$\frac{P_u}{\gamma B^2 L} = N_c \frac{c}{\gamma B} f_1\left(\frac{L}{B}\right) + N_q \frac{\gamma d}{\gamma B} f_2\left(\frac{L}{B}\right) + \frac{1}{2} N_\gamma f_3\left(\frac{L}{B}\right) \qquad (5.29)$$

where N_c, N_q, N_γ are all functions of ϕ and the depth of embedment d has been represented by a surcharge γd. The functions f_1, f_2, f_3 are shape factors. This result is thus entirely in accordance with our dimensional analysis. In fact, of course, the dimensional analysis places no constraints on the form of dependency that we should expect so that the additive result that is traditionally assumed (5.29) is only one acceptable possibility and certainly not the only one. This traditional result comes from empirical expediency and not from theoretical rigour.

Theoretical results for the analysis of bearing capacity are made much easier if certain simplifications are made (§7.3.4, §7.3.6). Thus, the footing might be assumed to be at the surface of the ground so that the effect of embedment is equivalent to the provision of a surcharge at foundation level. That appears to be a reasonable but possibly very conservative assumption: *strength* of soil above the level of the footing is entirely neglected. For a frictional soil the shear strength that can be mobilised at any point in the material depends on the stress level at that point. The heavier the soil in the region that is being brought to failure the greater the stress level and hence the greater the available strength. The incorporation of heavy soil into the theoretical analysis is tricky and tends to be, artificially, kept separate from the effect of the surcharge— hence the separate bearing capacity factors N_q and N_γ. In reality the soil unit weight above the footing level will certainly influence the way in which the surcharge is amplified to contribute to the footing load and a more multiplicative combination of effects would be implied.

5.2.7 Soil nonlinearity

We have looked at regimes of footing behaviour which are entirely governed by the elastic response of the soil or by the strength of the soil. In practice we are interested in the response of footings under working loads on a material which we know has an extremely limited elastic region: in the light of discussion in earlier chapters we can confidently assert that truly elastic, recoverable behaviour of the soil is unlikely to be dominant. We now need to introduce into our dimensional analysis some means of characterising the nonlinear material response of the soil. We need to have some prior understanding or assumption concerning the soil behaviour in order to be able to proceed: we need some soil constitutive model. The more complex this model is assumed to be the more difficult it will be to satisfy requirements of physical modelling.

Then we expect the settlement ρ of a rigid footing to be dependent on geometry (Fig 5.8) (radius a for a circular footing), load P, elastic properties (the elastic-plastic (?) soil is still supposed to possess underpinning elastic properties: shear and bulk moduli, G and K) and strength properties (cohesion c and friction ϕ) of the soil. The unit weight γ of the soil will characterise the rate of stress increase with depth which will influence the variation of stress-strain response with depth.

$$\rho = f(a, P, G, K, \gamma, c, \phi) \tag{5.30}$$

Applying our ideas of dimensional analysis we can deduce

$$\frac{\rho}{a} = f\left(\frac{P}{Ga^2}, \frac{G}{K}, \frac{c}{G}, \frac{\gamma a}{c}, \phi\right) \tag{5.31}$$

For an elastic-perfectly plastic model with shear strength c and shear stiffness G the ratio of strength to stiffness is the shear strain required to reach the strength of the soil which can be regarded as some sort of characteristic strain ϵ_c for the soil

$$\epsilon_c = \frac{c}{G} \tag{5.32}$$

The idea of a characteristic strain can be introduced more generally: for a hardening plastic model defined in terms of mobilised friction or stress ratio it

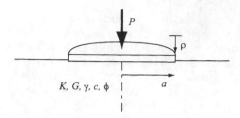

Figure 5.8: Rigid circular footing on elastic-plastic soil

could be the strain required to reach 50% of the frictional strength. We can then suggest that several of the dimensionless groups are actually strain-like quantities and that our modelling should try to ensure that ratios of strains are maintained as far as possible:

$$\frac{\rho}{a\epsilon_c} = f\left(\frac{P}{a^2 G\epsilon_c}, \frac{\gamma a}{G\epsilon_c}, \frac{G}{K}, \phi\right) \tag{5.33}$$

The characteristic strain ϵ_c typifies a class of soil models and does not need to be considered separately provided that all implied ratios of strains are maintained identical in the model and the prototype. Recall, however, that choosing to consider only a certain class of soil models does not constrain real soils to behave in the way that has been assumed.

The dimensionless quantity, angle of friction ϕ, has been included in the relationship and is left as a free dimensionless variable. If we are lucky then our class of soil models may be sufficiently characterised by ϵ_c, so that separate consideration of angle of friction may be unnecessary. However, our prior knowledge of soil behaviour would have to remind us that friction is not only a first order function of strain but is also a (possibly second order) function of mean stress and soil density.

5.3 Scaling laws revisited

We are now in a position to attack the scaling laws with rather more rigour and to discover the implications for physical geotechnical modelling. We are interested both in modelling on the laboratory floor at single gravity and modelling on a geotechnical centrifuge at multiple gravities so we will specifically allow for both these possibilities within a general framework. The mechanics of centrifuge modelling will be considered in the next chapter.

A summary of scale factors is given in Table 5.4[1]. The factors listed symbolically under the heading 'general' indicate the fundamental linkage between the various modelling decisions that might be taken: these are the ratios of model and prototype values. The particular factors listed under '1*g* (laboratory)' and '*ng* (centrifuge)' are the result of typical modelling choices—but other choices could be made. The table indicates aspirational guidance.

---

[1]Of course the list cannot be complete.

Table 5.4: Scale factors

quantity	general	scale factors	
		$1g$ (laboratory)	ng (centrifuge)
length	n_ℓ	$1/n$	$1/n$
mass density	n_ρ	1	1
acceleration	n_g	1	n
stiffness	n_G	$1/n^\alpha$	1
stress	$n_\rho n_g n_\ell$	$1/n$	1
force	$n_\rho n_g n_\ell^3$	$1/n^3$	$1/n^2$
force/unit length	$n_\rho n_g n_\ell^2$	$1/n^2$	$1/n$
strain	$n_\rho n_g n_\ell/n_G$	$1/n^{1-\alpha}$	1
displacement	$n_\rho n_g n_\ell{}^2/n_G$	$1/n^{2-\alpha}$	$1/n$
pore fluid viscosity	n_μ	1 or* $n^{1-\alpha/2}$	1 or* n
pore fluid density	$n_{\rho f}$	1	1
permeability (Darcy's Law)	$n_{\rho f} n_g/n_\mu$	1 or* $1/n^{1-\alpha/2}$	n or* 1
hydraulic gradient	$n_\rho/n_{\rho f}$	1	1
time (diffusion)	$n_\mu n_\ell^2/n_G$	$1/n^{2-\alpha}$ or* $1/n^{1-\alpha/2}$	$1/n^2$ or* $1/n$
time (creep)	1	1	1
time (dynamic)	$n_\ell (n_\rho/n_G)^{1/2}$	$1/n^{1-\alpha/2}$	$1/n$
velocity	$n_g n_\ell (n_\rho/n_G)^{1/2}$	$1/n^{1-\alpha/2}$	1
frequency	$(n_G/n_\rho)^{1/2}/n_\ell$	$n^{1-\alpha/2}$	n
shear wave velocity	$(n_G/n_\rho)^{1/2}$	$1/n^{\alpha/2}$	1

*scaling of pore fluid viscosity introduced in order to force identity of scale factors for diffusion time and dynamic time

A true model is obtained when all the governing laws of similitude are in place. However, often for geotechnical modelling it will be necessary to make do with an adequate model which maintains 'first order' similarity (Harris and Sabnis, 1999)—by arguing, from a proper consideration of the likely mechanisms of response, that some of the constraints imposed by dimensional analysis are of second order importance. This will often be the case for shaking table models where it is generally not possible to properly scale gravitational loads—and this will have other implications for material and system behaviour. Many authors have discussed scaling factors for models in general and geotechnical models in particular (eg Harris and Sabnis, 1999; Krawinkler, 1979; Iai, 1989; Schofield and Steedman, 1988)—some of these have been primarily concerned with the factors that are relevant to centrifuge modelling. A brief discussion is warranted here: often stiffness is not identified as a factor that needs to be considered separately and it seems to be helpful to use stiffness rather than strain as an independent quantity (although in practice the degree of independence may be somewhat illusory).

There are various scaling factors that need to be chosen and there may be discussion about the most fundamental set to use. Some of the factors that are introduced will be treated as independent factors though it will be shown that the soil will often manage to take control of the apparent (and possibly desirable) independence.

5.3.1 Length

We naturally start with a scale factor for length n_ℓ

$$\frac{(\text{length})_{\text{model}}}{(\text{length})_{\text{prototype}}} = n_\ell \qquad (5.34)$$

because the reduction of dimension is usually the primary objective in performing physical modelling.

5.3.2 Density

We have seen in our earlier discussion of dimensional analysis applied to geotechnical systems that the unit weight of the material—which is actually a product of density and gravitational acceleration—will often control response. Although we will usually try to make use of real soils in our physical models we can retain for the moment a scale factor for density n_ρ

$$\frac{(\text{density})_{\text{model}}}{(\text{density})_{\text{prototype}}} = n_\rho \qquad (5.35)$$

Even if we use the same material in model and prototype there may be (second order?) density differences arising from different stress levels and consequently different densities of packing. We could reckon to create an artificial granular material by using heavy particles of, say, iron ore instead of the

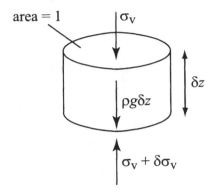

Figure 5.9: Gravitational variation of vertical stress in soil

usual silica or feldspar—but the uncertainties of effects on resulting mechanical response (from particle shape and interaction) would *probably* outweigh any benefit that might be obtained from an increased density.

5.3.3 Acceleration

We naturally introduce next a scale factor for gravitational acceleration n_g

$$\frac{(\text{acceleration})_{\text{model}}}{(\text{acceleration})_{\text{prototype}}} = n_g \tag{5.36}$$

which will be 1 for single gravity models and will take some chosen higher value n for models tested on a geotechnical centrifuge. There is an implicit assumption that dynamic accelerations must (will?) be scaled with the model equivalent gravitational acceleration so that there is a single scale factor for all accelerations. This seems logical because if dynamic vertical accelerations are being studied then the response of the soil or of elements of a geotechnical system will certainly depend on the proportion of gravitational acceleration that the dynamic accelerations take up.

Stresses build up in the ground because of the density of the soil and the gravitational acceleration. We know (Fig 5.9) that

$$\delta\sigma_v = \rho g \delta z \tag{5.37}$$

and hence the scale for stresses is $n_\rho n_g n_\ell$.

We make the assumption that we are interested in physical modelling of geotechnical problems in which autogenous—gravitational—stress gradients are important. If there are external anthropogenic forces (such as foundation loads) then we will expect to scale them in the same way as the internal forces: thus force on an isolated footing or pile will scale as $n_\rho n_g n_\ell^3$ while force per unit length of a plane strain system—such as prop loads for a long wall—will scale as $n_\rho n_g n_\ell^2$. If the external loads are large by comparison with the *in-situ* stresses

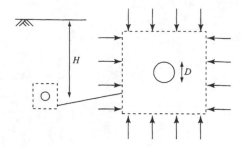

Figure 5.10: Deep tunnel: $H \gg D$

then the effect of the internal stress gradients may be negligible. Equally, there are geotechnical problems such as the deformation around a small very deep tunnel (Fig 5.10) where the gravitational variation of stress across the model may be negligible by comparison with the average level of pre-existing stress around the tunnel.

5.3.4   Stiffness

The concept of stiffness is difficult for a nonlinear material such as soil and in general we have to permit the possibility that stiffness of our geotechnical material may or may not be under our control to some extent. We therefore leave as independent a scale factor for shear stiffness N_G.

$$\frac{(\text{stiffness})_{\text{model}}}{(\text{stiffness})_{\text{prototype}}} = n_G \qquad (5.38)$$

We have introduced stiffness as an independent scaling factor. In many cases it will not really be independent because the deformation properties of soils are not easily controlled. The stiffness term might be considered to include two elements: the very small strain stiffness, which will in many situations control the dynamic response and propagation of waves through the model ground; and the nonlinear medium to large strain deformation properties of the soil. The small strain stiffness might be reckoned to be, to first order, dependent on the effective stress level, σ, according to a relationship of the form:

$$G \propto \sigma^\alpha \qquad (5.39)$$

Experimental experience suggests that the exponent α might be of the order of 0.5 for sands but of the order of 1 for clays. Evidently a value $\alpha = 0$ implies that the stiffness is independent of stress level. Table 5.4 shows scaling factors for single gravity modelling introducing this dependence together with factors for centrifuge models, in each case with linear scale $1/n$. For all models it is assumed that the same soil material has been used in prototype and model so that the scale factor for density is, to first order, unity.

For medium strain deformation response, the optimum approach to ensuring that model and prototype bear some resemblance is to invoke critical state soil mechanics and argue for similar values of state variable (or equivalent—section §2.6.1) between the two cases (Roscoe and Poorooshasb, 1963; Been and Jefferies, 1985). For a material with *local* critical state line slope λ in a conventional semi-logarithmic compression plane, then, if the value of density relative to the critical state density, allowing for stress change (this relative density *is* the state variable), is to be retained, a scale factor m on stresses implies a necessary change in initial specific volume or void ratio of

$$\Delta v = \Delta e = \lambda \ln m \qquad (5.40)$$

For a typical sand with λ of the order of 0.03, and with a linear model scale of, say 20, the necessary increase in specific volume or void ratio from prototype to model to maintain the similarity (in fact identity) of state variable would be around 0.1. For a typical sand the range of void ratio from maximum to minimum density determined by standard procedures might be around 0.4. The reduction in relative density required to maintain similarity is high: pluviation procedures commonly adopted for reliable and repeatable preparation of sand samples will probably not be able to achieve the necessary high initial void ratios. If the sand bed were being prepared for dynamic testing then, even if the initial void ratios could be achieved, the resulting extremely loose samples would not survive beyond the first shake of a model earthquake. Bolton and Steedman (1985) similarly note that no soil element will be able to mobilise more than the critical state strength once significant seismic shearing has started no matter what density and hence strength the element thought that it had a few milliseconds earlier.

For a typical clay with λ of the order of 0.2 the required increase in specific volume for a linear model scale of 50 is about 0.8 and, following typical correlations, this implies an increase in liquidity index (§1.8) also around 0.8. Again, the resulting samples may be initially rather soft and difficult to handle without disturbance.

So while one can see n_G as encapsulating the constitutive response of the soil (and not just the strictly elastic properties), we can deduce that we will not usually have much freedom to choose what might otherwise be considered the optimum or most desirable value for n_G. In writing down the particular scaling factors for $1g$ and ng models in Table 5.4 we have assumed that the stiffness has simply been allowed to take the value that corresponds to the stress level in the model according to (5.39). Thus in $1g$ models the stiffness scales as $1/n^\alpha$.

5.3.5 Strain

Strain results from changes in stress relative to the stiffness of the material. The chosen or imposed scale factor for stiffness will control the scale of strains and also various other quantities which are implicitly dependent on strains.

It is often suggested that the scale for strains should be unity in order to ensure geometric similarity and similar mobilisation of soil stiffness at all times but, desirable though this may be, modelling choices may often preclude it.

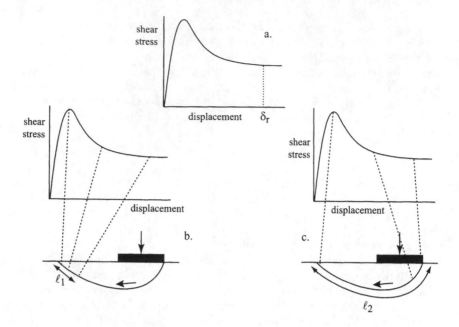

Figure 5.11: (a) Softening relationship between shear stress and relative displacement across an interface; (b) prototype footing; (c) small model footing

5.3.6 Displacement

Displacement results from the integration of a field of strains over distances within the soil and hence the scale for displacement is the product of the scales for strain and length. Although in general, because of the usual nonlinearity of the constitutive response of soils, it would be desirable for the strains at corresponding points in model and prototype to be the same, so that the scale for strain would be unity, this may be difficult in single gravity models because of the implicitly different scale factors for stress and stiffness. If strains are identical in model and prototype then the displacements will scale with the linear scale of the model n_ℓ.

If we are concerned about concepts such as stresses and strains then we are assuming that the soil is behaving as a continuous material so that such concepts have some clear meaning and relevance. If the geotechnical system under study leads to relative movement on interfaces—either between separate blocks of soil forming part of a failure mechanism, or between the soil and a structural element such as a pile or section of reinforcement—then the interface behaviour will be controlled by relative displacement across the interface and a small model may have difficulty in correctly reproducing the system response.

Consider a material for which the shear stress transmitted across the interface varies nonlinearly with displacement, with a residual stress beyond some relative displacement δ_r (Fig 5.11a). For a prototype system (Fig 5.11b), displacements at failure will be sufficiently large that only a small proportion of the failure surface, of length ℓ_1 is mobilising interface stress above the residual value at the moment that the soil at the emerging end of the failure surface just attains the peak of the shear stress:displacement relationship. In a small model the length ℓ_2 over which the stresses are above the residual value may make up a much higher proportion of the failure surface—in fact it is possible that no part of the failure surface will have fully lost strength as far as the residual value (Fig 5.11c). The prototype collapse load would then be overestimated from the small model.

Actually one might expect the interface behaviour to be controlled not by displacement *per se* but (non-dimensionally) by displacement scaled with particle diameter. Then, *if* the constitutive response of the soil could be left unaffected as the particle size were reduced, some of the problems associated with reduced model displacement might be overcome. Possibilities are limited: particle shape does not automatically remain unchanged as particle size falls; abrasion of asperities on large particles may occur more readily than for small particles; if particles become too small then surface forces (Van der Waals attractions) become significant in proportion to mechanical forces and the character of the particle interactions will change.

We conclude that care is necessary in extrapolating discontinuous phenomena where displacements rather than strains drive the response.

5.3.7 Permeability

In geotechnical engineering we are familiar with a coefficient of permeability k introduced through Darcy's law which says that the velocity v of fluid flow through the pores of a soil is proportional to hydraulic gradient i

$$v = ki \qquad (5.41)$$

The concept of hydraulic gradient is useful because hydraulic head is readily visualised as the height of a fluid column. However, what this form of Darcy's law conceals is that flow is actually driven by pressure gradient $\Delta p / \ell$:

$$v = \frac{k}{\gamma_w} \frac{\Delta p}{\ell} \qquad (5.42)$$

where $\gamma_w = \rho_w g$ is the unit weight of the water and ρ_w is its density.

Theoretical expressions for permeability of soils have been deduced by applying Poiseuille's law for flow through capillaries as a result of a pressure gradient, which does not need to be gravitationally driven (whereas the concept of unit weight implies gravity). A typical result produced by Taylor (1948) is

$$k = C d_s^2 \frac{\gamma_w}{\mu} \frac{e^3}{1+e} \qquad (5.43)$$

where d_s is a typical average grain size, e is void ratio, μ is the viscosity of the permeating fluid and C is a composite shape factor which somehow characterises

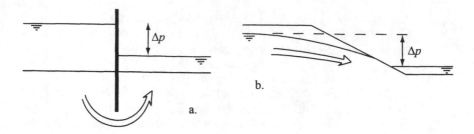

Figure 5.12: External hydraulic pressure difference Δp driving seepage(a)under sheet pile wall; and (b) through slope

the shape of typical void spaces within the soil and the tortuosity of the flow path. For a given soil there is an absolute or specific permeability K

$$K = \frac{k\mu}{\gamma_w} = \frac{k\mu}{\rho_w g} = Cd_s^2 \frac{e^3}{1+e} \tag{5.44}$$

which, neglecting changes in void ratio, is more or less a soil constant, independent of permeant and external conditions (such as temperature). The specific permeability has units of length^2 and is also expressed in darcys, where 1 darcy $= 0.987 \times 10^{-14}$ m^2. Since K is essentially independent of soil and permeant, the scale factor for permeability as conventionally used is the ratio of scale factors of fluid unit weight and viscosity.

$$k = \frac{K\rho_w g}{\mu} \tag{5.45}$$

and Darcy's law (5.42) becomes

$$v = \frac{K}{\mu} \frac{\Delta p}{\ell} \tag{5.46}$$

If we have a scale of pore fluid density $n_{\rho f}$ and a scale of pore fluid viscosity n_μ, then the scale for permeability is $n_{\rho f} n_g / n_\mu$ as shown in Table 5.4. If the same soil is used in the prototype and the model then $n_\rho = 1$ for the soil but there might be good reasons, as noted below, to use a pore fluid other than water so that $n_\mu \neq 1$ and $n_{\rho f} \neq 1$.

5.3.8   Hydraulic gradient

Hydraulic gradient i (5.41) appears to be a dimensionless quantity—a ratio of lengths—which therefore should automatically remain unchanged as we go from the prototype to the model. However, as we have seen in rewriting the flow equation (5.42), hydraulic gradient is actually a special way of presenting a pore pressure gradient in the soil.

Pore pressure gradients can emerge for two different reasons. There may be some overall steady seepage regime driven by some difference in elevation of free fluid surfaces across the soil (Fig 5.12). Pore pressures within the model will be controlled by this externally imposed pressure difference Δp and hydraulic gradients will be controlled by $\Delta p / \rho_w g \ell$. The scale factor for hydraulic gradient is then $n_p / n_{\rho f} n_g n_\ell$ where n_p is the scale factor for the external pressure difference.

However, pore pressures—often transient—will also develop as a result simply of total and effective stress change in the soil. If the soil permeability is low then any total stress change which would tend to produce a volume change of the soil will instead produce a change in mean effective stress (and hence pore pressure) in order to keep the volume of each soil element constant. The changes in total and effective stress scale with $n_\rho n_g n_\ell$ and hence the hydraulic gradient resulting from these pore pressures scales with $n_\rho / n_{\rho f}$.

In the same way that we concluded that external forces should be scaled with internal stresses, we deduce that, if we are trying to maintain similarity of seepage pore pressures and stress-induced pore pressures in prototype and model, we must scale the pressure Δp such that $n_p = n_\rho n_g n_\ell$ and we will then have consistent scaling of hydraulic gradient $n_\rho / n_{\rho f}$.

5.3.9 Time scales

There are three time scales that may be relevant for interpretation of the results of model tests. Many soils show some time dependent response in the form of creep, or strain rate or relaxation effects, or weathering or chemical decay or diagenetic bond growth. Insofar as these are real aspects of material response (rather than a misinterpretation of transient consolidation effects associated with migration of pore water down gradients of pore water pressure) then they will occur at the same rate, if the soil is subjected to the same stress conditions, whether in the prototype or in a model. Such creep effects are possibly driven by some law such as Arrhenius' law of reaction rates which will show an exponential influence of temperature on reaction rate. In principle, then, it might be possible to change the time scale for creep events by changing the temperature from prototype to model but this will not usually be feasible. In the absence of such stratagems the time scale for creep will be unity.

For many geotechnical situations the time scale that is of greatest interest will be that which controls the rate at which consolidation can occur. We have already seen that the normalisation of the equation of consolidation introduces a dimensionless time

$$T = \frac{c_v t}{H^2} \tag{5.47}$$

and recalling the definition of coefficient of consolidation and the definition of specific permeability this becomes

$$T = \frac{kt}{m_v \gamma_w H^2} = K \frac{t E_{oed}}{\mu H^2} \tag{5.48}$$

where the coefficient of volume compressibility m_v is a volumetric compliance, the inverse of E_{oed} the one-dimensional soil stiffness. The scale factor for diffusion time can therefore be deduced from the scale factors for fluid viscosity n_μ, soil stiffness n_G, and length n_ℓ giving an overall factor $n_\mu n_\ell^2 / n_G$.

An alternative argument deduces this time scale n_t for diffusion (consolidation) by insisting on similarity of two volumes: a flow volume (permeability $\times$ hydraulic gradient $\times$ area $\times$ time) and a volume change related to change in stress (strain $\times$ volume). Then

$$\frac{n_{\rho f} n_g}{n_\mu} \frac{n_\rho}{n_{\rho f}} n_\ell^2 n_t = n_\ell^3 \times \frac{n_\rho n_g n_\ell}{n_G} \qquad (5.49)$$

and hence, as before,

$$n_t = \frac{n_\mu n_\ell^2}{n_G} \qquad (5.50)$$

On a centrifuge with $n_\ell = 1/n$, $n_G = n_\mu = 1$, diffusion time scales as $1/n^2$ and there is an enormous benefit obtained from reduced consolidation times. Thus, a centrifuge model constructed at a linear scale of 1:100 and tested at 100g will consolidate 10^4 times faster than the prototype. One year of prototype time becomes about 53 minutes of model time. The advantage of centrifuge modelling for studying phenomena where consolidation or other diffusion effects dominate is very clear.

In a single gravity model with $n_\ell = 1/n$, $n_\mu = 1$, $n_G = 1/n^\alpha$, diffusion time scales as $1/n^{2-\alpha}$ and the benefit is not so great. For example, with $\alpha = 1$ (stiffness proportional to stress—typical for clays), time scales only as $1/n$. Although the drainage path is reduced by the factor n, the flow volume required to produce an equivalent change of scaled stress has increased by the same factor n because of the reduction in stiffness.

The third time scale of interest is that which governs dynamic events. A scale factor for velocity n_{vel} can be proposed from the need to maintain similarity of potential and kinetic energies in prototype and model. Change in potential energy per unit volume is unit weight $\times$ displacement. Kinetic energy per unit volume is density $\times$ velocity^2. Hence

$$n_\rho n_g \times \frac{n_\rho n_g n_\ell^2}{n_G} = n_\rho \times n_{vel}^2$$

and

$$n_{vel} = n_g n_\ell \sqrt{\frac{n_\rho}{n_G}} \qquad (5.51)$$

Then dynamic time scales as the ratio of displacement to velocity giving a dynamic time scale $n_\ell(n_\rho/n_G)^{1/2}$. Dynamic frequency scales as the inverse of this: $(n_G/n_\rho)^{1/2}/n_\ell$.

The time scales for dynamic events and diffusion events are different. This has significant implications for modelling of situations where both dynamic events and diffusion events are significant. For example, earthquake shaking of fine granular soils may induce liquefaction through pore pressure increase. However, there are likely to be vertical and horizontal gradients of pore pressure created during the earthquake and some dissipation of pore pressures will be able to occur (Fig 5.13). Whether or not dissipation can occur fast enough to remove the possibility of liquefaction will crucially control the overall geotechnical response.

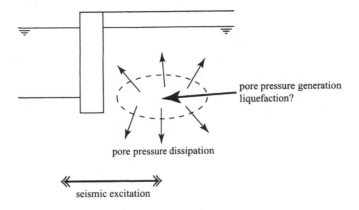

Figure 5.13: Simultaneous pore pressure generation and dissipation in soil around retaining structure under seismic excitation

Study of the scaling relationships for these two aspects of time suggests a route for solving the problem and bringing the two time scales into alignment. We require

$$\frac{n_\mu n_\ell^2}{n_G} = n_\ell \sqrt{\frac{n_\rho}{n_G}} \tag{5.52}$$

and hence

$$n_\mu = \frac{1}{n_\ell} \left(n_\rho n_G \right)^{1/2} \tag{5.53}$$

In other words we can force these two time scales to coalesce by changing the properties of the permeant. In the centrifuge, with $n_\ell = 1/n, n_\rho = n_G = 1$ we need to change the viscosity of the permeant such that $n_\mu = n$. Typically in centrifuge modelling a silicone fluid with viscosity 100 times that of water could be used in a model tested at 100g. The density of the silicone fluid is almost the same as that of water so the scaling laws are exactly satisfied and hydraulic gradients are not affected. Of course we then have to assume that there are no other changes in the mechanical behaviour of the soils which arise because of the change in pore fluid. For single gravity modelling the viscosity of the fluid to be used in a model at linear scale $n_\ell = 1/n$ would need to be increased by a factor of $n^{1-\alpha/2}$ (because of the inclusion of the stiffness scale n_G in (5.53)). Then for $n = 100$ and $\alpha = 1/2$ (typical of sand), $n_\mu = 31.6$.

5.3.10   Shear wave velocity

A further quantity of interest to us in considering dynamic modelling is the shear wave velocity which describes the speed with which shear disturbances propagate through the soil. (Compression wave velocity describes the propagation of longitudinal compression waves and an exactly similar argument can be used.) The shear wave velocity, v_s, can be shown theoretically (see §3.2.3) to be given by

$$v_s = \sqrt{\frac{G}{\rho}} \tag{5.54}$$

and consequently it scales according to the square root of the ratio of the scales of stiffness and density: $(n_G/n_\rho)^{1/2}$. The dynamic time scale can be recovered by considering the time taken for a shear wave to propagate through a model: (= typical dimension/shear wave velocity : $n_\ell(n_\rho/n_G)^{1/2}$). Both natural frequency of a soil layer (= shear wave velocity/typical dimension) and frequency of dynamic events scale in the same way—which may seem curious given that dynamic velocity and shear wave velocity have quite different scales which only align when

$$\left(\frac{n_G}{n_\rho}\right)^{1/2} = n_g n_\ell \left(\frac{n_\rho}{n_G}\right)^{1/2}$$

or, in other words

$$\frac{n_g n_\ell n_\rho}{n_G} = 1 \tag{5.55}$$

and the scale factor for strain is unity. However, the two frequencies emerge from different calculations. The natural frequency of the soil layer comes from consideration of the overall dimensions of the layer; the dynamic frequency comes from consideration of displacements within the soil. Thus it may be possible to preserve characteristics of dynamic *system* response without insisting on similarity of strains.

5.4   Soil-structure interaction

5.4.1   Footing

Modelling of soil-structure interaction requires further scale factors for structural elements. We have already seen in section 5.2.5 that to obtain similarity of settlements and stresses around a flexible footing we need to scale footing flexural rigidity D_r (5.20) with the stiffness of the soil through a dimensionless group Ga^3/D_r where a is a typical dimension (5.22). Thus the scale factor n_D for D_r should be (see also §8.4)

$$n_D = n_G n_\ell^3 \tag{5.56}$$

and if $n_G \neq 1$ then we will need to adjust the thickness and/or Young's modulus of the material from which the model footing is made in order to satisfy (5.56) and it will not be sufficient merely to scale the footing thickness according to the linear scale of the model.

For example, suppose that we have a prototype footing with thickness $h_p = 0.5$ m made of concrete with Young's modulus $E_p = 25$ GPa and Poisson's ratio 0.3 which we wish to model at $1g$ at a scale $n_\ell = 1/100$. The foundation soil is a sandy material for which $\alpha = 1/2$. We choose to model this with a footing made of aluminium alloy for which $E_m = 70$ GPa. We need to choose a value for the thickness h_m of the model footing. Thus, introducing scale factors n_E and n_h for footing Young's modulus and thickness respectively, we require

$$n_E n_h^3 = n_G n_\ell^3$$

or

$$n_h = n_\ell \left(\frac{n_G}{n_E}\right)^{1/3}$$

and substituting

$$n_h = \frac{1}{100}\left(\frac{1/100^{1/2}}{70/25}\right)^{1/3} = 3.29 \times 10^{-3} = 1/304$$

so that the model footing thickness is $500 \times n_h = 1.65$ mm. (We have assumed that the difference (if any) between the values of Poisson's ratio for prototype and model footing materials is negligible but this, if known reliably, could obviously be included in the calculation.)

5.4.2 Pile under lateral loading

We are interested in modelling the load transfer between a pile and the surrounding ground (Fig 5.14) (see also §8.2.2). We will assume that the pile is not being so heavily loaded that it is stressed by axial load or in bending beyond its elastic range. We assume also that we are not concerned with the ultimate lateral load capacity of the pile moving relative to the soil. Let us consider initially simply the response of the pile to lateral loading which will be governed primarily by the flexural rigidity of the pile EI—its actual physical dimensions and shape may be less important. The pile can be considered as a beam with certain loads applied both by loading at the ground surface or at the head of the pile and by the resistance of the ground to relative movement of pile and soil. If the soil responds elastically to this relative movement then the resisting force will be proportional to relative displacement according to some coefficient of subgrade reaction k and the equation governing the deformation of the pile will be of the form:

$$EI\frac{d^4y}{dz^4} = -ky \tag{5.57}$$

where z is the distance measured down the pile and y is the horizontal deflection of the pile. The coefficient k will be expected to be proportional to the shear modulus G of the soil $k = \beta G$ (although the pile:soil interaction is not strictly a process of pure shear).

Following the same procedure that we used in our interpretation of the consolidation equation we observe that this equation can be normalised by defining a dimensionless depth ζ

$$\zeta = \frac{z}{\ell} \tag{5.58}$$

where ℓ is the length of the pile, and a dimensionless pile deflection λ

$$\lambda = \frac{y}{y_o} \tag{5.59}$$

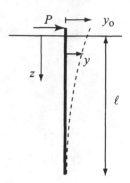

Figure 5.14: Pile under lateral loading

where y_o is the lateral deflection of the pile at its top (say). The equation then becomes

$$\frac{EI}{\ell^4}\frac{\mathrm{d}^4\lambda}{\mathrm{d}\zeta^4} = -k\lambda \tag{5.60}$$

and, since $k = \beta G$, a natural dimensionless group to characterise the problem is $G\ell^4/EI$ which describes relative pile:soil stiffness.

Alternatively, we can observe that terms in the solution of (5.57) involve $\mu\ell$ where $\mu^4 = k/EI = \beta G/EI$ and hence deduce again that $G\ell^4/EI$ is an appropriate dimensionless group to describe relative pile:soil stiffness (§8.2.2).

The soil quantity $G\ell^4$ somehow has an equivalence to the flexural rigidity EI of the pile. Then we might suppose that correct physical modelling will be obtained if we maintain the dimensionless ratio Φ_1

$$\Phi_1 = \frac{G\ell^4}{EI} \tag{5.61}$$

identical in the model and the prototype. If we have scale factors n_E and n_I for Young's modulus E and second moment of area I of the pile then we deduce that

$$n_E n_I = n_G n_\ell^4 \tag{5.62}$$

With a length scale $n_\ell = 1/n$ this leads to $n_E n_I = 1/n^{4+\alpha}$ for single gravity testing and $n_E n_I = 1/n^4$ for modelling on a geotechnical centrifuge with an acceleration of ng.

Let us consider different possibilities for modelling a prototype tubular pile which is 20m long, 0.5m diameter, with 25mm wall thickness, and made from steel with Young's modulus $E = 210$ GPa. For the sake of argument we will suppose that a length scale of $n_\ell = 1/100$ has been selected for the physical modelling.

Soil stiffness G identical in prototype and model: $\alpha = 0$

If $\alpha = 0$ and the stiffness is the same in the prototype and the model then the design of the model is the same whether the model is destined for testing at $1g$ or $100g$. For $1g$ testing, if we are studying the behaviour of piles in overconsolidated clay, then we might choose to prepare a model block of clay with a consolidation history which we reckon matches the typical history of the prototype soil—or even to sample the actual prototype material—and we might suppose that to a first order the stiffness properties of the ground are rather independent of depth over the length of the pile. The model soil thus has the same stiffness as the prototype soil and in a single gravity model where the stresses are reduced by a factor $1/n$ we would expect the strains also to be reduced by $1/n$.

We have a length scale $n_\ell = 1/n = 1/100$ so that our typical dimension ℓ, which is the length of the model pile, is 0.2 m. We need to reduce the flexural rigidity of the pile by $1/n^4$ in order to maintain the value of Φ_1. We could achieve this by making the model pile from steel (the same material as the prototype) with all dimensions reduced by $1/n$: 5 mm diameter, 0.25 mm wall thickness.

However, we might decide that such a model pile was rather delicate to manufacture and choose to replace the tubular prototype pile with a solid model pile. For the prototype tubular pile of diameter d_p and wall thickness t_p the second moment of area is $I_p = \frac{\pi}{8} t_p d_p^3$. For a solid model pile of diameter d_m the second moment of area is $I_m = \frac{\pi}{64} d_m^4$ and to maintain similarity we require $I_m/I_p = 1/n^4$ giving a model pile diameter $d_p = 3.98$ mm.

Alternatively we might choose to make the model pile out of a less stiff material such as aluminium alloy with $E_m = 70$ GPa and increase I_m and d_m to compensate. For a solid pile this would imply a model pile diameter of 5.23 mm.

Evidently the governing equation (5.57) is only concerned with elastic interaction between the pile and the ground. If we are concerned with the development of inelasticity or even eventual lateral failure with significant relative movement between soil and pile, then there may be other considerations (§8.9). For example, it may be seen as rather important to maintain the scaled geometrical profile of the pile so that the diameter should be correctly scaled to 5 mm (the solid aluminium model pile might be attractive). Failure suggests that behaviour at the pile-soil interface will be of concern. Modelling of surface effects might require deliberately roughened model piles to match rough—corroded?— prototype piles. 'Roughness' starts to introduce additional significant lengths— both absolute (the scale of asperities) and relative (the height of asperities in relation to particle size)—and the detail of the modelling of the pile surface in relation to the surrounding soil may not be straightforward.

Soil stiffness G dependent on stress: eg $\alpha = 0.5$

In a single gravity model created from a soil with $\alpha \neq 0$ we should take some account of the scaling of soil stiffness when we select the dimensions of our model pile. With $\alpha = 0.5$ we now require $n_E n_I = 1/n^{9/2}$.

With a length scale $n_\ell = 1/n = 1/100$ the diameter d_m of a solid model pile of material with Young's modulus E_m is given by:

$$\frac{\pi}{64} d_m^4 E_m = \frac{1}{n^{9/2}} \frac{\pi}{8} d_p^3 t_p E_p$$

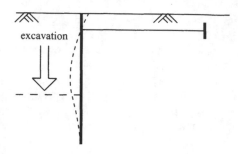

Figure 5.15: Anchored flexible retaining wall

leading to a diameter d_m = 2.94 mm for a solid aluminium model pile or
2.23 mm for a solid steel model pile.

If $\alpha = 1$, so that soil stiffness scales directly with stress then the flexural rigidity of the pile scales with $1/n^5$ leading to diameters of 1.65 mm and
1.25 mm for aluminium and steel model piles respectively. The departure from
proper scaling of the geometrical aspect ratio of the pile is now such that an alternative much less stiff material such as polypropylene ($E_m = 0.9$ GPa) might
be considered as an alternative (assuming that surface roughness can be properly modelled). With $\alpha = 1$ this would lead to a diameter of solid model pile
$d_m = 4.9$ mm which is geometrically more attractive.

5.4.3   Flexible retaining wall

A flexible retaining wall is an example of a geotechnical structure which deforms
under conditions of plane strain (as opposed to the pile in the previous example
which is an isolated structural element). A typical desirable output from modelling of such a structure would be the bending moment that is generated in the
wall as a result of excavation in front of the wall (Fig 5.15). As a plane structure,
the moment would be quoted as a moment per unit width of the structure M/b
and the structural property that will influence the bending will be the flexural
rigidity per unit width of the structure EI/b. The interaction between the wall
and the soil will be influenced by the stiffness G of the soil at a typical depth H.
For similarity we need to maintain the value of the dimensionless group (§8.7):

$$\Phi_2 = \frac{EI/b}{GH^3} \tag{5.63}$$

The quantity b is merely a notional unit width of model or prototype which
does not scale so that the consequence of considering plane models such as this
is that, in single gravity models with length scale $n_\ell = 1/n$, the flexural rigidity
scales with $1/n^{3+\alpha}$ instead of $1/n^{4+\alpha}$ as we found for the single pile.

We can now calculate the wall thickness required for our model. Suppose
that the prototype is made from concrete with Young's modulus $E_p = 20$ GPa
and thickness $t_p = 0.3$ m. We wish to construct our model wall at a linear scale
of 1/100 out of aluminium with Young's modulus $E_m = 70$ GPa. We assume
that $\alpha = 1/2$. Then for a model wall of thickness t_m we find that:

Table 5.5: Model wall thicknesses for flexible retaining wall

material	E_m GPa	t_m mm
steel	210	0.64
aluminium	70	0.9
microconcrete	10	1.75
polypropylene	0.9	3.9

$$\frac{\frac{1}{12}E_m t_m^3}{\frac{1}{12}E_p t_p^3} = 1/n^{7/2}$$

and hence

$$t_m = t_p \left(\frac{E_p}{E_m} \frac{1}{n^{7/2}} \right)^{1/3} = 0.3 \times \left(\frac{20}{70} \times \frac{1}{100^{7/2}} \right)^{1/3} = 0.9 \text{ mm}$$

We might alternatively reckon that interface effects could be more satisfactorily reproduced by making the wall using a modelling microconcrete with fine aggregate. With $E_m = 10$ GPa, for example, the wall dimension becomes $t_m = 1.75$ mm. Alternatively again, using polypropylene ($E_m = 0.9$ GPa), $t_m = 3.9$ mm which might in the end be more manageable. Various modelling possibilities are summarised in Table 5.5.

If the retaining wall is an anchored wall, as shown in Fig 5.15, then the anchor system will also need to be modelled. The ties can probably be regarded as essentially rigid so their properties may not matter in detail (but in some cases the interaction of flexible ties with consolidating or settling backfill might be important). However, the anchor system—perhaps a series of short individual anchor piles—is evidently important. Insofar as their role is to provide a fixed point of support for the wall, it is unlikely that we will be particularly concerned about their flexibility. We are primarily concerned about the rapid mobilisation of sufficient passive resistance as the anchor ties try to pull the anchors towards the wall. Mobilisation of passive resistance will potentially be affected by interaction with the wall itself. Correct geometrical modelling of the location and proportions (and interface properties) of the anchor piles will be key.

5.4.4 Buried flexible culvert

A buried flexible culvert (Fig 5.16) is another example of a geotechnical structure which deforms under conditions of plane strain. A typical desirable output from modelling of such a structure would be the bending moment per unit width M/b that is generated in the wall of the culvert as a result of construction procedures and surface traffic loading. The structural property that will influence the bending will again be a flexural rigidity per unit width of the structure EI/b. In addition, the moments will be influenced by the geometry of the culvert (characterised by a typical diameter D and a typical depth to mid-diameter from the

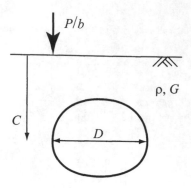

Figure 5.16: Buried flexible culvert

free ground surface C), by the unit weight of the soil ρg and the applied surface load per unit width (for example a line load P/b) and its line of action.

The interaction between the culvert and the soil will be influenced by the stiffness of the soil G at the typical depth C.

$$\frac{M}{b} = f\left(C, D, \frac{P}{b}, \rho g, \frac{EI}{b}, G\right) \tag{5.64}$$

which becomes[2]:

$$\frac{M/b}{\rho g D^3} = f\left(\frac{C}{D}, \frac{P/b}{\rho g C D}, \frac{EI/b}{G D^3}, \frac{PD^2}{EI}\right) \tag{5.65}$$

There is a geometric characteristic C/D which we need to retain to preserve similarity. Then there are various stress and stiffness related quantities in the loading spread over the cross section of the culvert P/bD and the stress in the ground at the mid-height of the culvert $\rho g C$; the flexural rigidity of the culvert EI/b and a corresponding soil stiffness GD^3. Evidently we need to retain similarity between these pairs of related quantities $(P/b)/\rho g CD$, $(EI/b)/GD^3$ and also maintain similarity between the loading and the structural stiffness $(P/b)D^2/(EI/b)$. Again, in single gravity models at linear scale $n_\ell = 1/n$, the flexural rigidity scales with $1/n^{3+\alpha}$.

We can follow the same procedure as before to calculate the wall thickness required for our model culvert. Suppose that the prototype culvert is made from steel with $E_p = 210$ GPa and has second moment of area $I/b = 6.25 \times 10^{-6}$ m^4/m. The prototype culvert will probably be made from steel formed into a corrugated section, and this section is reflected in the quoted second moment of area. We may well consider it acceptable to model the culvert with a curved sheet of thickness t_m. If we are modelling at a linear scale of $1/100$, and $\alpha = 1/2$, then for a model culvert made from steel sheet we find that:

---

[2]But rearrangement of the several groups shows that the dimensionless moment could just as well be $(M/b)/GD^2$.

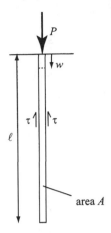

Figure 5.17: Pile under axial loading

$$\frac{1}{12}t_m^3 = 6.25 \times 10^{-6}\left(\frac{1}{100}\right)^{7/2} \tag{5.66}$$

and hence $t_m = 0.19$ mm.

Alternatively we might find that plastic sheet was more readily available and, so long as local plastic moment failure of the steel culvert is not initially reckoned to be a modelling issue, we can then reproduce all deformation effects with this different material. With $E_m = 0.9$ GPa we find $t_m = 1.2$ mm.

5.4.5 Piles under axial load

It is usually implicitly assumed that piles can be treated as axially rigid inclusions so that axial deformation of the pile is not immediately a constraint on physical modelling. However, a long pile, shedding its load by shaft friction along its length, may compress by an amount which is comparable with the relative movements between the pile and the ground (Fig 5.17). The possible need to model axial stiffness correctly needs to be borne in mind.

An appropriate dimensionless group can be constructed by comparing two characteristic forces (Nuñez and Randolph, 1984). Suppose that we have a pile which is sufficiently long and compressible that virtually no force reaches the toe of the pile when it is loaded at its top—all the load is shed to the soil in shaft resistance. For a pile of Young's modulus E, cross-sectional area A and length ℓ, the force generated by a settlement w_t at the top will be related to EAw_t/ℓ since w_t/ℓ provides some indication of the level of axial strain in the pile. If the pile of radius r_o is moving in elastic soil, then it is shown by Fleming *et al.* (1985) that the shear stress at the interface between pile and soil is related to the relative movement by (compare §8.2.3):

$$\frac{\tau}{w/r_o} \approx kG \tag{5.67}$$

where $k \approx 1/4$. So a characteristic shaft friction force is $2\pi r_o \ell (G/4)(w_t/r_o)$ and the dimensionless ratio of forces is

$$\frac{1}{2} \left(\frac{\ell}{r_o}\right)^2 \left(\frac{G}{E^*}\right)$$

where E^* is the Young's modulus for an equivalent solid pile so that $E^* \pi r_o^2 = EA$. Similitude in modelling then requires selection of model dimensions and stiffnesses so that $(\ell/r_o)\sqrt{(G/E^*)}$ is identical in model and prototype.

If we are concerned about development of limiting shear strength c_u at the pile-soil interface, as well as elastic transfer of stress, then (5.67) tells us that we will need to retain similarity of the ratio w/r_o. If we are modelling on a centrifuge with G, and probably c_u, unchanged from prototype to model, we will want to maintain full geometric similarity so that ℓ, r_o (and we hope w) scale directly with n_ℓ and the only modelling choice left to us will be the selection of the material and cross-section of the model pile so that EA scales with n_ℓ^2. Interface behaviour and surface roughness effects may also be important and we may be concerned with the nonlinearities of the pile:ground interaction as the relative movements develop. The constraints imposed to maintain complete similarity may be quite tight.

5.4.6   Dynamic soil-structure interaction

The response of a geotechnical structure under dynamic loading will depend, broadly, on the comparison of the frequencies over which the input energy (for example from vibrating machinery or from an earthquake) is distributed with the natural frequencies of the geotechnical structure. Deducing in advance what the natural frequencies of the combination of soil and structural elements will be may not be straightforward but we can assess the natural frequencies of the soil bed on its own and of the structure on its own. Understanding the way in which we should extrapolate from model observation to prototype expectation then requires us to think about maintaining similarity of dimensionless groups which characterise not only the static soil-structure interaction but also the natural frequencies.

We have already noted that dynamic frequencies scale as $(n_G/n_\rho)^{1/2}/n_\ell$ and that the natural frequency of a soil layer will scale with the same group of scaling factors. The natural frequencies of a laterally loaded pile of length ℓ, flexural rigidity EI, cross-sectional area A and density ρ_s, treated as a cantilever (in air), are multiples of $(\pi/\ell)^2 \sqrt{(EI/\rho_s A)}$. The scale factors for these natural frequencies n_{fs} will then be

$$n_{fs} = \frac{1}{n_\ell^2} \left(\frac{n_{EI}}{n_{\rho s} n_A}\right)^{1/2}$$

where n_{EI}, $n_{\rho s}$ and n_A are scale factors for flexural rigidity, density and area for the pile. Now we suppose that we have already decided to ensure that our pile satisfies similarity rules for static soil-structure interaction as outlined in section 5.4.2. Then $n_{EI} = n_G n_\ell^4$ and we require

$$\left(\frac{n_G}{n_{\rho s} n_A} \right)^{1/2} = \left(\frac{n_G}{n_\rho} \right)^{1/2} \frac{1}{n_\ell}$$

or

$$n_{\rho s} n_A = n_\rho n_\ell^2 \tag{5.68}$$

which seems reassuringly plausible but allows us to vary the material of the pile and the cross sectional area if necessary. For example, while it might be desirable to scale the perimeter of the pile correctly if shaft:soil interaction is important, we could still replace a solid section by a hollow section if this were more convenient experimentally to help us satisfy (5.62) *and* (5.68)—and at the same time we might replace the prototype material with some more convenient model material provided we could guarantee that we were going to remain in a range of either linear response or reproducible nonlinear material response. With linear scale $1/n$ on a centrifuge, or at $1g$ in a material having $\alpha = 0$, all frequencies scale as n. This is perhaps a little glib for, if we are not working on a centrifuge at prototype stress levels, then the need to scale flexural rigidity to cope with the changed soil stiffness may conflict with the scaling of cross-section and density.

This discussion may appear arcane but it should indicate that for complex situations there may well be several separate but interconnected features of response that are regarded as critical for correct modelling of the prototype. These features may well place conflicting and possibly unresolvable constraints on the selection of materials, dimensions and loadings for the model tests. Nevertheless, if the results of the model tests are to have any value and to be capable of being applied to a prototype, then understanding of these features and the physical characteristics that underlie the behaviour of the geotechnical structure is vital.

5.5 Single gravity modelling

The intention in this chapter has been to indicate how we can deduce the scaling factors that apply to various quantities that we may control or measure in our model tests. We need to know the scaling factors in order that we may be able correctly to extrapolate from observations made in model tests in order to predict behaviour at prototype scale. The links between apparently independent scaling choices have been demonstrated and it has then been shown what these composite scaling factors will look like when typical modelling decisions are made for single gravity and multiple gravity situations. Where some of these composite factors appear to run into difficulties it may be necessary to make alternative choices but this must always be done with care. For example, if we consider that soil strain is a first order indicator of mobilisation of strength for nonlinear soils then it will be wise to try to ensure that the scale factor for strains is always kept at unity.

It is obviously often harder to satisfy similarity constraints for single gravity modelling than for modelling on a geotechnical centrifuge at increased acceleration levels. The assumption that all aspects of stress:strain response can be encapsulated in a single soil stiffness G and, by implication, that *all* aspects of soil nonlinearity vary with stress$^\alpha$ may well be questioned. One is left with a probable expectation that it will be difficult to rely on a small single gravity model to provide an accurate representation of the response of a prototype—and this is of course the principal justification of centrifuge modelling which forms the subject of Chapter 6.

The single gravity laboratory model retains three attractions.

- As with other laboratory modelling the boundary conditions are, in principle, well defined and well controlled. The physical modelling provides a source of reliable data for supporting numerical modelling and back analysis.

- The size of the models can be quite large—limited only by available space and loading devices—so that the linear model scale from typical prototypes may be low. For dynamic modelling (Muir Wood *et al.*, 2002), shaking tables are typically used to test the seismic response of quarter scale model buildings, for example. The degree of extrapolation required of the supporting numerical modelling may then be low—and other undesirable effects that may be associated with small models (for example, particle size problems) may be somewhat avoided.

- Because the models are large the space available for instrumentation and actuators will be greater and more subtlety in loading, control and observation will be feasible. The disturbance to the soil arising from the finite size of instruments will be correspondingly lower.

6

Centrifuge modelling

6.1 Introduction

In discussing the choice of scaling factors required to maintain similitude between prototypes and small models it has been indicated in Chapter 5 that many (but not all) of the difficulties associated with scaling can be avoided if the stresses at corresponding points in the model and the prototype are the same. One way that this can be achieved is by using a geotechnical centrifuge to increase the local equivalent gravitational field in order to balance the decrease in stresses that would otherwise result from the chosen linear scale. In this chapter we will present the underpinning mechanics of centrifuge modelling, some of the machines that have been developed to create the artificial acceleration fields, and some of the ancillary devices that have been developed to perform centrifuge models of geotechnical problems and to observe what is happening in the soil. The intention is to give a flavour of the possibilities of geotechnical centrifuge modelling. More detail of particular techniques and applications can be found in conferences such as Kimura *et al.* (1998), Phillips *et al.* (2002) and books such as Taylor (1995). Developments in robotics, control, electronics and miniaturisation seem to be occurring so rapidly that any description of instrumentation or of techniques for modelling geotechnical processes at small scale must rapidly go out of date: we can provide only a dated snapshot here.

6.2 Mechanics of centrifuge modelling

The mechanical principle that underpins centrifuge modelling is simple: if a body of mass m is rotating at constant radius r about an axis with steady speed v (Fig 6.1) then in order to keep it in that circular orbit it must be subjected to a constant radial centripetal acceleration v^2/r or $r\omega^2$ where ω is the swept angular velocity. In order to produce this acceleration the body must experience a radial force $mr\omega^2$ directed towards the axis. We can normalise the centripetal acceleration with earth's gravity g and state that the body is being subjected to an acceleration of ng where $n = r\omega^2/g$.

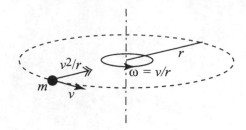

Figure 6.1: Object moving in steady circular orbit

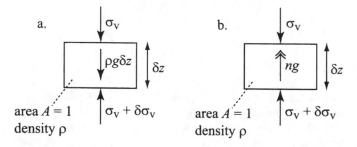

Figure 6.2: Element of soil (a) at surface of the earth and (b) on centrifuge

If we consider the equilibrium of an element of unit cross-sectional area and of thickness δz taken from a column of soil at the surface of the earth (Fig 6.2a) then we know that the increase of stress through the element must balance the weight of the element (which itself comes from the gravitational pull of the earth) in order to *prevent* any acceleration of the element:

$$\delta\sigma_v = \rho g \delta z \tag{6.1}$$

and, with constant density, at a depth z below the free surface

$$\sigma_v = \int_0^z \rho g \mathrm{d}z = \rho g z \tag{6.2}$$

On the centrifuge, if we consider the equilibrium of an element of unit cross-sectional area and of thickness δz (Fig 6.2b), then we see that the stress increase must provide the force necessary to *generate* the centripetal acceleration. The equation of motion becomes:

$$\delta\sigma_v = \rho n g \delta z \tag{6.3}$$

and at depth z/n below the free surface (assuming constant density)

$$\sigma_v = \int_0^{z/n} n\rho g \mathrm{d}z = \rho g z \tag{6.4}$$

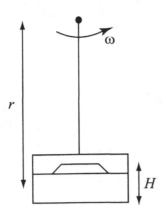

Figure 6.3: Finite dimensions of two-dimensional centrifuge model of embankment

Thus stresses are identical at geometrically equivalent points in the prototype and in the centrifuge model, provided the linear scale in the model is the inverse of the acceleration scale $n_g = n = 1/n_\ell$ (Table 5.4). Consequently we can expect that if mechanical behaviour of the soil is strongly dependent on stress level then such behaviour should be correctly reproduced in our centrifuge model.

There are a number of caveats that we need to insert. The value of n depends on r. Any model that we create—such as the model embankment on a soft clay foundation shown schematically in Fig 6.3—will have a finite radial dimension, and the value of n will vary through the model. The integration in (6.4) is thus not strictly correct. It is usually assumed that, provided the height H of the model is less than about $0.1r$ then the variation in the acceleration field is acceptable.

The acceleration field of earth's gravity is parallel (at the scale of civil engineering systems with dimensions small by comparison with the radius of the earth); the acceleration field on a centrifuge is radial. Model containers are usually fabricated for convenience with parallel sides (Fig 6.4a). Again it might be proposed that this difference will be negligible provided $B/r < 0.1$. A free water surface in a centrifuge model will adopt a cylindrical profile (Fig 6.4a) and in principle the soil surface should also follow this cylindrical profile—which would lead to obvious (but not necessarily intractable) problems in model preparation. If the model is, however, conveniently created with a flat surface then, in the radial acceleration field, this is equivalent to a gently curved surface in a parallel field (Fig 6.4b). A very soft soil may not be able to survive this implied somewhat hilly profile (Stone and Muir Wood, 1988).

Of course, one of the reasons for performing any physical modelling is to provide data of behaviour of boundary value problems against which numerical modelling, and hence by implication constitutive models, can be validated. Such numerical modelling can, in principle, take account of the actual variable and radial nature of the acceleration field.

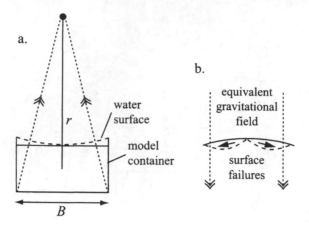

Figure 6.4: (a) Radial acceleration field on centrifuge; (b) flat surface 'feels' curved: soft soil may suffer 'slope' instability

For modelling of dynamic geotechnical problems there is a further detail. If an element of soil within the model is moving with velocity v^* as a result, for example, of some simulated earthquake event (Fig 6.5) (or underground explosion) then, because the soil is also rotating with an angular velocity ω, the element will experience a Coriolis acceleration $2\omega v^*$ and stresses must develop within the soil in order to produce this acceleration. The ratio of Coriolis to centripetal acceleration is $2v^*/r\omega = 2v^*/\sqrt{rng}$ and perhaps we should try to design our dynamic experiments in order to ensure that this ratio too remains below about 0.1.

Such limitations will be relevant to modelling of dynamic penetration problems. At the other extreme, if we wish to model explosions and the interaction of ejecta with the ground then we could try to ensure that velocities are so high that the trajectories of flying objects relative to the model are close to the absolute trajectories. For modelling of seismic response of geotechnical systems, where the prototype is being perturbed by horizontal base shaking, we could avoid these parasitic Coriolis accelerations by mounting our model on the centrifuge in such a way that the model 'horizontal' is parallel with the vertical axis of rotation of the machine.

It is tempting to think of the centrifuge model as having its own local gravitational acceleration field but really it should be thought of as simply having its own rotational velocity and hence radial acceleration field. Gravitational accelerations arise because of the attractive forces exerted on each other by two masses. The acceleration field on the centrifuge is entirely mechanical in origin. If an object is released within the centrifuge model container (Fig 6.6) then, if it is not in contact with the model, it will experience no forces apart from earth's gravity and will retain any initial absolute velocity that it possesses. *Relative* to the model it will appear to move because the model is itself moving in space

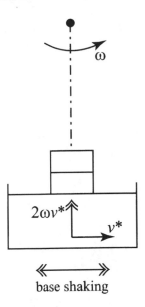

Figure 6.5: Coriolis acceleration experienced by particle moving with velocity $v*$ within model on centrifuge with angular velocity ω

(and the air in the model container will tend to drag the object with it) but the object will *not* accelerate towards the surface of the model (the modelled surface of the earth) with an acceleration of ng.

If an object (for example, a sand grain) is released above a centrifuge model at radius r with radial velocity $kr\omega$ then the position in the model, relative to the point of release (Fig 6.7a), is given parametrically by:

$$\frac{x}{r} = (1 + k\omega t)\sin \omega t \qquad (6.5)$$

$$\frac{y}{r} = (1 + k\omega t)\cos \omega t - 1 \qquad (6.6)$$

Typical trajectories are shown in Fig 6.7b.

If the object is released with velocity $kr\omega$ at an angle θ to the radius, measured positive in the direction of centrifuge motion (Fig 6.7c) then the parametric equations become:

$$\frac{x}{r} = \sin \omega t + k\omega t \sin(\omega t - \theta) \qquad (6.7)$$

$$\frac{y}{r} = (\cos \omega t - 1) + k\omega t \cos(\omega t - \theta) \qquad (6.8)$$

and it is evident (Fig 6.7d) that a combination of positive ejection together with directional vanes can help to give a somewhat more nearly 'vertical' path towards the surface of the centrifuge model. An illustration of such a path followed in

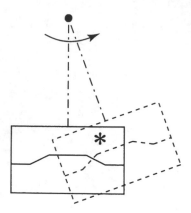

Figure 6.6: Object stationary in space appears to move relative to rotating centrifuge model

sand pluviation is given by Randolph *et al.* (1991). An exactly similar analysis can be used to follow the trajectory of objects ejected from the soil surface by modelled underground explosions (Steedman and Zeng, 1995): while *in* the soil such accelerated objects will of course experience forces generated by the Coriolis component of acceleration.

6.3 Centrifuges

There are two types of centrifuge that are in common use: beam and drum.

In a beam centrifuge (Figs 6.8, 6.9, 6.10), the model is rotated about a vertical axis at the end of a strong beam which at its other end carries some sort of a balancing mass or counterweight in order to prevent damaging out-of-balance rotatory forces on the centrifuge bearings. In many beam centrifuges the model is placed on a swinging platform (Figs 6.8, 6.9, 6.10) so that the local 'gravitational' acceleration field in the model is always coincident with the model vertical as the centrifuge speed is increased. This has obvious advantages for preparation of models.   Typical statistics for beam centrifuges are given in Table 6.1. The statistics of interest are evidently the physical dimensions of the model and the mean radius at which the model is moving, but also the acceleration capability of the centrifuge. There is an inevitable trade-off betwen the permitted mass of the model and the maximum acceleration that can be achieved. The 'power' of the centrifuge is usually quoted in g-tonnes—a given device may be able to tolerate a larger model but with lower permissible maximum acceleration. Fig 6.11 shows the three regimes of a typical limiting centrifuge performance envelope: in region *A* the performance is limited by the balancing capabilities of the counterweight; in region *B* the performance is limited by the mechanical stresses in the structural elements—there is a more or less inverse link between acceleration and total accelerated mass including the

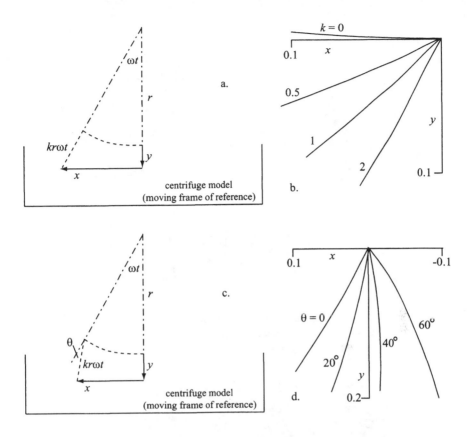

a.

centrifuge model
(moving frame of reference)

b.

c.

centrifuge model
(moving frame of reference)

d.

Figure 6.7: Object released above centrifuge model (a) with radial velocity $kr\omega$ giving (b) typical trajectories; (c) with velocity $kr\omega$ at angle θ giving (d) typical trajectories $k = 2$

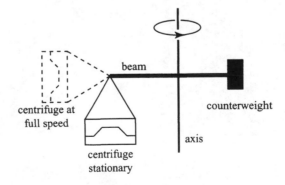

Figure 6.8: Schematic diagram of beam centrifuge: model on swinging platform

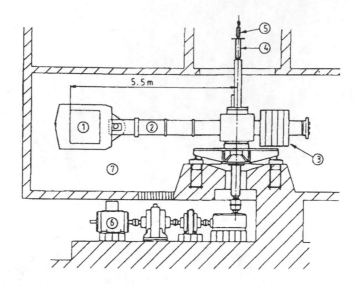

Figure 6.9: Diagram of Acutronic 680 beam centrifuge (Nicolas-Font, 1988) (1: swinging basket; 2: centrifuge beam; 3: counterweight; 4: slip-ring assembly; 5: rotary-joint assembly; 6: drive system; 7: aerodynamic enclosure)

Figure 6.10: Beam centrifuge at Hong Kong University of Science and Technology (photograph reproduced by kind permission of CWW Ng)

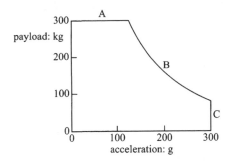

Figure 6.11: Beam centrifuge performance envelope (after de Souza, 2002)

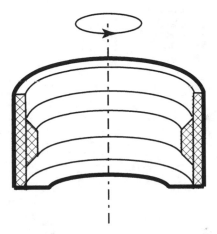

Figure 6.12: Schematic section through a drum centrifuge: continuous model of embankment

geotechnical modelling payload; in region C the performance is limited by the maximum rotational speed of the motor or its maximum rated power (Nicolas-Font, 1988).

In a drum centrifuge (Figs 6.12, 6.13) the model is created on the inside of a drum which typically rotates about a vertical axis. The model soil bed has no ends and can either be used for modelling a single long geotechnical structure or can be used as a model-scale test site on which various different structures can be created with a common geotechnical history of the foundation materials. Statistics for drum centrifuges are given in Table 6.2. The physical dimensions are of interest since they govern the intricacy of model construction processes. The acceleration levels attainable in drum centrifuges are typically much greater than for beam centrifuges but the physical dimensions are in general much lower.

A beam centrifuge might typically have a model depth of the order of 0.5 m and be accelerated to $100g$—equivalent to a prototype depth of 50 m. On a drum

Table 6.1: Beam centrifuges (based on http://geo.citg.tudelft.nl/allersma/tc2/cents.htm with additional information collected by the author)

country	owner	nom rad m	max accn g	max payload Mg	capacity g-tonnes	year
Australia	UWA	1.8	200	0.2	40	1989
Brazil	COPPE, UFRJ	0.5	450	0.2	90	1997
Brazil	Inst Tech Res Sao Paulo	0.75	200	0.05	10	1996
Canada	C'Core	5	200	2.2	220	1993
Canada	Queen's U	2.25	120	0.28	33.3	1997
Canada	Queen's U	0.9	60	0.03	1	1994
Canada	Queen's U	2.65	300	0.3	30	
Canada	U New Brunswick	1.6	200	0.11	22	1993
Colombia	U de los Andes	1.7	200	0.4	40	2000
Denmark	Danish Eng Acad	2.3	80	1.2	96	
France	CESTA	10	100	1	100	1956
France	LCPC, Nantes	5.5	200	2	200	1985
Germany	Ruhr U Bochum	4.125	250	2	500	1987
Germany	Ruhr U Bochum	1.8	200	0.4	40	
India	IIT, Bombay	4.5	200	0.625	125	2000
Italy	ISMES	2	600	0.4	240	
Japan	Aichi Inst Tech	1.36	200	0.075	16	1993
Japan	Chuo U	3.05	150	0.66	100	1988
Japan	Chuo U	1.18	270	0.15	30	
Japan	Fish Agy	3	150	0.25	37.5	1994
Japan	Hokkaido Devel Agy	2.5	200	0.3	60	1994
Japan	Japan Def Agy	2	100	0.15	15	
Japan	Kajima Co	2.63	200	1	100	1990
Japan	Kanazawa U	0.5	10000			1998
Japan	Kanto Gakuin U	0.4	500			
Japan	Kumamoto U	1.25	250	0.04	10	1996
Japan	Kyoto U	2.5	200	0.12	24	1988
Japan	Kyushu Inst Tech	1.27	150	0.18	27	1998
Japan	Kyushu U	0.75	200	0.005	1	1990
Japan	Min of Ag, For, Fish	1.3	200	0.07	14	
Japan	Min of Const	2	200	0.25	20	1987
Japan	Min of Const, PWRI	6.6	150	5	400	1997
Japan	Min of Labour	2.31	200	0.5	100	1988
Japan	Min of Trans, PARI	3.8	113	2.769	312	1980
Japan	Musashi Inst Tech	0.4	500			
Japan	Nagasaki U	1.5	200	0.06	12	1997
Japan	Nat Inst Ind Safe	2.1	200	0.5	100	
Japan	Nippon Koei Co	2.6	250	1	100	1996
Japan	Nishimatsu Co	3.8	150	1.3	200	1998
Japan	Nikken Sekkei Nakase Geot Inst	2.7	200	1	100	1992
Japan	Obayashi	7.01	120	7	700	2000
Japan	Ohita Tech Coll	0.8	200	0.04	8	1996
Japan	Osaka City U	2.56	200	0.12	24	1964
Japan	Saga U	0.75	200			
Japan	Science U Tokyo	0.27	420			1992
Japan	Shimizu Co	3.35	100	0.75	75	1991
Japan	Taisei Co	2.65	200	0.4	80	1990
Japan	Takenaka Co	6.5	200	5	500	
Japan	Tokyo Inst Tech	2.3	230		50	1995
Japan	Tokyo Inst Tech	1.25	150	0.25	38	
Japan	Toyo Co	2.2	250	0.3		1984
Japan	Utsunomiya U	1.18	120	0.15		

Table 6.1: *(continued)*

country	owner	nom rad m	max accn g	max payload Mg	capacity g-tonnes	year
Korea	Daewoo Inst Const Tech	2.7	200	1.2	120	1997
Netherlands	Delft Geot	6	400	5.5		1989
Netherlands	Delft Tech U	1.3	300	0.03	10	1989
PRChina	Chengdu Sci Tech U	1.5	250	0.1	25	1991
PRChina	China Inst Wat Res	5.03	300	1.5	450	1991
PRChina	Hehai U	2	250	0.1	25	
PRChina	Hong Kong UST	4	150		400	
PRChina	Inst Wat Cons Res	4.5	300	1.5	450	
PRChina	Nanjing Hydr Res Inst	5	200	2	400	1992
PRChina	Nanjing Hydr Res Inst	2.2	250	0.2	50	
PRChina	Nanjing Hydr Res Inst	1	500	0.01	5	
PRChina	Nanjing Hydr Res Inst	2.1	250	0.2	50	1989
PRChina	Shanghai Inst Rail Tech	1.55	200	0.1	20	
PRChina	Tsinghua U	2.2	250	0.2	50	1993
PRChina	Yangtze Riv Res Inst	3.5	300		180	
Portugal	LNEC	1.8	200	0.4	40	
Russia	Moscow Inst Rail Eng	2.5	322	0.17		1960
Singapore	Nat U Singapore	1.87	200	0.4	40	1991
Taiwan	Nat Cent U	3	200	1	100	1995
UK	Cambridge U	4.125	150	1	150	1973
UK	City U	1.6	200	0.2	40	1989
UK	Dundee U	0.325	400	0.001	0.4	
UK	Liverpool U	1.1	200	0.2	13	1978
UK	Manchester U	3.2	130	4.5	600	1971
UK	UMIST	1.5	150	0.75	100	
USA	USAF Eng Serv Cent	1.83	100	0.225	13	
USA	CalTech	1.3	175	0.035	7.5	
USA	Case Western Reserve U	1.37	200	0.182	20	1997
USA	Idaho Nat Eng Envir Lab	1.7	145	0.5	50	2002
USA	MIT	1.07	200	0.0681	13.6	1985
USA	New Mexico Eng Res Inst	1.8	100	0.227	25	
USA	Princeton U	1.3	200	0.076	10	
USA	Rensselaer Poly Inst	3	200	1	100	1989
USA	Sandia Lab	7.62	150	1.814	300	
USA	Sandia Lab	2.1	150	0.227	15	
USA	Sandia Lab	7.62	240	7.257	800	
USA	U Calif, Davis	1	175	0.09	9	1976
USA	U Calif, Davis	9.14	300	3.6	1080	1988
USA	U Calif, Davis	1	100	0.027		1985
USA	U Colorado, Boulder	1.5	220		15	1978
USA	U Colorado, Boulder	6	200	2	400	1988
USA	U Florida	1	100	0.023	2.5	
USA	U Florida	2	160	0.084		
USA	U Maryland	1.5	200	0.07	15	1983
USA	US Army Corps Eng, WES	6.5	350	8	1256	

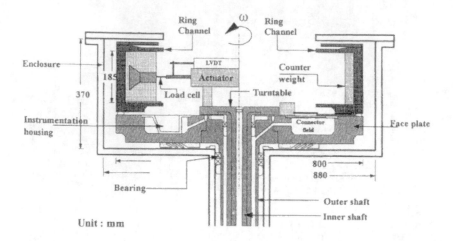

Figure 6.13: Diagrammatic section through drum centrifuge at Tokyo Institute of Technology with actuator arranged for pull-out test of enlarged base model footing (from Gurung *et al.*, 1998)

Table 6.2: Drum centrifuges (after Springman *et al.* (2001))

country	owner	$H \times d \times W$ m	payload Mg	max accn g	capacity g-t	year
Australia	UWA	$0.15 \times 1.2 \times 0.3$	0.6	484	290	1997
Brazil	COPPE	$0.17 \times 1 \times 0.25$	0.2	450	90	1996
Japan	Hiroshima U	$0.115 \times 0.74 \times 0.185$	0.13	416	50	1995
Japan	Kiso Jiban Co	$0.15 \times 1.2 \times 0.3$	0.6	484	290	1997
Japan	Tokyo Inst Tech	$0.15 \times 1.2 \times 0.3$	0.6	484	290	1997
Japan	Toyo Co	$0.3 \times 2.2 \times 0.8$	3.7	440	1600	1998
Japan	Utsunomiya U	$0.1 \times 0.8 \times 0.3$	0.195	150	43	1986
Switzerland	ETH Zurich	$0.3 \times 2.2 \times 0.7$	2	440	880	1999
UK	Cambridge U	$0.15 \times 2 \times 1$	1.7	400	675	1988
UK	Cambridge U	$0.12 \times 0.74 \times 0.18$	0.13	400	50	1995
UK	UMIST	$0.025 \times 0.25 \times 0.12$	0.006	1000	6.1	1971
USA	UC Davis	$0.13 \times 1.2 \times 0.23$	0.2	650	145	1979

Dimensions indicate depth H m × diameter d m × width W m.

centrifuge the soil bed might have a depth of about 0.2 m and be accelerated to 400g—equivalent to a prototype depth of about 80 m. The plan area of the test site for a drum of radius 1 m and height 0.5 m would then be around 2.4 km × 200 m at prototype scale. The high acceleration level for drum centrifuges leads to even more dramatic compression of diffusion time scales (see section §5.3.9): one year becomes 53 minutes at 100g but only 3.3 minutes at 400g.

Since we want to model geotechnical processes and to observe the response of our models, we need to make provision for hydraulic and electrical connections through slip-rings on the axis of the centrifuge (Fig 6.9) in order to provide power and control for actuators and other loading or process initiation devices. In early centrifuges all instrumentation signals were also led through high quality slip-rings to be logged away from the centrifuge. However, such signals may be of very low amplitude and these days it is usually preferred to undertake some signal conditioning and amplification, and analogue to digital conversion, and perhaps even logging of data, on the centrifuge itself using on-board computers placed close to the axis where acceleration levels are low. Downloading of digital information is much less sensitive to slip-ring noise and can take place in parallel with the logging itself.

6.4 Model preparation

As for any modelling of geotechnical elements or systems it is essential to know— and to be able to control—the past, the present and the future stress changes to which the soils are subjected. The first stage, covering the past and the present, represents the formation of the soil test bed and the establishment of some initial condition from which the effects of subsequent perturbation can be studied.

The first possibility might be to take a sample of real soil—particularly if it is intended to model an actual prototype. This is unlikely to be feasible or appropriate for a drum centrifuge but sizeable blocks of clay can be cut in the field and trimmed to size for a beam centrifuge strong-box. A period of steady centrifuge acceleration is then needed to reestablish pore pressure equilibrium and to establish some combination of vertical and lateral stresses which may not quite match the *in-situ* stress state. A block sample from one particular depth is being transformed into a complete soil layer (its stress state is being 'stretched') for the centrifuge model. Use of block samples in this way will be appropriate if the aim is to study behavioural characteristics that are linked to features of the natural structure and fabric of the soil—particle alignment, interparticle bonding, depositional details such as inclusion of seasonal varves of slightly different grain size and hence permeability. Of course, the scale of such natural features needs still to be small by comparison with typical dimensions of the model, otherwise similarity of prototype and model will be vitiated.

Use of block samples will not be helpful if such samples cannot be obtained without disturbance. There are techniques for *in-situ* freezing of block samples of sand so even for such materials use of undisturbed soil might not be completely ruled out. However, more often, when the behaviour of general classes of

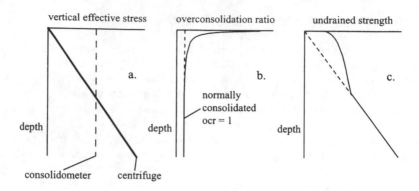

Figure 6.14: Preparation of clay sample by consolidation in consolidometer prior to establishing stress equilibrium on centrifuge: (a) vertical stresses; (b) resulting profile of overconsolidation ratio; (c) resulting profile of undrained strength

geotechnical systems is to be studied, the soil beds will deliberately be created from disturbed, remoulded, reconstituted or artificial materials.

Establishment of equilibrium effective stress states in clays requires consolidation and consolidation requires time. There is some advantage in performing as much as possible of the model preparation off the centrifuge itself, in order to avoid unnecessary machine occupancy. For beam centrifuge models it is standard to make use of free-standing consolidometers to prepare clay samples from slurry. Typically, the clay might be subjected to progressively higher stresses in the consolidometer (Fig 6.14) in order to give it at least enough strength to be handled and formed into a model which can be mounted on the centrifuge. (Any desired consolidation or overconsolidation history can be imposed on the *entire* clay block in the consolidometer.) When the centrifuge is brought up to speed, a stress field varying more or less linearly with radius is established (Fig 6.14a) so that the clay has an overconsolidation ratio which falls with depth (Fig 6.14b)—and the clay may indeed be normally consolidated below a certain depth. The strength of the clay will reflect this profile of overconsolidation (Fig 6.14c)—typical real soft clays have a somewhat stronger surface layer because of water table variation and other effects.

Sand samples can be prepared by direct pluviation into the model container (§6.2, §6.5): this provides a reasonably close approximation to the process by which sand deposits are formed in nature. Ideally pluviation should take place simultaneously across the whole area of the model in order to reproduce natural deposition over areas of great lateral extent.

On a drum centrifuge there is of course no separate model container and the soil layers have to be formed in the drum itself. Techniques have been developed (Laue *et al.*, 2002) for spraying clay slurry onto the drum. The centrifuge then has to be used for the consolidation process. It is necessary to be a little careful with the control of water in the drum: the clay needs to be kept saturated so the

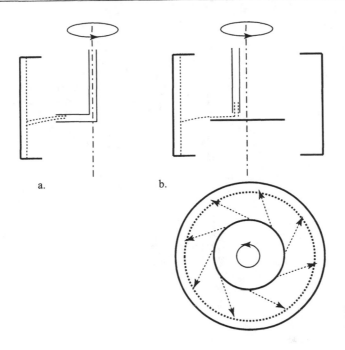

Figure 6.15: Preparation of sand sample in drum centrifuge (a) by feeding sand through nozzle; (b) by feeding sand onto spinning disc

water level should, during deposition, be always above the surface of the clay. However, the effective stress in the clay—and hence the undrained strength of the clay—at the surface is zero. If it is necessary to stop the centrifuge for a subsequent stage of model preparation, then the surface layer—hanging on a vertical face in the drum—may tend to slough off with any overlying water.

Sand beds can similarly be prepared on the drum by feeding sand from a nozzle or from a spinning disc, either of which can be raised and lowered to provide full coverage over the inside surface of the drum (Fig 6.15) (Laue *et al.*, 2002). We have noted that there is no reason for the sand to accelerate towards the inside surface of the drum—it will continue to travel at its velocity of delivery from the nozzle or disc with a trajectory slightly distorted by earth's gravity (and by air resistance inside the drum). Reproducibility and repeatability of the sand fabric are important. The detailed arrangement of sand particles will influence the eventual response of the soil—especially for low levels of deformation—so that the sand particle structures created by such a spraying technique (expedient though it is) will differ from those obtained by simultaneous deposition across the entire model cross-section.

If it is necessary to shape the surface of the sand model once it has been prepared then this can best be done with some tool operating within the spinning drum (Figs 6.13, 6.16, 6.17). If it is, however, necessary to stop the centrifuge in order to proceed to a subsequent stage of model preparation, then some strength

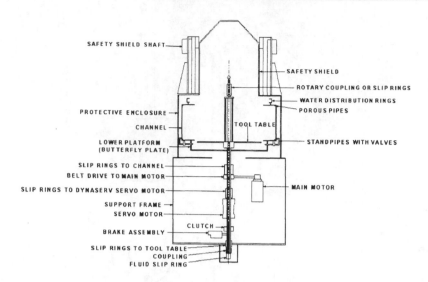

Figure 6.16: Diagrammatic section through drum centrifuge at ETH Zurich (Springman
et al., 2001)

can be developed by raising and lowering the water table in the soil layer round
the circumference of the drum, first more or less saturating and then leaving the
sand in a slightly damp state with surface tension between the sand particles
sustaining suction which then provides an effective confining pressure.

6.5   Geotechnical processes

The possibilities of modelling geotechnical processes on a centrifuge are limited
only by the ingenuity of centrifuge users. The sorts of things that we might wish
to do include: formation of cut slopes; creation of embankments; installation
and loading of shallow or deep foundations; construction of retaining structures;
formation of tunnels; and so on. Ideally we want to do as much as possible while
the centrifuge is 'in flight' but this will not always be feasible because of the
need to set in place and commission adequate instrumentation to monitor the
subsequent performance of our systems.

In the drum there is the possibility of using tools to manipulate the model.
These tools can either be held stationary in space and used, for example, to
'machine' the profile of the model—and literally create a continuous cut slope,
for example—or be rotated synchronously with the drum so that robotic opera-
tions can be performed at specific locations on the test bed—driving individual
piles or loading a single footing (Figs 6.13, 6.16, 6.17).

We are trying to follow the stages of a real construction process as closely as
possible, even if we are not trying to model a particular prototype. Often we will
need to create a stable geotechnical system and then perturb it in some way—
perhaps to bring it to failure. For example, we might want to study the effects

Figure 6.17: Pair of actuators mounted on the tool table of drum centrifuge at ETH Zurich (Springman *et al.*, 2001)

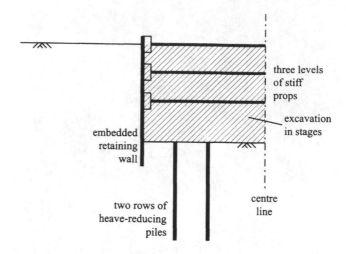

Figure 6.18: Model of propped retaining wall in overconsolidated clay with ground movements controlled using heave-reducing piles (inspired by McNamara and Taylor, 2002).

of using heave-reducing piles to limit the ground movements occurring behind an excavation (Fig 6.18, McNamara and Taylor, 2002). There are a number of stages involved in this modelling which should, as far as possible, match the stages of prototype construction. We need to start with the overconsolidated ground in pore pressure equilibrium—we will have to make decisions about the nature of the 'geological' history to impose. We have somehow to introduce the cantilever wall, and the heave-reducing piles, and the props and then to create the excavation. It will inevitably be necessary to make compromises.

An embankment can be prepared on the surface of the soil (which has been preconsolidated on the centrifuge) either at $1g$ by direct placement of a shaped pile of model fill or, preferably, at full speed by some process of controlled deposition from hoppers or nozzles over the model soil surface (noting again the non-ideal nature of the trajectory that the model fill will traverse between being released and coming to rest on the soil surface). The shapes of the resulting piles of model fill tend not to follow regular profiles (Fig 6.19) but so long as we can see what is happening then the actual shapes can be reproduced in parallel numerical modelling.

A surface footing is probably the simplest geotechnical system to model on the centrifuge. All that is required is some hydraulic or electro-mechanical device to lower the footing to the surface of the soil layer and then proceed to load it or cause penetration of the soil. Devices such as the LCPC robot (Fig 6.20) are evidently capable of this and other manoeuvres. If the footing is not intended to be quite at the surface of the soil then some initial preparation and positioning may be necessary at $1g$ before the centrifuge is rotating, leaving the loading stage to take place at the intended g level and stress level.

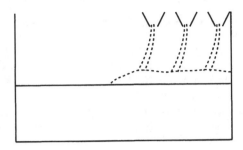

Figure 6.19: Preparation of model slope or embankment by deposition of sand fill in flight from hoppers mounted in model strong box

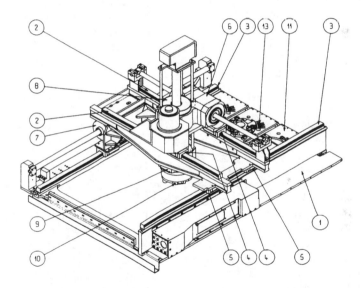

Figure 6.20: On-board centrifuge robot at LCPC, Nantes (Derkx *et al.*, 1998) (1: support beams; 2, 3: linear guide rails; 4, 5: sliders; 6, 7: brushless motors for x and y translation; 8: z axis translation by screw/nut system and brushless motor; 9: rotating robot arm controlled by ring gear and DC motor; 10: tool holder with built-in video camera; 11: three tool magazines; 13: proximity sensors used to identify tools)

Other geotechnical processes may require more compromises—some of which are more acceptable than others. Installation of individual model piles can be achieved using an on-board robot, by driving or steady jacking. The method of installation turns out to be rather important. The effect of development of lateral pressure on model piles installed in different ways is shown in Fig 6.21a. The dramatic difference in subsequent resistance (axial or lateral) of piles driven at $1g$ and at multiple gravities can be understood by considering the stress paths of typical elements around the pile (Fig 6.21c). The paths are shown in terms of a plane strain mean stress $s' = (\sigma'_v + \sigma'_h)/2$ and a shear stress $t = (\sigma'_v - \sigma'_h)/2$, where σ'_v and σ'_h are vertical and horizontal effective stresses respectively which are assumed to be, near enough, principal stresses. The driving of a pile will tend to increase the horizontal stress, probably without much drainage in a clay (we neglect the interface effects in the very disturbed zone immediately adjacent to the pile). The horizontal stress may possibly increase even above the vertical stress so that t falls below zero (BC at $1g$, PQ at ng in Fig 6.21c). If this occurs at low stress level (BC at $1g$ in Fig 6.21c), then subsequent consolidation (CD) will seek to reestablish a generally one-dimensional compression regime and the stresses may end up close to the K_o line. If pore pressure equilibration is required after pile installation at high stress level, however, (QR at ng in Fig 6.21c) then the total horizontal stresses may perhaps not change significantly and the soil will be left with *in-situ* radial stresses, before loading of the pile takes place, greater than the vertical stresses.

A similar effect is reported by Ng *et al.* (1998) in modelling sand compaction piles installed in soft clay at $1g$ or at multiple gravity. The technology that has had to be developed for in-flight installation of sand piles is rather more elaborate: first, a steel casing is jacked into the model ground; then sand is injected from a hopper using a hydraulically driven screw (Fig 6.22); the casing is steadily withdrawn as the sand is injected. This is fiddly but feasible: obviously it is desirable to do as much as possible under the multiple gravity environment.

Retaining structures are (simplistically) required either to support ground as excavation occurs in front of them, or to support new fill progressively placed behind them. Evidently the latter construction process might be modelled in the same way as the construction of an embankment by depositing material in flight behind a model wall. But in reality compaction of each layer would be required and the slightly uncontrolled nature of such a deposition process has usually led centrifuge modellers to prepare the fill—carefully—at $1g$ and then to load it by bringing the centrifuge to the desired acceleration—and then perhaps apply some surcharge or footing loading as an additional perturbation behind the wall (Fig 6.23). Although the *general* direction of the stress changes may not be too different, in detail there may be some difference between the desirable and the modelled stress paths (Fig 6.23b). In the context of the kinematic nature of soil stiffness (§2.5.3) (and the kinematic hardening models briefly introduced in section §3.5), however, the direction of the stress path immediately before the application of the footing load may be almost completely opposite (inset in Fig 6.23b) and this would have a major effect on the initial stiffness of the footing. These conclusions will certainly be dependent on the flexibility of the

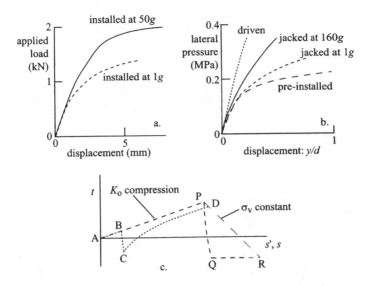

Figure 6.21: (a) Effect of pile installation procedure on subsequent development of axial resistance (after Yet *et al.*, 1994); (b) effect of pile installation procedure on development of lateral pressure (after Dyson and Randolph, 1998); (c) effective stress paths followed by elements close to pile for installation at 1*g* (ABCD) and *ng* (APQR)

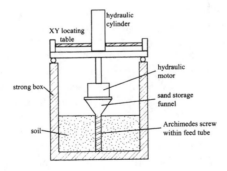

Figure 6.22: In-flight installation of sand compaction piles (after Ng *et al.*, 1998)

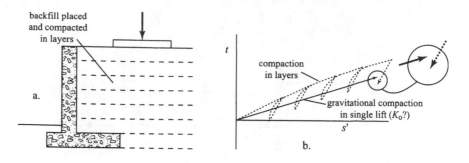

Figure 6.23: (a) Retaining wall supporting layers of backfill; (b) comparison of stress paths for typical element in backfill

wall which may permit much greater lateral deformation (and departure from a simple K_o stress path) as a result of compaction loads applied at prototype stress levels than would occur using typical $1g$ model preparation procedures.

Excavation in front of a pre-installed wall once again poses robotic challenges (Gaudin *et al.*, 2002). The problem is simultaneously to remove soil-like horizontal stresses *and* vertical stresses—and, desirably, to do this in stages leaving real soil below each excavation level (Fig 6.24a). The ratio of horizontal and vertical effective stresses before excavation (K_o) will depend on the soil type and the consolidation history of the soil. In a sandy soil it might be as low as 0.3, in a stiff clay as high as 3. Practically, the easiest way to apply a varying load over a deforming surface is to use fluid pressure (Fig 6.24b). The unit weight of soil is greater than the unit weight of water so one strategy is to use a heavy fluid such as an aqueous solution of zinc chloride. In a fluid, horizontal and vertical stresses are of course always the same at any level so an assumption might be made that it is more important to maintain the correct horizontal stresses on the pre-installed wall than to maintain the correct vertical stresses on the ground remaining in front of the wall. (Alternatively, McNamara and Taylor (2002) use a combination of heavy fluid with air pressure at the base of the eventually excavated soil in order to be able to provide separate control of horizontal and vertical stresses.) There will anyway be some uncertainty about the horizontal stress state that would remain after the installation of a prototype wall: a driven wall will increase lateral stresses; a concrete diaphragm wall poured under bentonite in a pre-cut trench will permit reduction in lateral stresses—so that the hydrostatic ($K_o = 1$) initial condition imposed by the fluid pressure might not be unreasonable. The detail of the stress paths sketched in Fig 6.24 can be disputed: the important message is to ponder the differences between prototype and model stress paths and to understand how those differences may affect the eventual geotechnical system response.

The modelling of tunnelling leads to a similar problem which may be slightly reduced if the diameter D of the tunnel is small by comparison with its depth C from the ground surface so that the gravitational variation of stress in the soil from crown to soffit of the tunnel is not great. Classic centrifuge tests by Mair

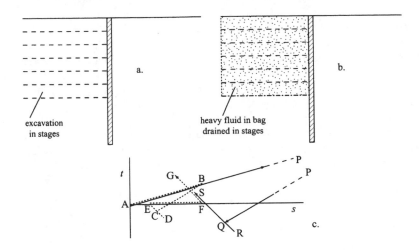

Figure 6.24: (a) Excavation in stages in front of pre-installed wall (APQRS); (b) 'excavation' by removal of fluid pressure (ABCDEFG); (c) comparison of stress paths for typical element in soil ([prototype: AP: consolidation; PQ: overconsolidation; QR: installation of wall; RS: excavation][model: AB: consolidation (1g); BC: overconsolidation (1g); CD: installation of wall (1g); DE: excavation and replacement by heavy fluid (1g); EF: centrifuge consolidation (1$g \to ng$); FG: drainage of heavy fluid]).

(1979) show normalised collapse conditions for long tunnels in soft clay tested in plane strain models(Fig 6.25a), compared with simple plasticity analyses (§7.3) (Fig 6.25b). Collapse was modelled by progressively reducing the pressure in a fluid filled bag. The occurrence of collapse is shown in terms of the tunnel support pressure σ_{Tc} normalised with undrained strength c_u as a function of tunnel geometry C/D. The spread of theoretical upper and lower bound results is modest and the data sit nicely in-between.

The three-dimensional problem of collapse of a partially unsupported tunnel heading was modelled using a preformed half-cylindrical tunnel (with a stiff brass model tunnel lining) with fluid pressure supporting the tunnel face and a length P of unlined perimeter (Fig 6.25d). These three-dimensional centrifuge tests have been interpreted using a non-dimensional stability number N_T:

$$N_T = \frac{n\rho g(C + D/2) - \sigma_{Tc}}{c_u} \tag{6.9}$$

comparing the difference between the tunnel support pressure σ_{Tc} at collapse and the mean vertical stress across the tunnel calculated from the soil density ρ, in a model being subjected to an acceleration ng on a centrifuge, with the undrained strength c_u. Tunnel heading collapse is capable of theoretical plasticity analysis for the two extremes of fully lined ($P/D = 0$) and fully unlined tunnels ($P/D = \infty$). Centrifuge model tests permit interpolation between these extremes, (Fig 6.25e), providing a valuable example of the integration of theoretical and physical modelling which provides reassuring support for both.

The fluid pressure which applies the same normal pressure on all parts of the unlined tunnel heading is now a very approximate replacement of the actual *in-situ* stress state which has different normal stresses in vertical, axial and transverse directions, and shear stresses on inclined surfaces. Robotic techniques are now available to achieve a more realistic modelling of the soil removal and lining placement (for example, Imamura *et al.*, 1998).

Surface settlements over a collapsing two-dimensional tunnel (Fig 6.25c) show that essentially the same profile of settlement is obtained at different model scales provided that the geometry of the collapsing tunnels is the same.

The capabilities of an on-board robot are to some extent limited by the proportion of the centrifuge payload that can be sacrificed for hardware as opposed to soil. Different approaches to the problem of simulating earthquake loading on a centrifuge have confronted this problem in different ways. On a $1g$ earthquake simulator—a shaking table—actuators are used to control all six translation and rotation movements. On a centrifuge, in general, researchers have limited themselves to one—horizontal—axis of shaking, relying on stiff bearings to prevent other uncontrolled parasitic modes of oscillation. Testing at $1g$ has shown how important it is to control—or at least monitor—all six degrees of freedom (even if some of the motions are *intended* to be zero) so that the detail of the motion to which a model has been subjected can be completely and correctly known. There may also be limitations on the types of simulated seismic motion that can be applied. It may be much easier to generate more or less sinusoidal motion—through conversion of rotary to linear motion—than the rather random excitation that characterises a typical earthquake. The

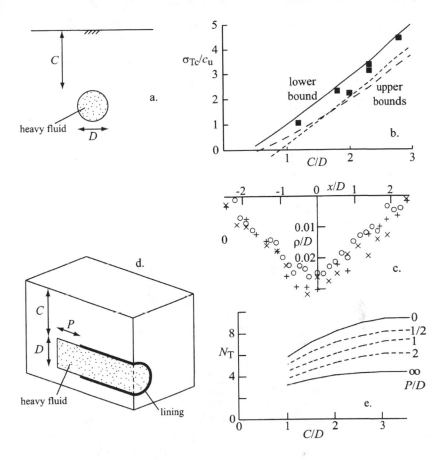

Figure 6.25: (a) Two-dimensional centrifuge model of tunnel collapse; (b) fluid pressure at tunnel collapse compared with theoretical analyses; (c) settlements over collapsing plane strain model tunnels with the same prototype diameter $D = 4.5$ m (o: 75g; ×: 125g; +: 125g); (d) three-dimensional model of partially lined tunnel heading; (e) collapse of partially lined tunnel headings (after Mair, 1979).

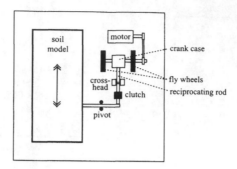

Figure 6.26: 'Stored angular momentum' model earthquake actuator (after Madabhushi *et al.*, 1998)

Cambridge 'stored angular momentum' shaker (Fig 6.26) applies bursts of some-what sinusoidal shaking to a single axis of excitation; the centrifuge at Hong Kong University of Science and Technology has a two axis shaking table mounted with actuators controlling the two (model) horizontal degrees of freedom (Fig 6.27). The VELACS project, which compared capabilities for physical and numerical modelling of liquefaction events caused by seismic loading, showed how impor-tant are the details of model preparation and the exact replication of imposed model shaking if reasonably similar results are to be obtained at different testing locations (Arulanandan *et al.*, 1994).

At the other extreme, Allersma (1998) shows how the melting of a block of ice can be used to produce loading of centrifuge models extremely economically (if slightly uncontrollably) and with minimal penalty in terms of use of payload capacity. For example, a prestressed spring restrained by a block of ice can form the loading system for the pull-out of a buried anchor (Fig 6.28). Obviously such imaginative devices are ideal for small centrifuges which are to be made readily available for student projects with limited technical support.

6.6   Pore fluid

We discovered in section §5.3.9 that there were separate rules governing the scaling of time for diffusion processes, such as consolidation or migration of pollutants, and dynamic processes, such as the generation of pore pressures during earthquakes. Where these time scales conflict it is necessary to take special action. The problem is likely to be most obvious in fine sands where the need to model correctly the degree of dissipation that can occur *during* a seismic event becomes important because this will have a major influence on the likelihood of liquefaction occurring. The usual way of satisfying the similarity requirements is to modify the viscosity of the pore fluid by adding glycerol to the pore water or by using silicone oil. By this means it is certainly possible to alter the permeability of the soil by a factor of 100. There may be some concern that the use of such a pore fluid may influence the mechanical properties of

Figure 6.27: On-board earthquake actuator for centrifuge at Hong Kong University of Science and Technology (see also Fig 6.10) (photograph reproduced by kind permission of CWW Ng).

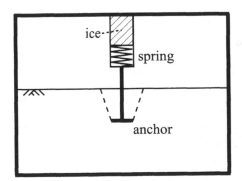

Figure 6.28: Use of melting block of ice to control pull-out loading of a buried anchor (after Allersma, 1998)

the soil itself by affecting the interparticle mechanics or changing the surface chemistry of clay particles.

There are a number of requirements that influence the choice of pore fluid for dynamic centrifuge modelling:

- a Newtonian fluid;

- same compressibility as water;

- chemically polar for use with all particle shapes and sizes;

- correct scaling of surface tension;

- non-toxic and environmentally suitable;

- made from readily obtainable constituents;

- easily accurately mixed for reliability and repeatability;

- stable with time; and

- non-corrosive.

The Delft Geotechnics pore fluid (Allard and Schenkenveld, 1994) has been presented as one candidate—a water based solution of undeclared chemistry—which satisfies all these requirements, and demonstrably produces unaltered mechanical response. It is shown that its viscosity can be varied over three orders of magnitude by varying the concentration of the mixture, while the density varies by only $< \pm 2\%$.

6.7   Site investigation

Whatever method is used to prepare the soil layers for subsequent perturbation on the centrifuge it is helpful to have some techniques that can be used to study the *in-situ* properties of the soil at the augmented acceleration levels at which they are to be employed. Such techniques can be used to explore spatial variability across a model as well as profiles with depth. Techniques that have been used at prototype scale can be adapted for use at small scale. Thus Garnier (2001, 2002) describes the use of miniature cone penetrometer, vane and even pressuremeter in centrifuge models.

The cone penetrometer is widely used for site investigation and site characterisation and for construction control at full scale (§1.2.3). There are limits to the miniaturisation of model cones for centrifuge application: for example, the diameter needs to be large in relation to the particle size (for example, d_{50}) in order to obtain reliable results; and the penetrometer needs to be strong enough not to buckle as it is pushed in. Gui et al. (1998) suggest a limit $D_{cone} > 20d_{50}$ and also note that it takes a penetration of about $5D_{cone}$ to mobilise cone resistance so that the precise detection of strength changes will be slightly smeared. (However, Foray et al. (1998) find that for model piles $D_{pile} > 200d_{50}$ is needed to avoid particle scale effects at the interface between pile and soil.) Centrifuge

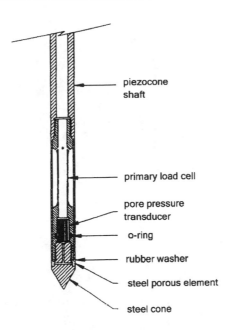

piezocone
shaft

primary load cell

pore pressure
transducer

o-ring

rubber washer

steel porous element

steel cone

Figure 6.29: Piezocone penetrometer for centrifuge site investigation (after Esquivel and Ko, 1994)

tests may be performed at different levels of excitation—one might as well use a single probing device in all models so that the scaled 'prototype' penetrometers are of different sizes. A miniature cone penetrometer with external diameter 8 mm is described by Almeida and Parry (1984). Scaled up (at 100g, say) such a cone is perhaps more like a small model pile. This device measured, at the top of the penetrometer, the separate loads on the tip and the shaft—somewhat like a standard full size cone (of diameter 35 mm). Esquivel and Ko (1994) describe a miniature piezocone (Fig 6.29) of diameter 12.7 mm, using two load cells, the tip resistance being measured just behind the point of the cone, and with a pore pressure transducer also being incorporated.

Miniature vane tests require a combination of penetration and rotation control and have the same applicabilty as a full size vane—though again it will not be feasible to think of scaling down a standard field vane by the linear scale of the centrifuge model.

The analytical intepretation of the output of any penetration device requires some assumed mechanism of deformation around the device—the tip of the cone or the blades of the vane. Theoretical analysis of cone penetration draws an analogy with the creation of a spherical or cylindrical cavity in the soil. This expansion process depends on both the undrained strength c_u and the shear stiffness G of the soil through a rigidity factor $I_R = G/c_u$—and therefore depends on effective stress level and history of overconsolidation. Interpretation of cone tests to give a profile of strength requires accompanying assumptions.

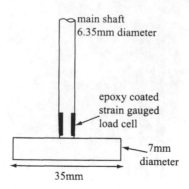

Figure 6.30: Sketch of T-bar penetrometer for centrifuge site investigation (after Stewart and Randolph, 1991)

Plastic flow *around* a cylindrical object is, however, capable of close theoretical analysis and, exploiting this, the T-bar (Fig 6.30) has been developed by Stewart and Randolph (1991) as a device which can be pushed or pulled through a clay layer to record the profile of resistance and hence undrained strength. All is never ideal: even for this cylindrical object the interpretation depends on the surface roughness. However, an average strength factor $N_t \approx 10.5$ is proposed to convert force P per unit length of a T-bar of diameter D to undrained strength c_u:

$$\frac{P}{c_u D} = N_t \qquad (6.10)$$

and this factor is independent of stress level and history of overconsolidation. This T-bar too will average the strength of the soil over a distance of some 30-50 mm from the penetrometer.

Geophysical techniques are used in the field not only to give general profiling of soil layers but also to give information about shear wave velocities and hence shear stiffnesses. Shear wave velocity can only be deduced by measuring travel time over a known distance so it gives a smeared out average soil property between transmitter and receiver (§3.2.6). Miniature piezoceramic devices—'bender elements'—are quite widely used for measurement of shear wave velocities in laboratory element tests and the same technology can be used to record shear wave velocities and, perhaps more importantly, *changes* in shear wave velocities, in model tests. Bender elements are formed from two pieces of piezoceramic (Fig 6.31) glued together, separated by a metal sheet. When subjected to appropriately chosen electrical excitation one side tends to become longer, the other side shorter with the result that the combined element bends and transmits a shear motion to the soil in which it is embedded. A similar element, used in passive mode as a receiver, generates voltages when it is deformed by the arriving shear wave. (By varying the connections is it possible to use the same element to send and receive shear and compression waves (Lings

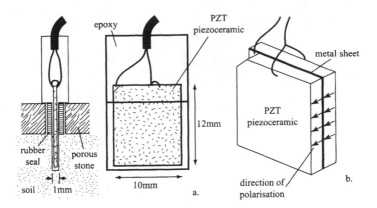

Figure 6.31: Bender element for laboratory geophysical studies: (a) mounted in triaxial end platen; (b) construction of element (after Lings and Greening, 2001)

and Greening, 2001).) In laboratory element tests such geophysical techniques can now be used to build an extensive picture of the evolving elastic anisotropy of the soil. It is quite likely that there will be continuing developments in their application to model tests.

6.8 Instrumentation

In full-scale geotechnical systems we might wish to monitor pore pressures, displacements, contact stresses, and structural resultants (such as bending moments) and in dynamic situations, accelerations. We will want to measure these things in our centrifuge models too. The two constraints that are encountered are the need to be able to operate in a high ambient acceleration field and the need for miniaturisation if the observations are to be regarded as plausibly point values.

Pore pressures are typically measured with Druck transducers (Fig 6.32)— 6.35 mm diameter—which use a silicone diaphragm as a differential pressure sensitive element. A porous stone is provided to separate the diaphragm from the soil and, as in any piezometer, this porous stone needs to be saturated before use and to be kept saturated during use. Where the scale of detail of a geotechnical system is small—one could imagine modelling sand drains which, in a prototype are at 1.5 m centres, at a scale of 1/100, or sand compaction piles of model diameter 20 mm at 40 mm centres (Lee, 2002)—the precise location of the pore pressure transducer between the drains may be rather important— or, turning this restriction round, the readings of pore pressure change and rates of dissipation need to be interpreted in the knowledge that the region over which the pressures are measured has a size which may be of a similar order of magnitude to the critical dimensions of the model.

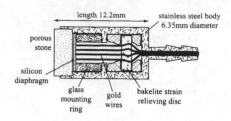

Figure 6.32: Druck pore pressure transducer (adapted from Taylor (1995) and König et al. (1994))

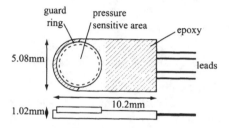

Figure 6.33: Miniature contact stress transducer (after Lee et al., 2002)

Contact stresses are difficult to measure reliably at any scale—primarily because of the effect that a stress transducer of stiffness different from the boundary within which it is embedded has on the local stress field. Contact stress transducers have been devised with miniature strain-gauged elements. Garnier (2001) describes very stiff miniature cells which have been used with model hopper flow tests which deflect less than 1 μm under a normal stress of 100 kPa. The total stress cell shown by Lee et al. (2002) is shown in Fig 6.33. Strain gauging of structural elements—piles, flexible walls, tunnel linings—may be more reliable and it may be possible to deduce interface stresses from gradients of bending moment or axial force. However, differentiation of experimental observations always introduces errors. The use of tactile pressure sensitive mats is described by Springman et al. (2002). These can measure local stresses over a grid of 1936 contact points over an area 56 mm × 56 mm. The sensitivity is not particularly good (range/256) and calibration is not straightforward but this is evidently a promising emerging technology.

Displacements at discrete points on a model can be measured with LVDTs (linear variable differential transformers) which use a core attached to the model moving relative to a fixed coil. If used to measure settlement of the soil surface, the core, accelerated on the centrifuge, behaves like a surface penetrometer and needs to be provided with a contact pad to spread the loading and reduce the contact pressure and possible penetration (Fig 6.34).

Non-contact laser techniques can be used to monitor displacements across surface profiles: the device has to be driven along a known course at a finite

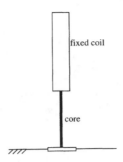

Figure 6.34: LVDT for surface displacement measurement with pad to reduce contact pressure

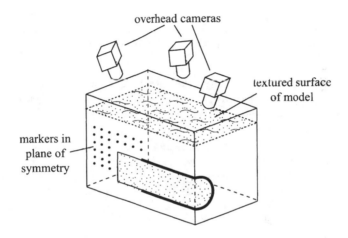

Figure 6.35: Close-range photogrammetry for recording three-dimensional surface displacements (inspired by Taylor *et al.*, 1998)

speed so these measurements are most useful for recording displacement patterns under rather steady state conditions. Close range photogrammetry (Figs 6.35, 6.36) can be used to give a three-dimensional instantaneous view of the surface displacements using two or three cameras mounted above the surface of the model. The procedure is essentially similar to that used in mapping from aerial photography except that the cameras are fixed and it is the actual movement of the ground surface that leads to differences in successive photographic images.

Plane models can be viewed through a lateral glass panel. Discrete markers can be placed in the lateral face of the model as it is being prepared and then monitored by flash photography while the centrifuge is in flight (Fig 6.37 and Fig 6.35). Measurement of the positions of these markers on successive photographs using automatic image analysis techniques can be used to deduce fields of strain increment during the model test. Any measurement of

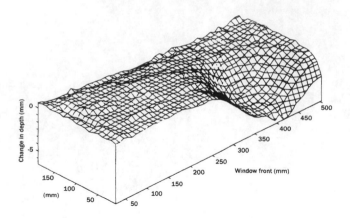

Figure 6.36: Three-dimensional surface displacements over tunnel heading determined using close-range photogrammetry (from Taylor *et al.*, 1998)

displacements seen through the transparent side of a plane model may be influenced by friction between the soil and the side boundary. Marker bands of coloured sand can be used to give visual evidence of formation of shear bands and also for general qualitative evidence of magnitudes of settlement (Fig 6.37).

Alternatively, if the soil used in the model has a clearly visible texture—or it is possible to apply some texture to the visible surface (Take and Bolton, 2002)—then close range digital photography can be used with subsequent Particle Image Velocimetry (PIV) to record the changing appearance of the fabric. An accuracy of 1/15000 of the field of view is estimated by White and Bolton (2002) with typical displacement precisions 4-15 μm depending on the size of the patch of observed fabric whose displacement is being followed. White and Bolton quote a typical patch being followed with dimensions 2-4 mm, with image analysis techniques being used to detect movements with a precision of 1/15 pixel, using a digital camera with pixel resolution 1760 × 1168. Fabric photography can also indicate other effects such as soil particle rotation and breakage which will not be detected by monitoring of individual markers (Fig 6.38). The great advantage of using such digital photographic techniques is that information concerning a very large number of points in the plane section of the model can be obtained extremely rapidly and analysed automatically with an objectivity that eliminates the human factor involved in many other techniques. We are essentially obtaining *field* rather than point information.

Some techniques that are appropriate to single gravity models can also be used, after a test, for centrifuge models. Radiography can be used to detect the location, after the test, of threads of bismuth or lead paste injected before the test. This can be helpful in detecting the position of failure surfaces within the soil (Fig 6.39) although it is of course not necessarily possible to determine at what stage during testing on the centrifuge these discontinuities actually developed. A similar result can be obtained by inserting coloured spaghetti into clay

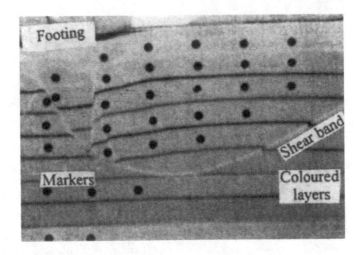

Figure 6.37: Observation of individual markers and marker bands of coloured sand in two-dimensional plane models of footing (from Bakir *et al.*, 1994)

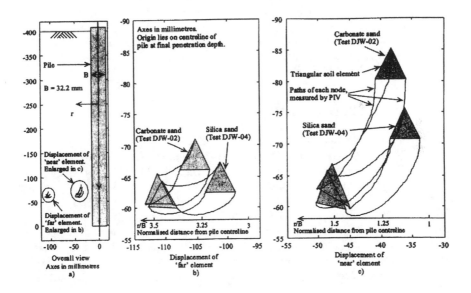

Figure 6.38: Close range digital photography used to detect displacements and particle rotations around a driven pile in sand (White and Bolton, 2002).

Figure 6.39: Radiographic observation of bismuth threads to detect location of failure surfaces under embankment on soft clay, after completion of centrifuge model test (Bassett and Craig, 1988).

models: the pasta takes up moisture from the clay and, thus softened, follows the movements of the clay without impediment (see, for example, Tamate and Takahashi, 2000). The spaghetti have to be located after the completion of the centrifuge test using careful exhumation techniques. It is obviously impossible to separate displacements which occurred while the model was being tested at full centrifuge acceleration and those which have developed during the slowing of the centrifuge to a halt.

An exhumation technique is also used by Muir Wood, Hu and Nash (2000) to detect the deformed shape of model stone columns (similar to sand compaction piles) in soft clay after a loading test on a model footing at single gravity (Fig 6.40). The sand in the model column/pile is carefully sucked out and plaster of Paris (gypsum) poured in around a wire armature. Removal of the surrounding clay then reveals the deformed shapes of the columns complete with internal bulging and shear planes.

6.9   Modelling and testing

When Andrew Schofield began promoting the use of geotechnical centrifuges in the west in the 1960s he had a vision of centrifuge modelling becoming a natural and inevitable tool in geotechnical design. Craig (1985)—who had, in his use of the Rowe centrifuge at Manchester, used centrifuge modelling in support of design of actual prototypes such as major embankment dams, foundations of offshore structures, and the Oosterschelde storm surge barrier—ended by writing: *If centrifuge work is to continue, it should have a positive role beyond phenomenological studies and the development of design rules by parametric variation in idealised, non-specific models.* Lee (2002) talks of *the philosophy of modelling versus testing*—modelling leading to predictions, testing leading to validation.

In practice, centrifuge modelling has probably been used more for study of generic problems than for reproduction of the response of particular prototypes. We have seen some of the possible reasons for this already: scale, boundaries, and processes.

Small features which may have a significant effect on system response—the presence of sand lenses, for example—cannot be directly modelled at small scale.

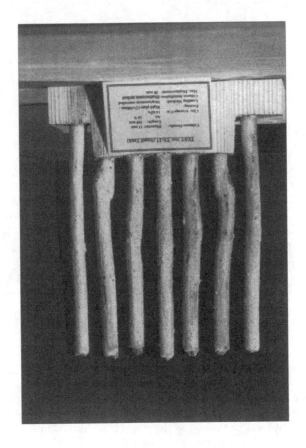

Figure 6.40: Exhumation to discover deformed shape of model stone columns (after Muir Wood *et al.*, 2000)

Key features of soil fabric or of soil behaviour (such as the formation of patterns of localised deformation or the concentration of displacement into a shear band (Muir Wood, 2002)) with a characteristic length that is small by comparison with typical problem dimensions at prototype scale may be of the same order of magnitude as the scaled problem dimensions if prototype soil is used in the small model.

As in any modelling—numerical or physical—boundaries have to be introduced with characteristics (complete absence of friction, for example) and locations which are appropriate to but will not unduly influence the system response that is being studied. Advantage can be taken of symmetry to insert a smooth centre-line boundary and study half the problem. Usually it will be desirable to make the active part of the geotechnical system—the foundation, wall, embankment, tunnel—as large as possible: this is where gradients of displacement will be greatest. However, there needs to be space beyond to the outer boundaries of the (usually rigid) container so that the active parts do not feel too constrained.

Inevitable compromises have to be made in reproducing geotechnical processes. Robotic possibilities are steadily increasing but the benefit of improved modelling of some details of construction, or of geological history has to be weighed against the cost of greater on-board complexity (and associated risk of malfunction)—and loss of pay-load capacity for the geotechnical elements of the model.

What centrifuge modelling does supremely well is to reveal mechanisms of geotechnical behaviour at prototype stress levels. Such behaviour can be revealed in problems covering a range of scales extending right up to neotectonics and mountain orogeny (Jeng et al., 1998). The art of successful geotechnical centrifuge modelling is to ensure that the simplicities of the modelling do not distort these mechanisms. As many as possible of the likely important effects must be included. A centrifuge model is a closely controlled boundary value problem, conducted with real soils incorporating all their constitutive vagaries—many of which are hidden from the numerical or constitutive modeller. Data from a well-designed centrifuge model can thus be used to validate numerical modelling which, validated, can be used to extrapolate to closer modelling of prototype details. You should have knowledge and close control over the preparation and history of the soils (supported by element testing and constitutive modelling) and exactly how they have been treated in the centrifuge modelling—including the time scales and acceleration histories. All these—possibly non-ideal—details can be incorporated in parallel numerical modelling.

It is as true for centrifuge modelling as for any other type of geotechnical modelling that you should always start out with a prediction of what you expect to happen. If the observation of the model manages to surprise and to confound these prior expectations, then reflection is required to develop deeper understanding and improve the next predictions (Fig 6.41). Indeed, a good model test is precisely one which surprises—this is the way in which scientific understanding advances—and we should try to design our model tests with this in mind[1]. Scientific conjectures cannot be proved (absence of evidence *so far*

---

[1] Although it is usually hoped that modelling intended to support prototype design decisions will not surprise.

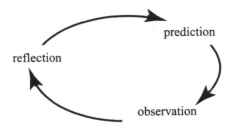

Figure 6.41: Reflective practice cycle

which conflicts with a conjecture does not indicate that such evidence must always be absent), but they can be refuted. Rival hypotheses can be sifted using carefully chosen testing.

6.10 Closure

Looking at the proceedings of the conferences on centrifuge modelling that have taken place every few years over the past couple of decades one is struck by the way that this technique has reached a maturity in this period. Centrifuge modelling is now regarded as a legitimate tool to support geotechnical research or design. Papers seem less concerned with descriptions of centrifuge hardware, more concerned with novelties in process modelling and observation or quite simply with the presentation of geotechnical phenomena. Indeed the most recent conference (Phillips *et al.*, 2002) is deliberately devoted to 'physical modelling' rather than specifically to centrifuge modelling to emphasise the general acceptance of the centrifuge as one of a number of techniques of physical modelling and not one that requires special attention and justification.

Kimura (2000) charts the growth in centrifuge usage in Japan which is perhaps the only country where it has been adopted as a matter of routine in the way that Andrew Schofield had hoped. From 3 centrifuges in 1980, Japan had some 32 beam centrifuges and 5 drum centrifuges by 1998. The number is now certainly higher. More interesting than the number is their distribution: of the 37 centrifuges only 50% were in universities, 25% were in national research institutes and the remaining 25% in private industry, 19% in general contractors. These include some extremely large machines.

One may debate the merits of different centrifuge sizes. Evidently the larger the model the less the concern about the accuracy of representation of detail, the less the concern about the effects of particle size, or the larger the geotechnical prototypes that can be modelled. Evidently too the infrastructure required to keep large machines operational and in full usage is extensive. There is obviously a role for small machines that can be safely operated by individual researchers with little technical support. The more widely that such machines can calmly penetrate the world of undergraduate teaching the greater the likelihood that the potential of centrifuge modelling will become generally understood and accepted even by sceptical practising engineers and the greater the chance of the acceptance of the vigorous refutation by Schofield (2000) of Terzaghi's assertion of the 'utter futility' of attempts to rely on the results of small-scale geotechnical models.

7

Theoretical modelling

7.1   Introduction

There is often advantage in being able to obtain rapid theoretical solutions to analyses of the behaviour of geotechnical systems without having to resort to the full complexity of numerical modelling. It has been emphasised several times, in the context both of numerical modelling and of physical modelling, that it should always be possible to obtain supporting 'back-of-the-envelope' estimates of response to give reassurance that the results of the numerical or physical modelling are secure and that the governing phenomena are understood. These estimates may not be particularly precise but should be of the correct order of magnitude. Simple calculations can also be used to perform rapid parametric studies in order to reveal the most influential parameters and hence make more efficient the subsequent detailed numerical or physical modelling.

In this chapter we will describe some of the tools that are available to support and perform such theoretical modelling. Some of these—such as the elastic stress distributions and the exact results of simple consolidation problems—are capable of exact theoretical closed-form analysis and can be used to provide a direct check on the implementation and programming of algorithms required for finite element or finite difference numerical modelling. Some of the other examples included in this chapter actually require some numerical strategy for their solution so that, as a check on full numerical analysis, with all its concomitant approximations, it appears that we are merely substituting another approximate numerical analysis. However, for the simple theoretical models that we are considering here the numerical solution algorithm, if one is required, will usually itself be relatively simple to implement and rather evident to the user. Moreover, it is likely that the numerical procedure will not be the same as that involved in the full finite element or finite difference analysis (at least the subroutines will be newly prepared) so that there will be a degree of independence in the comparisons that are made.

309

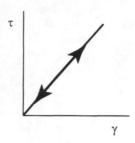

Figure 7.1: Linear elastic stress-strain relationship

7.2   Elastic stress distributions

Two of the examples of empirical modelling that were introduced in Chapter 1 (§1.2.2, §1.2.3) made more or less explicit reference to a requirement to estimate the stress distribution in the ground resulting from a loaded footing. If the ground can be considered to be isotropic *and* homogeneous *and* elastic *and* linear (Fig 7.1) (these are not insignificant assumptions), then combination of equations of equilibrium, kinematic compatibility (definition of strain in terms of gradients of displacements), and an elastic constitutive law (Hooke's law) leads to a set of partial differential equations which is capable of exact explicit solution for certain sets of boundary conditions and which is capable of numerical solution for more general boundary conditions. If a particular problem can be approximated by an idealised problem for which the boundary conditions match those of a standard elastic solution then the stress state and deformation pattern can be directly stated.

A particular advantage of assuming linear elastic soil response is that stress resultants deduced as effects of different applied loadings can be superimposed. A particularly useful building block for geotechnical problems in the present context is the stress state produced by a vertical point load P acting at the surface of a semi-infinite elastic half space (Fig 7.2): this is the Boussinesq problem. The resulting stresses, referred to cylindrical coordinates, are

$$\sigma_z = \frac{3Pz^3}{2\pi R^5} \tag{7.1}$$

$$\sigma_r = -\frac{P}{2\pi R^2}\left[-\frac{3r^2 z}{R^3} + \frac{(1-2\nu)R}{R+z}\right] \tag{7.2}$$

$$\sigma_\theta = -\frac{(1-2\nu)P}{2\pi R^2}\left[\frac{z}{R} - \frac{R}{R+z}\right] \tag{7.3}$$

$$\tau_{rz} = \frac{3Prz^2}{2\pi R^5} \tag{7.4}$$

where $R = \sqrt{r^2 + z^2}$ as shown in Fig 7.2 and ν is Poisson's ratio.

Any other vertical surface loading can be considered as a series of vertical pressures acting on infinitesimal areas (in other words, a series of equivalent

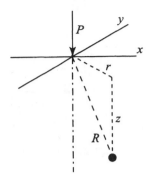

Figure 7.2: Vertical point load on surface of elastic half-space

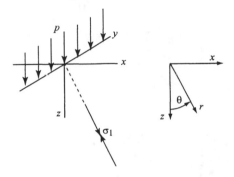

Figure 7.3: Line load on surface of elastic half-space

point loads) and the combined effect can be obtained by integration. Exact closed form integration is possible for certain convenient regular geometries of the loaded area; numerical integration is required for more irregular geometries.

Some convenient building block results are given here. For a line load of intensity *p per unit length* on the surface of the infinite half space (Fig 7.3) the resulting stress field is deduced from integration of the stress resultants for the point load and is radial from the line of application of the load[1]. The major principal stress is

$$\sigma_1 = \sigma_r = \frac{2pz}{\pi r^2} \qquad (7.5)$$

The minor principal stress is

$$\sigma_3 = \sigma_\theta = 0 \qquad (7.6)$$

The intermediate principal stress in the direction parallel to the applied load is

$$\sigma_2 = \sigma_y = \nu \frac{2pz}{\pi r^2} \qquad (7.7)$$

[1]Note the coordinate system defined in Fig 7.3.

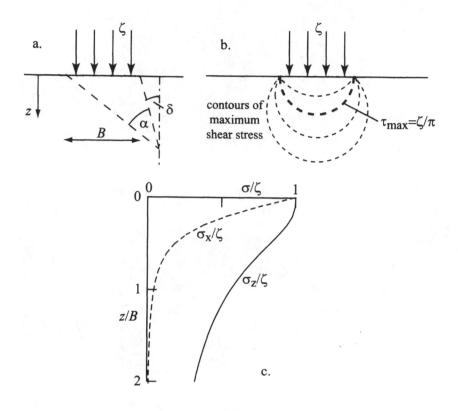

Figure 7.4: (a) Uniform strip load; (b) contours of maximum shear stress; (c) vertical and in-plane horizontal stresses beneath centre-line

The line load can then be integrated to give the stresses beneath a uniformly loaded strip (Fig 7.4) with load of intensity ζ *per unit area*. These are most conveniently defined in terms of auxiliary angles shown in Fig 7.4a. Referred to rectangular Cartesian axes they are

$$\sigma_z = \frac{\zeta}{\pi} \left[\alpha + \sin\alpha \cos(\alpha + 2\delta) \right] \tag{7.8}$$

$$\sigma_x = \frac{\zeta}{\pi} \left[\alpha - \sin\alpha \cos(\alpha + 2\delta) \right] \tag{7.9}$$

$$\sigma_y = 2\nu\alpha\frac{\zeta}{\pi} \tag{7.10}$$

$$\tau_{xz} = \frac{\zeta}{\pi} \sin\alpha \sin(\alpha + 2\delta) \tag{7.11}$$

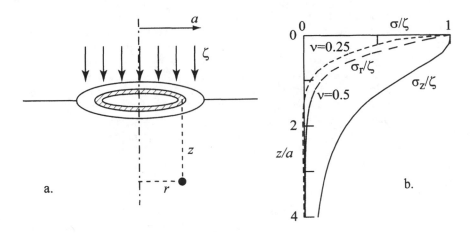

Figure 7.5: (a) Uniformly loaded circular area as sum of set of loaded rings; (b) stresses on centre-line

The major and minor principal stresses are given by

$$\sigma_1 = \frac{\zeta}{\pi} [\alpha + \sin \alpha] \tag{7.12}$$

$$\sigma_3 = \frac{\zeta}{\pi} [\alpha - \sin \alpha] \tag{7.13}$$

Loci of constant principal stress and of maximum shear stress (dependent only on α) are circles passing through the edges of the loaded area (Fig 7.4b). The greatest value of maximum shear stress, equal to ζ/π, occurs for $\alpha = \pi/2$, thus tracing a semicircle through the ends of the loaded area (Fig 7.4b). The vertical and horizontal in-plane stresses on the centreline of the strip are shown in Fig 7.4c.

Under conditions of axial symmetry, the stresses for the point load can be integrated to give the stresses *on the centreline* of a ring loading and thence integrated again to give the stresses *on the centreline* under a uniformly loaded circular area of radius a (Fig 7.5) with intensity of loading ζ *per unit area*. The stress resultants in cylindrical coordinates are

$$\sigma_z = \zeta \left\{ 1 - \left[1 + \left(\frac{a}{z} \right)^2 \right]^{-\frac{3}{2}} \right\} \tag{7.14}$$

$$\sigma_r = \sigma_\theta = \frac{\zeta}{2} \left\{ (1 + 2\nu) - 2(1 + \nu) \left[1 + \left(\frac{a}{z} \right)^2 \right]^{-\frac{1}{2}} + \left[1 + \left(\frac{a}{z} \right)^2 \right]^{-\frac{3}{2}} \right\} \tag{7.15}$$

The variations of these vertical and horizontal (radial) stresses *on the centreline* of the loaded area are shown in Fig 7.5b.

As the problem being analysed becomes more complicated so the calculations of the stresses and strains become more intricate. Fortunately, a comprehensive compendium of elastic solutions for problems of relevance in geotechnical engineering has been produced by Poulos and Davis (1974), scouring the literature for solutions usually obtained by numerical integration of the equations and presented in the form of tabulated results. If the elastic material description is thought to be plausible and if the boundary conditions can be made to fit a standard form then the solution for the distribution of stresses and displacements is immediately available and no further numerical analysis is required. A huge computational economy is possible. Results are available for finite layers, multiple layers, horizontal surface loads, internal loads within the soil mass, rigid and flexible loaded areas of different shapes, etc.

It is of interest to note, as might be expected from the dimensional analysis of section §5.2.5, that the stiffness of the material plays no part in the stress distributions for a homogeneous linear elastic material that have been presented here. Further, for a line or strip loading, which are both plane strain loadings, the value of Poisson's ratio has no influence on the stresses in the plane perpendicular to the long direction of the load and influences only the out-of-plane stress. For the circular loaded area the value of Poisson's ratio does not affect the vertical stress. The variation of vertical stress reveals the combination of vertical equilibrium with load spreading beneath the loaded area: the stress spreads out more rapidly beneath the circular load than beneath the strip load. Poisson's ratio controls the lateral push generated by the vertical load. Use of an elastic analysis to estimate vertical stresses as part of an empirical procedure for calculation of settlement (§1.2.2, §1.2.3) has a certain logic: these stresses are not affected by the value of Poisson's ratio. We can expect the details of constitutive modelling (for example, the occurrence of plasticity) to have a much greater effect on the lateral stresses.

7.3   Plastic failure analysis

At the other extreme from the elastic model we look at some of the calculations that can be made using a perfectly plastic material model (Fig 7.6). It does not matter whether the material is rigid before it reaches the condition of perfect plasticity or whether it behaves elastically: if we imagine that our geotechnical system is in the process of failing then the strains that will be occurring are sufficiently large to dwarf any prefailure elastic strains. The independence of plastic collapse loads both from any elastic properties and from any particular state of initial stress is thus intuitively correct—and it can also be proved theoretically (see, for example, Calladine, 1985).

All the examples considered here are plane strain problems. In all cases it will be assumed that the out-of-plane normal stress is the intermediate principal stress so that limiting stress states are controlled entirely by the in-plane major and minor principal stresses.

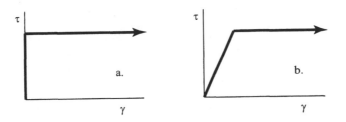

Figure 7.6: (a) Rigid-perfectly plastic stress-strain response; (b) elastic-perfectly plastic stress-strain response

7.3.1 Bound theorems

Our exploration of plastic collapse loads of geotechnical systems is aided by the two bound theorems of plasticity which again are intuitively correct but can be proved subject to the condition that the plastic material obeys an associated flow rule—so that strain increments are always aligned in the direction of the outward normal to the yield surface at the current stress state (Calladine, 1985).

Collapse is concerned with deformations continuing indefinitely in order to form a failure mechanism. Our search for appropriate failure mechanisms—combining information about kinematics with the yield criterion for the soil—endeavours to discover the mechanism that produces the lowest collapse load. The *upper bound theorem* says that the soil will always be cleverer than we are in spotting the easiest mechanism by which to fail so that our estimates of collapse loads made by studying mechanisms of failure will in general be *unsafe*.

Elements of soil within a geotechnical system may yield—which, for the perfectly plastic materials considered in this section, means fail—without the entire system failing. As further loads are applied the stresses must redistribute somewhat from their elastic pattern in order to make best use of the available strength of the material. The *lower bound theorem* says that any system of internal stresses which is in equilibrium with the applied loads and which nowhere violates the yield condition for the soil will lead to a lower bound estimate of the collapse load for the system. The soil will always be cleverer than we are in finding ways in which to redistribute the stresses. Estimates of collapse loads obtained by seeking admissible internal stress fields will be *safe*.

7.3.2 Structural example

The plastic analysis of steel structures makes use of these bound theorems to home in on more or less exact estimates of collapse loads. For simple structures exact results can usually be obtained with coincident upper and lower bound values—admissible systems of internal moments corresponding to kinematically acceptable mechanisms of collapse. For more complicated structures we may well have to make do with a gap between our best upper and lower bounds.

For moment collapse of steel beams the yield criterion is one-dimensional (Fig 7.7): the moment cannot have a magnitude greater than M_p, the full plastic moment of the beam:

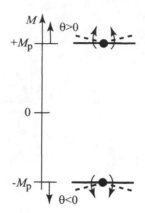

Figure 7.7: Yield criterion for steel beam in bending and associated plastic rotations

$$-M_p \leq M \leq +M_p \tag{7.16}$$

and the plastic deformations are associated with the plastic moments (Fig 7.7) in the sense that if $M = +M_p$ then the plastic rotations θ must be positive and if $M = -M_p$ then the plastic rotations θ must be negative (Fig 7.7). This is usually so obviously necessary that we do not bother to check that it is so for any assumed mechanism.

Consider a beam of length ℓ built in at both ends and subjected to a uniformly distributed total load W (Fig 7.8). Collapse occurs with hinges at the centre and at the two ends (yield first occurs at the ends). Sketching the distribution of moments (Fig 7.8b) gives a lower bound estimate of the collapse load:

$$W = 16\frac{M_p}{\ell} \tag{7.17}$$

(The free moment diagram indicates the bending moments in the beam under its actual loading but with the end fixities released, the nett moments are the difference between these free moments and the moments imposed by the end constraints.) Studying the mechanism of collapse, we equate the work done by the descending load and the work absorbed by plastic rotation in the hinges (Fig 7.8c) and obtain the same result. We conclude that we have in fact found the exact collapse load for the beam.

Next consider a beam of length ℓ built in at one end and simply supported at the other end and again subjected to a uniformly distributed total load W (Fig 7.9). We might guess that this would fail with the same mechanism (Fig 7.9c), with a central hinge. Consideration of the work balance then gives an upper bound to the collapse load:

$$W = 12\frac{M_p}{\ell} \tag{7.18}$$

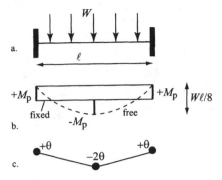

Figure 7.8: (a) Beam built in at both ends; (b) internal moments at collapse; (c) mechanism of collapse

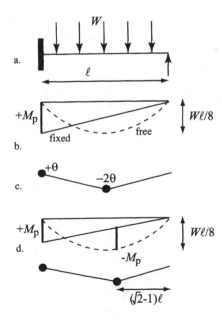

Figure 7.9: (a) Beam built in at one end; (b) statically admissible distribution of internal moments; (c) collapse mechanism with central hinge; (d) most critical collapse mechanism and corresponding distribution of internal moments at collapse

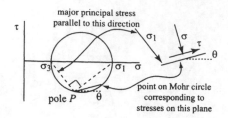

Figure 7.10: Mohr's circle of stress and pole construction

A simple statically admissible distribution of internal moments with the root moment equal to M_p and to the maximum free moment in the beam $W\ell/8$ (Fig 7.9b) gives a lower bound but certainly safe estimate of the collapse load:

$$W = 8\frac{M_p}{\ell} \tag{7.19}$$

We are left with a rather wide gap between the upper and lower bounds on the collapse load, but the lower bound is indeed safe.

In fact, if we study the general class of mechanism with one hinge somewhere along the beam and optimise this mechanism to find the lowest upper bound (Fig 7.9d), then we find that this lowest collapse load occurs when the hinge is at $(\sqrt{2} - 1)\ell = 0.41\ell$ from the simply supported end and:

$$W = 2\left(3 + 2\sqrt{2}\right)\frac{M_p}{\ell} = 11.66\frac{M_p}{\ell} \tag{7.20}$$

We can find a corresponding statically admissible system of internal moments (Fig 7.9d) and once again obtain coincident upper and lower bounds for the collapse load for which we now deduce that we have an exact value.

7.3.3   Continuum problems—Mohr's circle

For continuum problems we replace the limiting moment in a structural element by a limiting shear stress in a continuum element. We replace the plastic hinge in a structural element by a sliding surface in a continuum element. We may expect that it will be more difficult to find exactly matching upper and lower bounds except for the simplest of systems.

Many of the plastic stress fields for two-dimensional problems can be most readily explored with reference to the geometry of Mohr's circle of stress. The *pole* construction is particularly helpful (Fig 7.10). Mohr's circle shows all the combinations of shear stress and normal stress for planes having different orientations through a particular point within the material. Suppose that we know the stress state σ, τ on one plane which makes an angle θ with the horizontal. We locate this stress point on the Mohr circle and draw a line through this point making an angle θ with the horizontal. The point where this line intersects the circle again is called the *pole P*. We can find the stresses on any other plane by

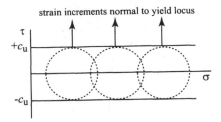

Figure 7.11: Cohesive soil: strength independent of stress level; normality of plastic strain increments

drawing a line through P parallel to that plane. In particular we can discover the orientations of the principal planes by drawing lines from P to the points of intersection of the Mohr circle with the normal stress axis (Fig 7.10).

Note that in this chapter we will take compression stresses as positive, following the usual soil mechanics convention. With compression positive we have to take *counter-clockwise* shear stresses as positive in plotting and interpreting Mohr's circles.

7.3.4   Bearing capacity of cohesive soil

A purely cohesive soil has strength independent of stress level (Fig 7.11). The maximum diameter of any Mohr circle which represents soil which is just reaching the yield/failure condition is $2c_u$. This material describes the undrained failure of clay: normality implies that plastic shearing occurs without change in volume, there is no plastic strain component linked with the normal stress σ.

Stress fields: lower bounds

In building up statically admissible stress fields we will usually start by dividing the soil into zones of uniform stress state separated by strong stress discontinuities or stress jumps. There are some expressions that we can derive to describe the relationship between the stress conditions on each side of a discontinuity which will be useful in helping us to develop candidate stress fields. Equilibrium tells us (Fig 7.12a) that the stresses on the plane of the discontinuity must be the same on each side of the discontinuity. However, the stresses on planes orthogonal to the discontinuity will be different on each side. We assume that there is a limiting stress state on each side of the discontinuity so that both Mohr circles have diameter $2c_u$ and that they intersect at a point corresponding to the plane of the discontinuity. We assume that the discontinuity is horizontal (the relative orientations of the discontinuity and the Mohr circles are arbitrary but choosing this orientation simplifies the geometry) and can therefore discover the location of the poles of the circles for each side of the discontinuity (P_A and P_B) by drawing a horizontal line through the point of intersection (Fig 7.12b).

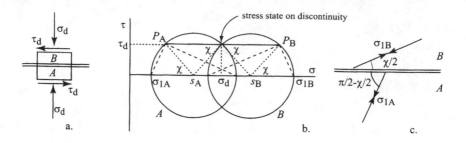

Figure 7.12: Cohesive soil: (a) stress discontinuity; (b) corresponding Mohr's circles of stress; (c) orientation of major principal stress

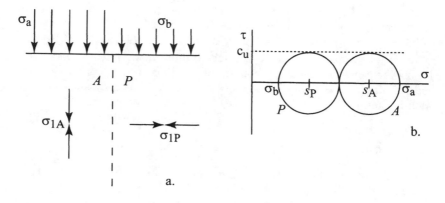

Figure 7.13: Bearing capacity of cohesive soil: single discontinuity

If the shear stress mobilised on the discontinuity is τ_d then we can define an angle χ (Fig 7.12b):

$$\sin \chi = \frac{\tau_d}{c_u} \tag{7.21}$$

so that χ can vary between 0 and $\pi/2$ and indicates the 'strength' of the discontinuity. The jump in mean stress across the discontinuity is

$$\Delta s = s_B - s_A = 2c_u \cos \chi \tag{7.22}$$

Study of the geometry of the two circles (Fig 7.12b) and the orientation of the discontinuity (Fig 7.12c) shows that the jump in direction of the major principal stress is

$$\Delta \theta = \frac{\pi}{2} - \chi \tag{7.23}$$

Let us suppose that we are trying to find the greatest value of the footing stress σ_a that can be placed on the surface of soil carrying a general surcharge σ_b (Fig 7.13a). Let us start by considering a very simple stress field with a

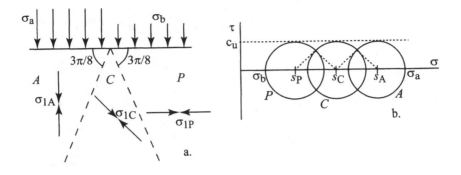

Figure 7.14: Bearing capacity of cohesive soil: two discontinuities

single vertical discontinuity (Fig 7.13a). The major principal stress is vertical under the footing and horizontal under the surcharge so that from (7.23) $\chi = 0$ and the two Mohr circles touch on the normal stress axis (Fig 7.13b). Although we can deduce the bearing capacity directly from the Mohr circles it is helpful to build it up from consideration of the shift in centre of the Mohr circles (7.22) and from the position of the boundary stresses, σ_a and σ_b, in relation to the centre of each circle. Then we can write:

$$\sigma_a - \sigma_b = c_u + \Delta s + c_u = c_u + 2c_u + c_u = 4c_u \qquad (7.24)$$

We can repeat this process using two discontinuities. It seems reasonable to suppose that we will wish to obtain equal rotations of major principal stress across each discontinuity and this enables us to deduce that $\chi = \pi/4$ for each discontinuity and that they should be symmetrically disposed with angle $3\pi/8$ to the horizontal (Fig 7.14a). The estimate of bearing capacity now becomes

$$\sigma_a - \sigma_b = c_u + 4c_u \cos\frac{\pi}{4} + c_u = 2\left(1 + \sqrt{2}\right) c_u = 4.83c_u \qquad (7.25)$$

We can repeat this for any number of stress discontinuities. The more discontinuities we have, the smaller the jump in direction of major principal stress across each one and the smoother the overall stress field will become. For n discontinuities $\chi = (\pi/2)(n-1)/n$ and this angle tends to $\pi/2$ as n increases (Fig 7.15a). The outer discontinuities make an angle $\pi/2 - \chi/2$ with the horizontal and this angle tends to $\pi/4$ as n increases. The bearing capacity is given by

$$\sigma_a - \sigma_b = 2c_u\left\{1 + n\cos\left[\frac{\pi}{2}\frac{(n-1)}{n}\right]\right\} \qquad (7.26)$$

The bearing capacity is also clearly tending towards some asymptotic value (Fig 7.15b)—as would be expected because we are steadily improving our estimate of the lower bound on the collapse load.

We can calculate this asymptote from study of equation (7.26) but it is easier simply to discover what happens when the jump in stress state across each

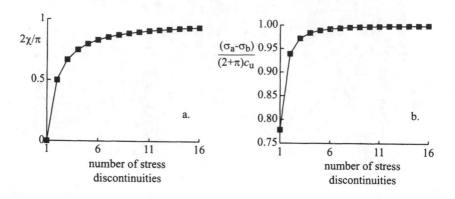

Figure 7.15: (a) Variation of angles in stress field and (b) variation of corresponding bearing capacity as number of discontinuities increases

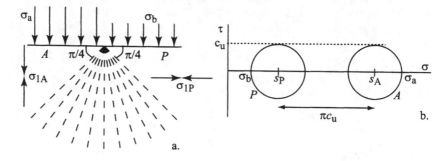

Figure 7.16: Bearing capacity of cohesive soil: infinite number of discontinuities = fan zone

discontinuity becomes infinitesimally small. With $\chi = \pi/2 - \delta\vartheta$ the infinitesimal jump in mean stress (7.22) becomes

$$\delta s = 2c_u \sin \delta\vartheta \approx 2c_u \delta\vartheta \tag{7.27}$$

If we have an infinite number of discontinuities making up a total rotation of major principal stress $\Delta\vartheta$ then the overall increase in mean stress is

$$\Delta s = 2c_u \Delta\vartheta \tag{7.28}$$

and we have a continuous *fan* zone of angle $\Delta\vartheta$. Such fan zones form a basic building block in constructing admissible stress fields for cohesive soils.

For our bearing capacity problem the total fan angle has to be $\Delta\vartheta = \pi/2$ (Fig 7.16) and the bearing capacity is calculated to be:

$$\sigma_a - \sigma_b = c_u + 2c_u \frac{\pi}{2} + c_u = (2 + \pi)\, c_u = 5.14 c_u \tag{7.29}$$

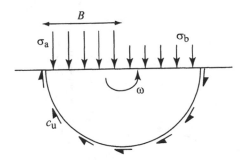

Figure 7.17: Bearing capacity of cohesive soil: semicircular failure mechanism

Mechanisms: upper bounds

We have optimised our *class* of lower bound equilibrium stress fields[2]. Next we should approach the problem from the other end and look for possible collapse mechanisms. This upper bound approach requires us to look at the balance of work done by moving loads and the energy absorbed by the deforming soil. Our background knowledge of the use of slip circles in analysis of geotechnical collapse (for example, in studying the stability of slopes) suggests that we might start by investigating the possibility of using circular failure surfaces for this bearing capacity problem too. Circular mechanisms are certainly kinematically compatible, since relative movement between adjacent blocks occurs with no separation or volume change.

The simplest mechanism will be one that includes a semicircular rigidly sliding block beneath the loaded area (Fig 7.17). If this block is given a small rotation ω then the work balance, matching the work done by the descending load σ_a with the work absorbed in sliding on the interface, and the work done in lifting the surcharge stress σ_b, is:

$$\sigma_a B\left(\frac{B\omega}{2}\right) = c_u \pi B^2 \omega + \sigma_b B\left(\frac{B\omega}{2}\right) \tag{7.30}$$

giving

$$\sigma_a - \sigma_b = 2\pi c_u \tag{7.31}$$

In fact we can improve this upper bound on the failure load by considering a more general circular failure mechanism in which the angle subtended by the failing block at the centre of the circle is 2β (Fig 7.18a). The work equation now becomes:

$$\frac{(\sigma_a - \sigma_b)B^2}{2}\omega = c_u 2\beta \left(\frac{B}{\sin\beta}\right)^2 \omega \tag{7.32}$$

---

[2]Though we cannot be certain that there are no other *classes* of possible stress fields that we should investigate.

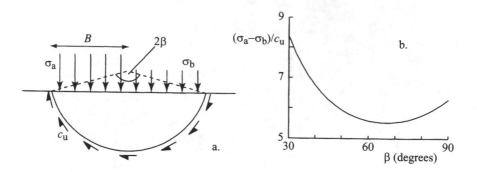

Figure 7.18: Bearing capacity of cohesive soil: (a) circular arc failure mechanism; (b) variation of bearing capacity with angle β

or

$$\sigma_a - \sigma_b = \frac{4c_u\beta}{\sin^2\beta} \tag{7.33}$$

We can now vary β and find the value of β which minimises the value of $(\sigma_a - \sigma_b)/c_u$. The variation of $(\sigma_a - \sigma_b)/c_u$ with β is shown in Fig 7.18b. The minimum value occurs when:

$$\tan\beta = 2\beta \tag{7.34}$$

which has a solution $\beta \approx 1.16^c = 66.8°$ and corresponding value $\sigma_a - \sigma_b = 5.52c_u$.

As an aside, this approach using circular mechanisms can be applied to the estimation of the bearing capacity of a soil with undrained strength which varies linearly with depth from a value c_o at the surface (Fig 7.19a):

$$c_u = c_o(1 + kz) \tag{7.35}$$

For a loaded area of width B, the bearing capacity is:

$$\frac{(\sigma_a - \sigma_b)}{c_o} = \frac{4}{\sin^2\beta}\left[\beta + kB\left(1 - \beta\cot\beta\right)\right] \tag{7.36}$$

This can be minimised (numerically) to find a dependence of $(\sigma_a - \sigma_b)/c_o$ on kB as shown in Fig 7.19b. The variation of the depth of the failure mechanism is shown in Fig 7.19c.

An alternative class of failure mechanisms makes use of rigid blocks sliding on planar failure surfaces. The mechanism may look unrealistic—it is clear that it will not be possible for large displacements to occur without problems at the corners—but for infinitesimal movements there can be no objection. Again, kinematic compatibility is ensured if sliding occurs between the blocks with no developing separation. The first mechanism, with two sliding blocks, is shown in

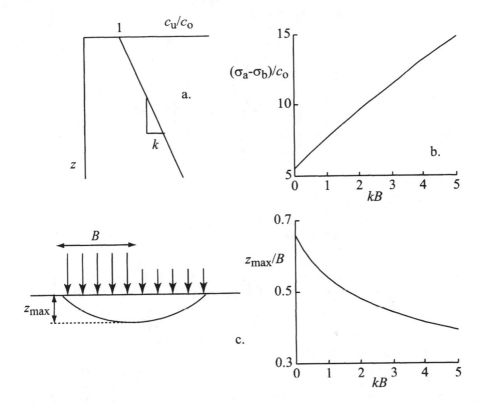

Figure 7.19: (a) Cohesive soil with strength varying linearly with depth; (b) bearing capacity calculated using circular arc failure mechanism; (c) depth of critical circular arc failure mechanism

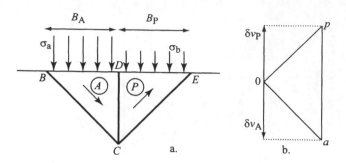

Figure 7.20: Bearing capacity of cohesive soil: (a) mechanism with two sliding blocks; (b) corresponding displacement diagram (a and p indicate displacements of blocks A and P respectively relative to origin 0)

Fig 7.20a. We will adopt the simple initial assumption that the blocks are both $45°$ triangles, though we could evidently check that this geometry did indeed give us the lowest upper bound.

The energy balance requires us to consider relative displacements on any sliding surface—we can construct a displacement diagram for the infinitesimal displacements as shown in Fig 7.20b. The 'active' block A can only move down in a direction at $\pi/4$ to the horizontal and the 'passive' block P must move up in a direction at $\pi/4$ to the horizontal. The interface between the two blocks is vertical so that *relative* displacements must occur in a vertical direction. The loaded area of width B_A moves down by an amount given by the vertical component of displacement δv_A of block A, and the surcharged area of width B_P moves up by an amount given by the vertical component of displacement δv_P of block P. In general, with uniform cohesion throughout the soil, the energy balance requires that

$$\sigma_a B_A \delta v_A - \sigma_b B_P \delta v_P = \sum_{i=1}^{n} c_u \Delta \ell_i \delta v_i \qquad (7.37)$$

where the mechanism consists of n interfaces of lengths $\Delta \ell_i$ between rigid blocks displacing relatively by δv_i. We can conveniently tabulate the contributions to this energy balance (Table 7.1): the footing width is $B (= B_A = B_P)$ and the vertical component of footing displacement is δ_v. The energy balance then gives:

$$(\sigma_a - \sigma_b) B \delta_v = 6 c_u B \delta_v \qquad (7.38)$$

or

$$\sigma_a - \sigma_b = 6 c_u \qquad (7.39)$$

A mechanism involving three sliding blocks is shown in Fig 7.21a. These blocks have been chosen with equal angles of $\pi/3$ but again these angles could be optimised to produce the lowest estimate of the collapse load for such a three

Table 7.1: Bearing capacity: two sliding blocks (Fig 7.20)

sliding surface	sliding length $\Delta \ell_i$	relative displacement δv_i	energy dissipated
BC	$B\sqrt{2}$	$\delta_v\sqrt{2}$	$2c_u B\delta_v$
DC	B	$2\delta_v$	$2c_u B\delta_v$
CE	$B\sqrt{2}$	$\delta_v\sqrt{2}$	$2c_u B\delta_v$
Total energy dissipated			$6c_u B\delta_v$

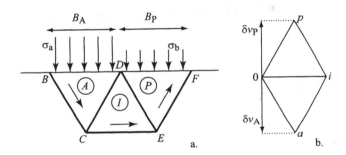

Figure 7.21: Bearing capacity of cohesive soil: (a) mechanism with three sliding blocks; (b) corresponding displacement diagram (a, i and p indicate displacements of blocks A, I and P respectively relative to origin 0)

Table 7.2: Bearing capacity: three sliding blocks (Fig 7.21)

sliding surface	sliding length $\Delta \ell_i$	relative displacement δv_i	energy dissipated
BC	B	$2\delta_v/\sqrt{3}$	$2c_uB\delta_v/\sqrt{3}$
DC	B	$2\delta_v/\sqrt{3}$	$2c_uB\delta_v/\sqrt{3}$
CE	B	$2\delta_v/\sqrt{3}$	$2c_uB\delta_v/\sqrt{3}$
DE	B	$2\delta_v/\sqrt{3}$	$2c_uB\delta_v/\sqrt{3}$
EF	B	$2\delta_v/\sqrt{3}$	$2c_uB\delta_v/\sqrt{3}$
Total energy dissipated			$10c_uB\delta_v/\sqrt{3}$

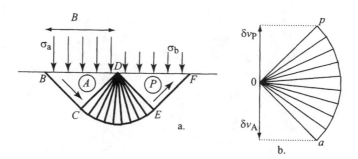

Figure 7.22: Bearing capacity of cohesive soil: (a) mechanism with large number of sliding blocks; (b) corresponding displacement diagram

block mechanism. The displacement diagram is shown in Fig 7.21b. The energy absorbed in the sliding between the blocks is tabulated in Table 7.2. The energy balance then gives:

$$\sigma_a - \sigma_b = \left(10/\sqrt{3}\right)c_u = 5.77c_u \qquad (7.40)$$

While this process can of course be repeated for increasing numbers of sliding blocks disposed in optimised orientations it is easier to jump directly to a large number of blocks sliding as shown in Fig 7.22. We imagine large 45° blocks—an 'active' block A, and a 'passive' block P—at each end, sliding down and up respectively at 45° to the horizontal, and then a large number of infinitesimally thin triangular blocks, each one sliding on what has now become an essentially circular boundary with the rigid undeforming region beyond. It is important to note that kinematic compatibility requires that each triangular block should *translate*, not rotate (rotation would not be compatible with the sliding of the active and passive blocks) and therefore that there is relative sliding motion on each of the radial interfaces between the blocks as well as on the circumferential boundary. The displacement diagram is shown in Fig 7.22b.

Table 7.3: Bearing capacity: large number of sliding blocks (Fig 7.22)

sliding surface	sliding length $\Delta \ell_i$	relative displacement δv_i	energy dissipated
BC	$B/\sqrt{2}$	$\delta_v \sqrt{2}$	$c_u B \delta_v$
circumferential sliding	$B \delta \theta / \sqrt{2}$	$\delta_v \sqrt{2}$	$\int_0^{\pi/2} c_u B \delta_v \mathrm{d}\theta$
radial sliding	$B/\sqrt{2}$	$\delta_v \sqrt{2} \delta \theta$	$\int_o^{\pi/2} c_u B \delta_v \mathrm{d}\theta$
EF	$B/\sqrt{2}$	$\delta_v \sqrt{2}$	$c_u B \delta_v$
Total energy dissipated			$(2+\pi) c_u B \delta_v$

For a footing of width B the radius of the 'fan' of sliding blocks is $B/\sqrt{2}$. For a footing displacement with vertical component δ_v the sliding displacement on the circumferential boundary is everywhere $\delta_v \sqrt{2}$ while the sliding displacement on each radial interface is $\delta_v \sqrt{2} \delta \theta$ where $\delta \theta$ is the angle each block subtends at the centre of the fan. The elements of the energy dissipated in the sliding blocks are compiled in Table 7.3. Then the energy balance tells us that

$$\sigma_a - \sigma_b = (2 + \pi) c_u \qquad (7.41)$$

and we discover (serendipitously?) that we have obtained coincident upper and lower bounds to our estimates of the bearing capacity of the cohesive soil. This single value must therefore be the exact and correct value of the collapse load.

There is an intuitive plausibility about this result since the final optimised stress field has obvious resemblance to the optmised mechanism. However, in fact, to be completely certain of the validity of this result, we should also demonstrate that the stress field can be extended into the whole of the soil while simultaneously satisfying equilibrium and never violating the yield/failure condition anywhere. In fact of course we expect that the soil will only be *at* failure just on the boundary of the failing region and we may well have some freedom in the way we complete the stress field. For this bearing capacity problem the completion of the stress field is not particularly difficult. For other problems it may be harder formally to demonstrate this completion.

Bearing capacity with inclined loading

Having seen how the building blocks of uniform stress state and fan zones of infinitesimal discontinuities of stress can be used we can rapidly produce results for some other simple configurations. Suppose that the applied loading is applied at an angle δ to the vertical so that it imposes shear stresses τ_a as well as normal stresses σ_a (Fig 7.23a), with $\tau_a/\sigma_a = \tan \delta$. How does the presence of shear stresses affect the bearing capacity? To answer this question we have to introduce an auxiliary angle Δ in Mohr's circle of stress (Fig 7.23b) to characterise the degree of mobilisation of shear strength on the horizontal boundary plane :

$$\sin \Delta = \frac{\tau_a}{c_u} \qquad (7.42)$$

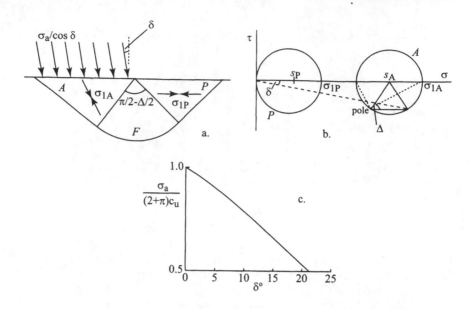

Figure 7.23: (a) Bearing capacity of cohesive soil with inclined loading; (b) Mohr circles; (c) variation of bearing capacity with inclination of load (Note that σ_a is the *normal* component of the applied stress)

We discover from the geometry of Mohr's circle that the major principal stress under the loaded area makes an angle $\Delta/2$ to the vertical and therefore that the angle of the fan of stress discontinuities must be $\pi/2 - \Delta/2$. Proceeding as before, the change in mean stress from the passive region under the surcharge to the 'active' region under the footing is $2c_u \left(\pi/2 - \Delta/2\right)$ and the bearing capacity can be found:

$$\sigma_a - \sigma_b = [1 + (\pi - \Delta) + \cos \Delta]c_u \qquad (7.43)$$

For the situation where there is no surcharge[3], we can present the result in terms of the variation of bearing capacity σ_a/c_u with angle δ of the applied loading (Fig 7.23c) noting that this angle δ is not the same as the angle of the major principal stress to the vertical (Fig 7.23a, b). Fig 7.23c has been drawn with the bearing capacity scaled with the bearing capacity for purely vertical loading to emphasise the reduction caused by the application of shear stresses.

In the limit, when full shear strength is mobilised on the horizontal surface, $\Delta = \pi/2$, $\delta = \cot^{-1}(1 + \pi/2) = 21.25°$ and

$$\sigma_a = \left(1 + \frac{\pi}{2}\right) c_u \qquad (7.44)$$

so that the bearing capacity is exactly halved by comparison with that calculated for purely vertical loading. We can understand the effect of spinning the wheels

---

[3]In the presence of surcharge, $\sigma_b \neq 0$ the value of δ depends on the value of σ_b.

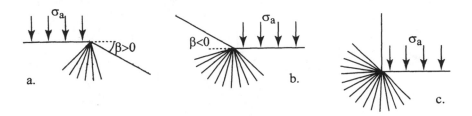

Figure 7.24: Bearing capacity of footing on edge of slope

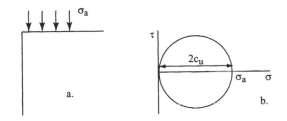

Figure 7.25: Bearing capacity of unsupported excavation

of a car sitting on soft ground: applying traction (shear stress) to the wheels merely encourages bearing failure and the car sinks further in.

7.3.5 Wall retaining cohesive soil

It is simple to extend the solution for the bearing capacity of a footing on the surface of a horizontal bed of soil to the problem of a footing on the edge of a slope (Fig 7.24), provided the soil is weightless. If the soil has self weight then the solution becomes somewhat more analytically challenging: Schofield and Wroth (1968) show one result; more extensive presentation of the required analytical armoury is given by Sokolovskii (1965). Without self weight, the angle β of the slope (Fig 7.24) merely reduces the angle of the fan of stress discontinuities so that the bearing capacity becomes, for a slope free of shear stress:

$$\sigma_a = (2 + \pi - 2\beta)\,c_u \qquad (7.45)$$

where the angle β can be positive (Fig 7.24a) or negative (Fig 7.24b). One special case is obtained when $\beta = -\pi/2$ (Fig 7.24c) and the bearing capacity of a strip load at the base of an excavation is calculated as $2\,(1 + \pi)\,c_u$.

Another specical case is found for $\beta = \pi/2$ (Fig 7.25): the bearing capacity of an unsupported excavation is $2c_u$. Alternatively we can deduce that the active pressure required to support a smooth retaining wall subjected to a surcharge σ_a (Fig 7.26) is $\sigma_h = \sigma_a - 2c_u$. Recall again that we are ignoring the self-weight of the soil. However, for this particular case, because there is essentially a single Mohr circle describing the stress state behind the wall, we can add a stress γz

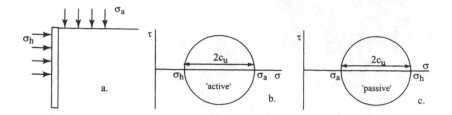

Figure 7.26: Cohesive soil: pressure on retaining wall

to horizontal and vertical stresses at all depths z below the horizontal surface without affecting the size of the Mohr circles and hence without violating the yield/failure criterion. Then at any depth in the limit:

$$\sigma_h = \sigma_a + \gamma z \pm 2c_u \qquad (7.46)$$

where the negative sign corresponds to active collapse of the wall with vertical stresses greater than horizontal stresses (Fig 7.26b) and the positive sign to passive collapse (Fig 7.26c). Tension cracks can thus extend to a depth $z_c = 2c_u/\gamma$ (assuming that the soil cannot sustain any tensile total stresses).

We can—for weightless soil—extend the analysis for the active collapse of a rough wall on which shear stresses as well as normal stresses are developed (Fig 7.27). As before we introduce an auxiliary angle Δ to indicate the mobilisation of available cohesion on the vertical surface (Fig 7.27b). The major principal stress then has angle $\Delta/2$ to the vertical just behind the wall and we must introduce a fan of stress discontinuities having the same angle (Fig 7.27a). If the inclined stress on the wall makes an angle δ to the horizontal (the normal to the wall) and has horizontal (normal) component σ_b, then

$$\sigma_b \tan \delta = \tau_w = c_u \sin \Delta \qquad (7.47)$$

where τ_w is the shear stress mobilised on the back of the wall. Then from consideration of the stress changes through the fan and the Mohr circles for the regions immediately behind the wall and under the surcharge σ_a

$$\sigma_b + c_u \cos \Delta + 2c_u \Delta/2 + c_u = \sigma_a \qquad (7.48)$$

or

$$\frac{\sigma_a - \sigma_b}{c_u} = 1 + \cos \Delta + \Delta \qquad (7.49)$$

There is again a limiting value for $\Delta = \pi/2$

$$\frac{\sigma_a - \sigma_b}{c_u} = 1 + \pi/2 \qquad (7.50)$$

The variation of $(\sigma_a - \sigma_b)/c_u$ with $\Delta = \sin^{-1}(\tau_w/c_u)$ is shown in (Fig 7.27c). The angle Δ is defined directly in terms of the shear stress τ_w (Fig 7.27c): whereas the value of angle of inclination of the wall stress δ always depends on the value of σ_b (7.47) (Fig 7.27b).

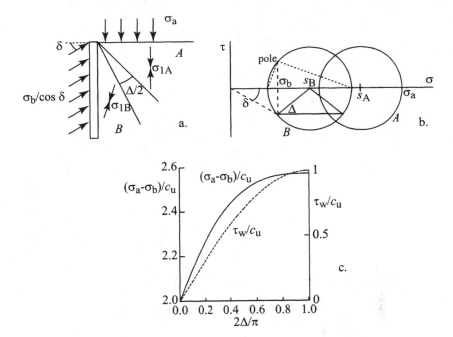

Figure 7.27: Cohesive soil: 'active' pressure on rough wall

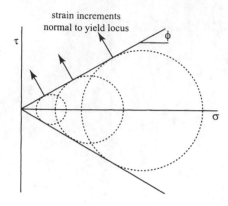

Figure 7.28: Strength of frictional soil; strain increments implied by associated flow

7.3.6   Bearing capacity of frictional soil

We next consider frictional soil in which the strength depends on the stress level (Fig 7.28). These analyses will be of relevance to drained behaviour of soils. As a purely frictional material the soil has *no* strength unless there is some non-zero mean stress holding the soil together. We will discover that, whereas the contributions to bearing capacity of cohesive soil are essentially *additive*, contributions to bearing capacity of frictional soil are *multiplicative*. Because of the utter dependence of strength on stress the existence of self-weight of the soil has a major effect on plasticity calculations. We will ignore it here. The mathematical consequences of including self weight in such calculations are explored by Sokolovskii (1965).

Stress fields: lower bounds

We can proceed as for the cohesive soil, looking at the consequences of introducing a strong discontinuity of stress state in the soil between two regions in which the soil is at yield/failure. The two Mohr circles are shown in Fig 7.29a and the stresses in the physical plane are shown in Fig 7.29b. The available friction angle of the soil is ϕ; the angle of friction mobilised on the plane of the discontinuity is ϕ_d. We introduce an auxiliary angle χ which indicates the degree of mobilisation of the available friction on the plane of the discontinuity:

$$\sin \chi = \frac{\sin \phi_d}{\sin \phi} \qquad (7.51)$$

By constructing the poles and considering the geometry of the two Mohr circles we find that the rotation of direction of major principal stress that occurs across the discontinuity is $\pi/2 - \chi$. The ratio of the mean stresses at the centres of the two Mohr circles can be found by calculating the shear stress on the discontinuity from the geometry of each circle.

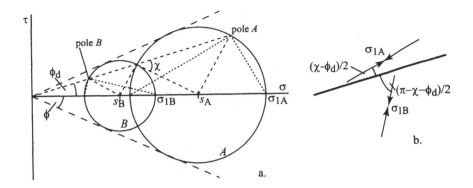

Figure 7.29: Stress discontinuity in frictional soil: (a) Mohr's circles of stress; (b) orientation of major principal stress

$$s_B \sin\phi \sin(\chi + \phi_d) = s_A \sin\phi \sin(\chi - \phi_d) \tag{7.52}$$

$$\frac{s_A}{s_B} = \frac{\sin(\chi + \phi_d)}{\sin(\chi - \phi_d)} \tag{7.53}$$

This ratio has the limiting value for $\chi = 0$:

$$\frac{s_A}{s_B} = \frac{1 + \sin\phi}{1 - \sin\phi} \tag{7.54}$$

which can of course be confirmed by the geometry of the Mohr circles. At the other extreme, for $\chi = \pi/2 - \delta\theta$,

$$\frac{s_A - s_B}{s_A + s_B} \approx \frac{\delta s}{2s} \approx \tan\phi \delta\theta \tag{7.55}$$

Since we are ignoring the self-weight of the soil the *only* way in which any strength can be generated in the soil is by having some finite surcharge (σ_b) on the area beyond the footing (Fig 7.30a). We start by considering the possible stress fields with one single stress discontinuity (Fig 7.30). The jump in direction of major principal stress across the discontinuity is then $\pi/2$ and the auxiliary angle is $\chi = 0$. We can build up the expression for the bearing capacity as a product of three terms:

$$\frac{\sigma_a}{\sigma_b} = \frac{\sigma_a}{s_A} \frac{s_A}{s_P} \frac{s_P}{\sigma_b} = (1 + \sin\phi)\frac{1 + \sin\phi}{1 - \sin\phi} \frac{1}{1 - \sin\phi} = \left(\frac{1 + \sin\phi}{1 - \sin\phi}\right)^2 \tag{7.56}$$

For $\phi = 30°$ this gives $\sigma_a/\sigma_b = 9$.

With two discontinuities (Fig 7.31) we may suppose that it will be advantageous to have an equal jump in direction of major principal stress, of $\pi/4$, across

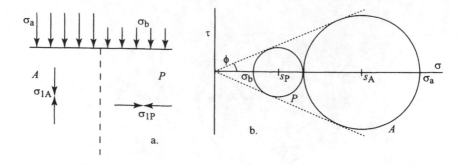

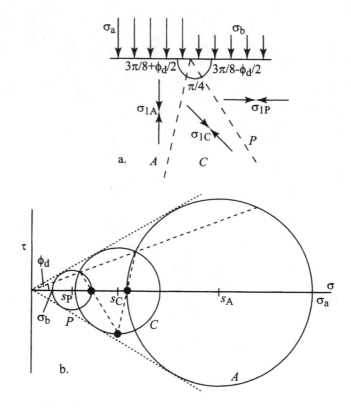

Figure 7.30: Bearing capacity of frictional soil: single stress discontinuity

Figure 7.31: Bearing capacity of frictional soil: two stress discontinuities (points ● indicate poles of Mohr circles)

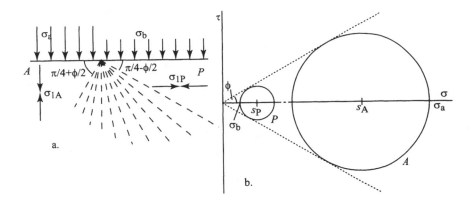

Figure 7.32: Bearing capacity of frictional soil: fan of infinitesimal stress discontinuities

each of them. Then $\chi = \pi/4$ and the mobilised friction on the discontinuities is given by $\sin \phi_d = \sin \phi/\sqrt{2}$. Comparison with Fig 7.29 shows that we have to place the discontinuities at angles $3\pi/8 \pm \phi_d/2$ to the horizontal (Fig 7.31). We can calculate the bearing capacity in stages again:

$$\frac{\sigma_a}{\sigma_b} = \frac{\sigma_a}{s_A} \frac{s_A}{s_C} \frac{s_C}{s_P} \frac{s_P}{\sigma_b} = (1 + \sin \phi) \left[\frac{\sin(\chi + \phi_d)}{\sin(\chi - \phi_d)} \right]^2 \frac{1}{1 - \sin \phi} \qquad (7.57)$$

For $\phi = 30°$ this gives $\sigma_a/\sigma_b = 14.72$.

Although this process could be continued, steadily adding additional discontinuities, the algebra becomes somewhat tedious—and, since we may suspect that the optimum result will be obtained with an infinite number of discontinuities, we might as well go there directly. An infinite number of discontinuities implies that $\chi \approx \pi/2$ and $\phi_d \approx \phi$ and hence, from Fig 7.29, that the limiting discontinuities must make angles $\pi/4 \pm \phi/2$ with the horizontal (Fig 7.32). The angle of the fan of discontinuities is $\pi/2$ so that integration of (7.55) gives:

$$\int_{s_A}^{s_P} \frac{ds}{s} = \int_0^{\pi/2} 2 \tan \phi d\theta \qquad (7.58)$$

$$\ln(s_P/s_A) = \pi \tan \phi \qquad (7.59)$$

whence we can calculate the bearing capacity:

$$\frac{\sigma_a}{\sigma_b} = \frac{\sigma_a}{s_A} \frac{s_A}{s_P} \frac{s_P}{\sigma_b} = \frac{1 + \sin \phi}{1 - \sin \phi} e^{\pi \tan \phi} \qquad (7.60)$$

For $\phi = 30°$ this gives $\sigma_a/\sigma_b = 18.40$.

Mechanisms: upper bounds

We recall that the bound theorems of plasticity require associated plastic flow, that is that the strain increment vectors must be normal to the yield surface.

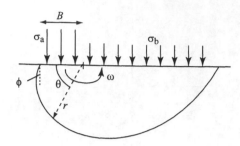

Figure 7.33: Bearing capacity of frictional soil: logarithmic spiral failure mechanism

That imposed no problem for the cohesive soil for which undrained failure was expected to occur at constant volume and be a purely shearing process. For a frictional material we have a difficulty: we have seen in sections §2.6, §3.3.4, §3.4.1 that in general soils show angles of dilation which are significantly lower than angles of friction. However, we are constrained in seeking true upper bounds to maintain the condition of normality for the frictional soil (Fig 7.28). This requires us to ensure that any mechanisms that we propose generate relative motion at an angle of dilation equal to the angle of friction on any sliding surface. However, there is another factor which in some ways eases our calculations: we noted in §3.3.4 that for an associated flow frictional material there was no energy dissipated in frictional shearing. Thus, so long as our mechanisms are kinematically admissible, we do not need to concern ourselves with internal dissipation. In fact we have only to match the work done by our footing load descending with the work done against the rising surcharge pressure. The geometry of the failure mechanism is thus crucial.

A circular failure mechanism is clearly not kinematically acceptable: the equivalent for the frictional soil is a logarithmic spiral (Fig 7.33) rotating about the edge of the footing of width B. This has equation

$$r = Be^{\theta \tan \phi} \tag{7.61}$$

where r and θ are defined in Fig 7.33. The length of the region over which the surcharge is lifted is therefore $Be^{\pi \tan \phi}$. For a small rotation ω of the logarithmic spiral the energy balance gives:

$$\sigma_a B \frac{B\omega}{2} = \sigma_b Be^{\pi \tan \phi} \frac{Be^{\pi \tan \phi} \omega}{2} \tag{7.62}$$

and hence

$$\frac{\sigma_a}{\sigma_b} = e^{2\pi \tan \phi} \tag{7.63}$$

For $\phi = 30°$ this gives $\sigma_a/\sigma_b = 37.62$.

Building up mechanisms with planar sliding surfaces is a bit fiddly: we can go directly to an infinite number of infinitesimal triangular sliding blocks (Fig 7.34) separating rigid triangular blocks with sides making angles $\pi/4 \pm \phi/2$ to

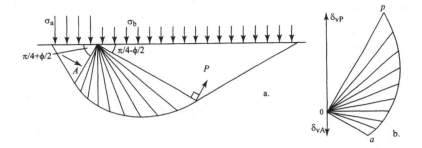

Figure 7.34: Bearing capacity of frictional soil: (a) mechanism with large number of sliding blocks; (b) corresponding displacement diagram (points a and p indicate displacements of blocks A and P relative to fixed point 0)

the horizontal as shown. The displacement diagram for this mechanism is shown in Fig 7.34b: it can be checked that this will guarantee that every sliding surface (recall that each triangle is *translating* not rotating) dilates at angle ϕ. From this we can deduce that, for a vertical component of displacement of the footing δ_v, the vertical component of displacement of the surcharge is

$$\delta_v \tan(\pi/4 + \phi/2)e^{(\pi/2)\tan\phi} \tag{7.64}$$

Geometry of the mechanism shows that the length of the region over which the surcharge is lifted is

$$B \tan(\pi/4 + \phi/2)e^{(\pi/2)\tan\phi} \tag{7.65}$$

Our energy balance then tells us that

$$\frac{\sigma_a}{\sigma_b} = \tan^2(\pi/4 + \phi/2)e^{\pi\tan\phi} = \frac{1 + \sin\phi}{1 - \sin\phi}e^{\pi\tan\phi} \tag{7.66}$$

and we have obtained coincident upper and lower bounds for the bearing capacity, albeit using a somewhat unrealistic material model. This ratio of stresses is the bearing capacity factor N_q for the frictional material and the expression that we have found is precisely that generally accepted for geotechnical design calculations (see, for example, Lancellotta, 1987).

Bearing capacity: inclined loading

If the applied footing load is inclined then we have to proceed in rather the same way as for the cohesive soil, introducing an auxiliary angle χ to characterise the friction mobilised by the applied load on the horizontal surface (Fig 7.35). For a load with inclination δ:

$$\sin\chi = \frac{\sin\delta}{\sin\phi} \tag{7.67}$$

and the major principal stress in the soil beneath the footing makes an angle $(\chi + \delta)/2$ with the vertical. Approaching the problem from the lower bound

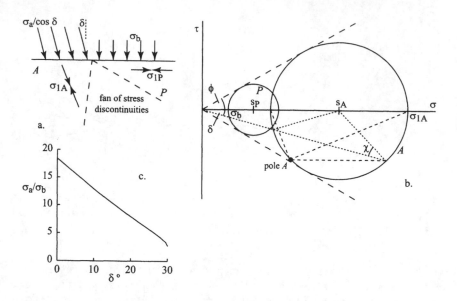

Figure 7.35: Bearing capacity of frictional soil with inclined surface loading ($\phi = 30°$)

direction, seeking admissible stress fields within the soil, we deduce that we can modify the optimum stress field for the vertical footing load by reducing the angle of the fan of infinitesimal discontinuities by this same amount. Then, as usual, we can calculate:

$$\frac{\sigma_a}{\sigma_b} = \frac{\sigma_a}{s_A}\frac{s_A}{s_P}\frac{s_P}{\sigma_b} = [1 + \sin\phi\cos(\chi + \delta)]\, e^{(\pi - \chi - \delta)\tan\phi}\frac{1}{1 - \sin\phi} \qquad (7.68)$$

with the limiting value

$$\frac{\sigma_a}{\sigma_b} = (1 + \sin\phi)\, e^{(\pi/2 - \phi)\tan\phi} \qquad (7.69)$$

for $\delta = \phi$ and $\chi = \pi/2$. The variation of the ratio σ_a/σ_b with δ is shown in Fig 7.35c for $\phi = 30°$.

Retaining wall

For a smooth wall the stress field is straightforward (Fig 7.36): we have a single Mohr circle to describe the stress state and obtain the Rankine active and passive limiting values:

$$\frac{\sigma_h}{\sigma_a} = \frac{1 \mp \sin\phi}{1 \pm \sin\phi} \qquad (7.70)$$

In fact for this simple situation, as we well know, the result can be interpreted also for a soil with self-weight as an indicator of the limiting ratio of horizontal and vertical stresses in the soil.

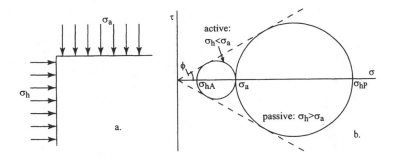

Figure 7.36: Smooth wall retaining frictional soil

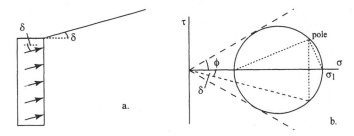

Figure 7.37: Rough wall retaining frictional soil: single state of stress requires inclined backfill

When the wall is rough the simple Rankine approach breaks down unless we start modifying the geometry of the wall. A single Mohr circle can only be sustained for heavy soil if the backfill behind the wall slopes at the same angle as the angle of mobilised friction on the back of a vertical wall (Fig 7.37)[4]. Alternatively, if the backfill is horizontal then the back of the rough wall must be inclined. These seem excessive constraints.

For weightless soil we can find a solution for the stress on the wall by introducing a fan of infinitesimal stress discontinuities separating one stress state immediately behind the wall and another stress state beneath the surcharged horizontal surface (Fig 7.38). We need an auxiliary angle χ to characterise the mobilisation of friction (through a wall friction angle δ) on the vertical surface (Fig 7.38b).

$$\sin \chi = \frac{\sin \delta}{\sin \phi} \tag{7.71}$$

Then the major principal stress for the 'active' case, with the soil moving down relative to the wall (this defines the direction of the shear stress), makes an angle $(\chi - \delta)/2$ to the vertical and this becomes the required angle of the fan of

[4]The same angle of friction is mobilised on all planes parallel to the slope as is mobilised on the back of the wall. All stresses are zero at the free surface, and the Mohr circle is there degenerate.

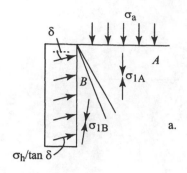
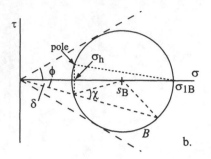

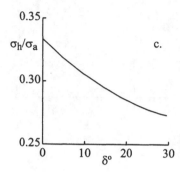

Figure 7.38: Rough wall retaining frictional soil: 'active' plastic failure; (b) shows Mohr circle for region B behind wall; (c) variation of σ_h/σ_a for $\phi = 30°$

stress discontinuities. As usual, we can inspect the geometry of the Mohr circles to deduce the ratio of horizontal wall pressure to surcharge stress.

$$\frac{\sigma_h}{\sigma_a} = \frac{\sigma_h}{s_B}\frac{s_B}{s_A}\frac{s_A}{\sigma_a} = [1 - \sin\phi\cos(\chi - \delta)]\,e^{-(\chi-\delta)\tan\phi}\frac{1}{1+\sin\phi} \qquad (7.72)$$

with limiting value

$$\frac{\sigma_h}{\sigma_a} = (1 - \sin\phi)\,e^{-(\pi/2-\phi)\tan\phi} \qquad (7.73)$$

The variation of σ_h/σ_a with δ is shown in Fig 7.38c for $\phi = 30°$.

The same procedure can be followed in reverse for a rough wall being pushed into the soil, developing passive pressures (Fig 7.39). The soil now moves up relative to the wall. The auxiliary angle χ is defined as before, and the angle of the fan of stress discontinuities is $(\chi + \delta)/2$. Now

$$\frac{\sigma_h}{\sigma_a} = \frac{\sigma_h}{s_B}\frac{s_B}{s_A}\frac{s_A}{\sigma_a} = [1 + \sin\phi\cos(\chi + \delta)]\,e^{(\chi+\delta)\tan\phi}\frac{1}{1-\sin\phi} \qquad (7.74)$$

with limiting value

$$\frac{\sigma_h}{\sigma_a} = (1 + \sin\phi)\,e^{(\pi/2+\phi)\tan\phi} \qquad (7.75)$$

The variation of σ_h/σ_a with δ is shown in Fig 7.39c for $\phi = 30°$.

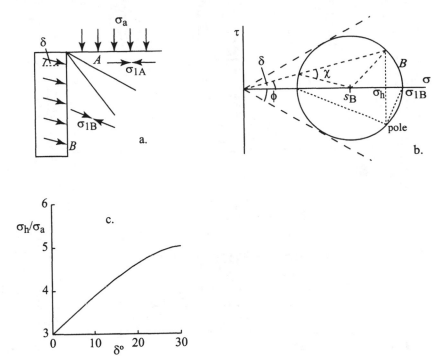

Figure 7.39: Rough wall retaining frictional soil: 'passive' plastic failure; (b) shows Mohr circle for region B behind wall; (c) variation of σ_h/σ_a for $\phi = 30°$

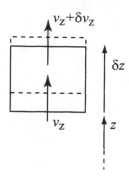

Figure 7.40: One-dimensional consolidation

7.4 One-dimensional consolidation

Consolidation is a transient process of unsteady flow in which there is coupling between flow and volume change as the soil gradually adjusts to a new effective stress regime. It is simplest to develop the consolidation equation for conditions of one-dimensional flow and deformation (Fig 7.40) (§4.2.1, §4.8, §5.2.3).

The equation for vertical flow with permeability k_z in the vertical direction tells us that the velocity of flow v_z is linked with gradient of pore pressure u and the rate of change of volumetric strain (which is the same as the vertical strain):

$$-\frac{\partial v_z}{\partial z} = \frac{k_z}{\gamma_w} \frac{\partial^2 u}{\partial z^2} = -\frac{\partial \epsilon_z}{\partial t} \tag{7.76}$$

where the negative sign is required because the volumetric strain for any element of the porous medium, which for the one-dimensional situation is equal to the vertical strain ϵ_z, is assumed to be *positive in compression*.

Vertical strain develops because of a change in vertical effective stress: the assumed one-dimensional deformation means that this is the only independent contribution to deformation that we need to consider. Let us suppose that the soil has a one-dimensional stiffness E_{oed} (which could be measured in an oedometer). Vertical strain is linked to change in vertical effective stress $\delta\sigma'_z$ (also positive in compression) and hence, from the principle of effective stress, to the difference between changes in vertical total stress $\delta\sigma_z$ and pore pressure δu:

$$\delta\epsilon_z = \frac{\delta\sigma'_z}{E_{oed}} = \frac{\delta\sigma_z - \delta u}{E_{oed}} \tag{7.77}$$

Then, combining (7.76) and (7.77)

$$\frac{k_z}{\gamma_w} \frac{\partial^2 u}{\partial z^2} = -\frac{1}{E_{oed}} \left(\frac{\partial\sigma_z}{\partial t} - \frac{\partial u}{\partial t} \right) \tag{7.78}$$

We substitute

$$c_v = \frac{k_z E_{oed}}{\gamma_w} \tag{7.79}$$

in order to define the coefficient of consolidation, c_v, which has dimensions of (length^2/time). The one-dimensional consolidation equation becomes

$$c_v \frac{\partial^2 u}{\partial z^2} = \frac{\partial u}{\partial t} - \frac{\partial \sigma_z}{\partial t} \tag{7.80}$$

and, for the particular situation where the total stress σ_z remains constant during the transient change of pore pressure,

$$c_v \frac{\partial^2 u}{\partial z^2} = \frac{\partial u}{\partial t} \tag{7.81}$$

which is the standard diffusion equation governing any gradient driven flow. Exactly similar equations can be written to describe transient diffusion of heat or concentration. Solutions developed for different physical situations are therefore directly interchangeable.

This one-dimensional consolidation equation is a linear equation which can be made non-dimensional by writing

$$U = u/u_i \tag{7.82}$$

$$Z = z/H \tag{7.83}$$

$$T = \frac{c_v t}{H^2} \tag{7.84}$$

where u_i is a reference pore pressure, H is a characteristic length, and T emerges as a dimensionless time factor. The equation then becomes

$$\frac{\partial^2 U}{\partial Z^2} = \frac{\partial U}{\partial T} \tag{7.85}$$

The assumptions that underpin this equation are material characteristics: incompressible pore fluid, incompressible soil particles, flow of pore fluid governed by Darcy's law, constant stiffness E_{oed} during the consolidation process; and boundary conditions: one dimensional deformation and flow. There is also an implicit assumption that the deformations that occur are sufficiently small that the geometry of the soil element for which the equation is derived does not change and the volume of this element for which flow and deformation are considered is referred always to an initial configuration which does not move in space. For soft clay soils, with high void ratios and low values of stiffness E_{oed} the change in geometry incurred during consolidation may be substantial—but the assumption of constant E_{oed} and permeability k_z during the consolidation process may also then become untenable.

Let us develop three aspects of the solution of the consolidation equation (7.85). We will consider the simple case of a layer of soil for which the vertical stress has been increased rapidly over an area of large lateral extent, for example by placing fill on the ground surface (Fig 7.41). Since the fill has been placed rapidly at normalised time $T = 0$, there is everywhere an initial excess pore pressure u_i above the static equilibrium value. In other words for $T = 0$, $U = 1$ for all Z. The overlying soil is assumed to be fully drained so that the problem is driven by the reduction of U to zero at $Z = 0$.

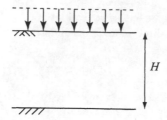

Figure 7.41: One-dimensional consolidation resulting from surcharging of soil layer underlain by impermeable rock

a. Semi-infinite layer

Initially there is a propagation process as the sensation of the reduction of pore pressure at the surface progressively spreads into the body of the clay. The problem is the same as that of suddenly changing the temperature at one end of a long conducting bar for which Carslaw and Jaeger (1959) show that the solution can be written

$$U = \mathrm{erf}\left(\frac{Z}{2\sqrt{T}}\right) \tag{7.86}$$

where $\mathrm{erf}(x)$ is the error function

$$\mathrm{erf}(x) = \frac{2}{\sqrt{\pi}} \int_0^x e^{-t^2}\, dt \tag{7.87}$$

which is tabulated in Carslaw and Jaeger and other standard texts. The variation of this normalised pore water pressure U is shown in Fig 7.42a. The error function is within 1% of 1 for values of the argument greater than about 2. From (7.86) therefore we can deduce that the normalised depth Z_p to which the consolidation front has penetrated at any time T is given by

$$Z_p \approx 4\sqrt{T} \tag{7.88}$$

It is normally more useful to describe the progress of consolidation through the developing settlement at the surface of the consolidating soil. A reference settlement is required: the settlement of a finite layer of characteristic thickness H ($Z = 1$) in which the pore pressure falls from u_i ($U = 1$) to zero and the effective stress increases by a corresponding amount. The reference settlement is then

$$\rho_o = \frac{u_i H}{E_{oed}} \tag{7.89}$$

Settlement of the layer occurs because the soil is becoming more compressed as water is squeezed out at the surface of the layer. The rate at which water leaves the soil is governed, through Darcy's law, by the pore pressure gradient at the surface of the layer. From (7.86) (and Carslaw and Jaeger) this is

$$\frac{u_i}{H}\left(\frac{\partial U}{\partial Z}\right)_{Z=0} = \frac{u_i}{H}\frac{1}{\sqrt{\pi T}} \tag{7.90}$$

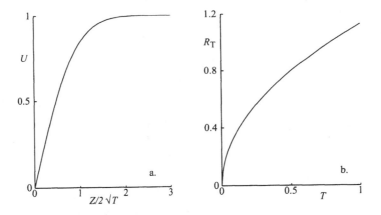

Figure 7.42: One-dimensional consolidation: (a) dimensionless variation of pore pressure with depth; (b) dimensionless variation of degree of consolidation with time

The rate of volume change of the layer, which is the rate of settlement, is then

$$\frac{d\rho}{dt} = \frac{k_z}{\gamma_w} \frac{u_i}{H} \left(\frac{\partial U}{\partial Z} \right)_{Z=0} = \frac{k_z u_i}{\gamma_w} \frac{1}{\sqrt{\pi c_v t}} \tag{7.91}$$

and the total settlement to time t, as a proportion of the settlement of this reference layer, is

$$R_T = \frac{\rho}{\rho_o} = \frac{k_z u_i}{\gamma_w} \frac{E_{oed}}{u_i H} \int_0^t \frac{1}{\sqrt{\pi c_v t}} \, dt = \sqrt{\frac{4T}{\pi}} \tag{7.92}$$

and this is plotted in Fig 7.42b: the degree of consolidation varies with the square root of time[5].

Evidently this analysis will be valid also for a finite clay layer, of thickness $Z = 1$, provided the distance that the consolidation front has penetrated is less than the thickness of the layer or, approximately, from (7.88) $T < 1/16$ or $t < H^2/16 c_v$.

b. Finite layer

The second case to be considered is in a sense the most realistic and for this reason leads to the most complex solution. A finite layer of compressible soil of thickness H is underlain by impermeable rock (Fig 7.41). The pore pressure throughout the layer is initially equal to u_i ($U = 1$ at $t = T = 0$). The governing partial differential equation can be solved using standard techniques to give a

---

[5]Note that U is a local variable describing the normalised pore pressure whereas R_T is a system variable describing the overall settlement of a particular consolidating system.

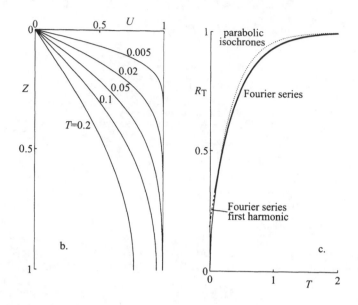

Figure 7.43: One-dimensional consolidation: (a) isochrones of pore pressure from Fourier series solution; (b) degree of consolidation from Fourier series and parabolic isochrone solution

general Fourier series solution:

$$U = \frac{4}{\pi} \sum_{m=0}^{\infty} \left\{ \frac{1}{2m+1} e^{-\pi^2(2m+1)^2 \frac{T}{4}} \sin\left[\frac{\pi}{2}(2m+1)Z\right] \right\} \qquad (7.93)$$

where m is an integer. The degree of consolidation at time T is given by

$$R_T = 1 - \frac{8}{\pi^2} \sum_{m=0}^{\infty} \left[\frac{1}{(2m+1)^2} e^{-\pi^2(2m+1)^2 \frac{T}{4}} \right] \qquad (7.94)$$

This Fourier series solution considers the entire clay layer throughout the analysis and is not really concerned with the propagation of disturbances into the layer from a boundary. It can be used to generate a family of isochrones showing the spatial variation of pore pressure at different times (Fig 7.43a). We observe that it is only during an initial phase that there is an agreement with the form of the general isochrone for the infinite layer that was shown in Fig 7.42a: but the Fourier series solution requires a very large number of terms to be evaluated in order to produce an accurate representation of this stage of the analysis because the series solution struggles to match the physical constraints.

As time goes by, however, the first harmonic becomes dominant (Fig 7.43a) and it is sufficient to consider just the first term of the Fourier series

$$U = \frac{4}{\pi} \sin\frac{\pi Z}{2} e^{-\pi^2 \frac{T}{4}} \qquad (7.95)$$

and

$$R_T = 1 - \frac{8}{\pi^2} e^{-\pi^2 \frac{T}{4}} \tag{7.96}$$

This latter relationship is also plotted in Fig 7.43b together with the exact result (7.94). The first harmonic approximation is obviously in error for low values of T (it suggests that $R_T = 0.189$ for $T = 0$) but is extremely close for T greater than about 0.1.

Thus, in addition to the exact Fourier series solution, which may be cumbersome to use, we have found two solutions which are quite accurate within certain ranges of the problem and which are much simpler to work with. For small times the theory for an infinite layer is appropriate ((7.86), (7.92)); for large times the finite layer theory with a single harmonic to describe the spatial variation of pore pressure is sufficient ((7.95), (7.96)). We learn that we can choose our solution technique and modelling simplification to be appropriate to the problem under consideration.

7.4.1 Parabolic isochrones

The observation that the isochrones have a characteristic shape both in the initial stage of consolidation (Fig 7.42a, eq (7.86)) and in the later stages (Fig 7.43a, eq (7.95)) suggests an alternative approximate approach to the analysis of the consolidation problem which preserves the overall physics of the problem while using simpler mathematical functions.

Let us assume that at all times the pore pressure isochrones have a common geometric shape. We could in principal assume any shape—we have seen that a sine function might be very suitable for the later stages of the problem—but it turns out that the mathematics become particularly simple if we assume that the isochrones are parabolic (Schofield and Wroth, 1968). The analysis then proceeds in two stages—in a sense, linking the infinite layer and the single harmonic solutions.

The pore pressure is assumed to vary parabolically with distance from the drainage boundary (Fig 7.44). The effective stresses are given by the difference between the applied total stress and this pore pressure so that volume change of the clay is linked with the area above the parabolic isochrone through the one-dimensional stiffness E_{oed}. The rate at which water flows out of the clay is controlled by the slope of the parabolic isochrone at the drainage boundary. The problem then reduces to a simple differential equation, deduced from the geometry of the parabola, linking the rate of change of volume of the soil with the rate at which water flows out of the soil.

There are two stages to the analysis of consolidation using parabolic isochrones: after the pore pressure falls to zero at the drainage boundary the effect of this drainage propagates steadily into the clay (stage 1: Fig 7.44a, c). The maximum pore pressure is equal to the applied total stress. Once the drainage 'shock' has reached the opposite side of the clay (assumed to be an impermeable surface—which could be a plane of symmetry in a doubly drained block of soil) the pore pressure at this impermeable boundary steadily reduces (stage 2: Fig 7.44b, c).

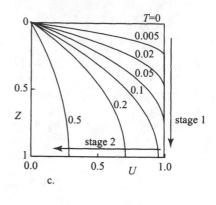

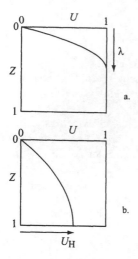

Figure 7.44: One-dimensional consolidation: parabolic isochrones: (a) stage 1: propagation of boundary drainage into layer; (b) stage 2: fall in pore pressure at undrained boundary; (c) isochrones for combined solution

During the first stage, the effect of the reduction of the pore pressure at the top boundary is allowed to propagate into the clay. The parabolic isochrone is always tangential to the line $U = 1$ (Fig 7.44a). We use our physical understanding of the overall consolidation process to derive the governing equation for the current location λH of the consolidation front. Just as for the calculation of the degree of consolidation for the finite layer analysis, the volume change of the clay, and hence the settlement of the clay, result from the change in effective stress

$$\rho = \frac{H u_i}{E_{oed}} \int_0^\lambda (1 - U)\, \mathrm{d}Z \tag{7.97}$$

From the geometry of a parabola this can be directly written

$$\rho = \frac{H u_i}{E_{oed}} \frac{\lambda}{3} \tag{7.98}$$

and the rate of change of settlement is therefore

$$\frac{\mathrm{d}\rho}{\mathrm{d}t} = \frac{H u_i}{3 E_{oed}} \frac{\mathrm{d}\lambda}{\mathrm{d}t} \tag{7.99}$$

From conservation of mass this rate of settlement must be equal to the rate at which water is leaving the top surface of the clay, which is given by Darcy's law applied to the exit gradient of the parabolic isochrone:

$$\frac{\mathrm{d}\rho}{\mathrm{d}t} = \frac{k_z}{\gamma_w} \frac{2 u_i}{H \lambda} \tag{7.100}$$

Equating (7.99) and (7.100), solving the differential equation, and substituting for T we find

$$\lambda = \sqrt{12T} \qquad (7.101)$$

which is directly comparable with the approximate expression (7.88), and

$$R_T = \sqrt{\frac{4T}{3}} \qquad (7.102)$$

which is remarkably close to the exact expression (7.92) for the infinite layer: it is also plotted in Fig 7.43b. This first stage of consolidation continues until $\lambda = 1$ (and $R_T = 1/3$) at $T = 1/12$.

Then a second regime takes over (Fig 7.44b, c). The geometry of the parabola remains the same but now it is the pore pressure U_H at the base of the layer where the isochrone is always vertical (implying zero flow at an impermeable boundary) that varies with time. Now the single sinusoidal harmonic of the Fourier series solution has been replaced by a single parabolic curved isochrone. The physical reasoning is exactly the same as before leading to the governing equation

$$\frac{dU_H}{dT} = -3U_H \qquad (7.103)$$

with the solution

$$U_H = e^{-3\left(T - \frac{1}{12}\right)} \qquad (7.104)$$

and

$$R_T = 1 - \frac{2}{3}e^{-3\left(T - \frac{1}{12}\right)} \qquad (7.105)$$

This variation of degree of consolidation with time is included in Fig 7.43b.

We conclude that we can capture the essence of the consolidation problem either by using an exact analysis and recognising the different regimes of response, or by standing one step back from the exact equation and adopting a simpler mathematical description which is still strongly physically based but which looks at the physics of the whole system rather than of the individual components: the overall physical process can then be followed rather clearly.

These stratagems that are adopted to ease the solution are of course quite separate from the assumptions that underlie the consolidation theory itself.

7.4.2 General power law approximate solution

Parabolic isochrones were chosen by Schofield and Wroth because of their geometric simplicity and because they match well the exact solution of (7.85). Let us explore the use of a general power law variation of pore pressure:

$$U = 1 - (1 - \frac{Z}{\lambda})^n \qquad (7.106)$$

or

$$u = u_i \left[1 - \left(\frac{\ell - z}{\ell}\right)^n\right] \qquad (7.107)$$

During the first stage of consolidation (Fig 7.44), U is scaled by a pore pressure u_i equal to the applied total stress σ_v and $\ell = \lambda H$ is the distance into the clay to which the drainage front has propagated. During the second stage (Fig 7.44), u_i is the pore pressure at the distant impermeable boundary ($= U_H \sigma_v$) and ℓ is equal to H, the full drainage path length or thickness of the clay. For all values of $n > 1$ the form of (7.107) preserves the physical requirement to have zero slope at $z = \ell$ where there is no flow and zero value at $z = 0$ where there is free drainage. The parabolic isochrones are given by $n = 2$.

During the first stage the solution of the mass balance equation gives:

$$\lambda = \sqrt{\frac{2n(n+1)c_v t}{H^2}} = \sqrt{2n(n+1)T} \qquad (7.108)$$

where $T = c_v t/H^2$ is the usual normalised non-dimensional time. The degree of consolidation R_T, defined as the ratio of current average effective stress to eventual average effective stress (or the proportion of long term volume change that has occurred in the clay) is:

$$R_T = \frac{1}{n+1} \lambda = \sqrt{\frac{2n}{n+1} T} \qquad (7.109)$$

These equations apply until $\lambda = 1$ and $T = 1/[2n(n+1)]$.

During the second stage it is the central pore pressure u_i that is changing. The solution is:

$$\frac{u_i}{\sigma_v} = U_H = e^{-[(n+1)T - 1/2n]} \qquad (7.110)$$

and the degree of consolidation is given by:

$$R_T = 1 - \frac{n}{n+1}\frac{u_i}{\sigma_v} = 1 - \frac{n}{n+1}e^{-[(n+1)T - 1/2n]} \qquad (7.111)$$

The effect of n on the rate of consolidation is shown in Fig 7.45. The higher n the faster the flow from the sample and the faster the consolidation. Closest agreement with the exact theoretical solution is in fact obtained with $n = 1.5$.

7.4.3   Pore pressures for stability analysis of embankment on soft clay

When embankments are constructed over soft deposits of cohesive soils, it is short term stability during construction that usually governs the rate of construction: in the long term, pore pressure dissipation leads to beneficial transfer of load to effective stress in the soft soils and hence increased strength.

Stability analyses can be performed using any one of very many commercial packages and can be assessed in terms of either total or effective stresses. In terms of total stresses, the current profile of undrained strength of the soil is required: to determine the stability at different stages during construction the extent of the transfer of embankment load to undrained strength is needed. It might be proposed that once the vertical effective stress exceeds the preconsolidation pressure a further increase in vertical effective stress of $\Delta\sigma'_v$ produces a

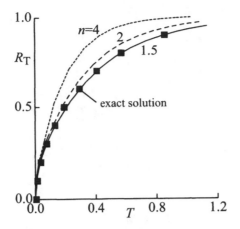

Figure 7.45: Effect of exponent n on rate of consolidation

consequent increase in undrained strength of about $0.2\Delta\sigma'_v$ (Muir Wood, 1990). Some means of estimating the increase in vertical effective stress is therefore required.

In terms of effective stresses, the strength properties do not change during construction (assuming that the construction process and subsequent consolidation do not produce any significant change in the fabric or cementation of the soil): it is the current distribution of pore pressure through the soft soil that controls the stability. Some means of estimating this distribution of pore pressure is required. We are not concerned here with the relative merits of total and effective stress analyses, nor with analysis of stability of embankments on soft clays. The intention is to explore rather simple ways in which the degree of consolidation and hence pore pressure generation and effective stress development in a soft clay foundation can be estimated for an embankment which is not simply wished into place in zero time but for which some drainage is expected and indeed required to occur during construction in order to ensure stability at all stages. The procedure that is developed could be very easily extended to stage construction in which phases of embankment construction are separated by rest periods in which consolidation occurs.

Up to now we have assumed that the total applied vertical stress σ_v is constant during consolidation. However, if we are concerned to estimate the pore pressures that develop under an embankment during construction then the vertical stress must be allowed to vary. We will assume that the construction rate is sufficiently slow that the clay is at all times in its 'stage two' state with the isochrone having reached the distant impermeable boundary (Fig 7.44b). The governing equation now becomes:

$$\frac{n}{n+1}\frac{du_i}{dt} + \frac{nc_v}{H^2}u_i = \frac{d\sigma_v}{dt} = \xi \tag{7.112}$$

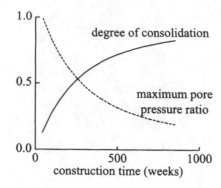

Figure 7.46: One-dimensional consolidation with increasing load

It will be assumed that the rate of increase of vertical stress ξ is maintained constant. Since in general the final height of the embankment—and final value of vertical stress σ_c—will be fixed, it will be convenient to work in terms of a construction time t_c:

$$t_c = \frac{\sigma_c}{\xi} \tag{7.113}$$

and corresponding dimensionless construction time:

$$T_c = \frac{c_v \sigma_c}{\xi H^2} \tag{7.114}$$

The solution for the pore pressure u_c at the end of construction is then:

$$\frac{u_c}{\sigma_c} = \frac{1}{nT_c} \left[1 - e^{-(n+1)T_c} \right] \tag{7.115}$$

with corresponding degree of consolidation

$$R_T = 1 - \frac{n}{n+1} \frac{u_c}{\sigma_c} \tag{7.116}$$

Take as an *example* construction of an embankment of height 8 m of fill with unit weight 22 kN/m^3 (so that $\sigma_c = 176$ kPa) on a layer of soft clay of thickness $H = 8$ m and with coefficient of consolidation $c_v = 10^{-7}$ m^2/s. The value of n is taken as 2 corresponding to parabolic isochrones. The variation of maximum pore pressure and degree of consolidation with construction time (shown in weeks) is shown in Fig 7.46. For any construction time less than about 42 weeks the maximum pore pressure will be equal to the applied total stress (the full total pressure applied by the embankment) and equation (7.112) is no longer appropriate. The degree of consolidation at the end of construction in this case is very low (less than 15%) and the remaining pore pressures—which will tend to reduce stability—high. If it were estimated, from separate slope stability analyses, that an average degree of consolidation of 70-80% (say)

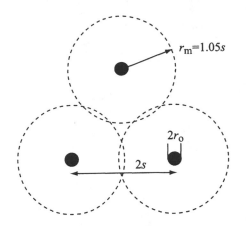

Figure 7.47: Radius of influence r_m of drains in triangular array

were required in order to bring the stability to an acceptable level then this would imply construction times of between 450 and 750 weeks which would be unlikely to be acceptable for typical road construction projects. (Of course the excess pore pressures will be lower and the effective stresses will be greater near the drainage boundary at the surface of the clay so that the effective degree of consolidation for shallow failure mechanisms will be locally higher than that implied by Fig 7.46.)

The one-dimensional analysis is not able to accommodate the variation of vertical stress that occurs under the side slopes of the embankment but serves to indicate that for these values of geometric and soil parameters it is likely that some other measures will be required in order to speed up construction. An obvious way of doing this is to provide an array of vertical drains under the embankment.

Assume that a triangular array of vertical drains has been installed. Each drain is working perfectly and radial drainage to drains dominates over vertical flow. The drains are of radius r_o and the radius of influence of each drain is r_m (Fig 7.47). Typical semi-empirical rules (Leroueil *et al.*, 1985) suggest that for a triangular array of vertical drains at spacing s the radius of influence for calculation of progress of consolidation should be $r_m \approx 1.05s$. The total vertical stress σ_v on the clay is uniform over the region of influence of each drain but it is assumed that the drainage is so effective that pore pressures in each cell around a particular drain are not influenced by the pore pressures around neighbouring drains.

The analysis assumes that the deformation of the soil is one-dimensional, with only vertical movement, but that drainage occurs radially to the vertical drains. This is obviously an approximation: more detailed analyses of related problems are reported by Al-Tabbaa and Muir Wood (1991).

The consolidation is governed by radial flow to the drain. As in §7.4.2 we consider a power law relationship for the spatial variation of excess pore pressure

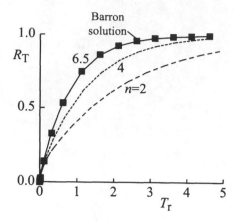

Figure 7.48: One-dimensional consolidation with radial flow: effect of exponent n $(r_m/r_o = 10.5)$

with radius and again consider the two stages of consolidation: in the first, the drainage 'shock' propagates into the clay from the drain; in the second, the pore pressure at the distant impermeable boundary progressively reduces. The pore pressure is now given by:

$$\left(\frac{u}{u_i}\right) = 1 - \left(\frac{\lambda r_m - r}{\lambda r_m - r_o}\right)^n \tag{7.117}$$

where r is the radius. In the first stage u_i is equal to the total stress σ_v and λ increases steadily from r_o/r_m to 1; in the second stage u_i is the pore pressure at the impermeable boundary and $\lambda = 1$.

For this problem it is convenient to define a dimensionless time T_r:

$$T_r = \frac{c_v t}{r_m^2} \tag{7.118}$$

Then, for the first stage, the solution is given with λ as linking variable:

$$T_r = \frac{[(\lambda r_m/r_o) - 1]^2 \, [4(\lambda r_m/r_o) + (3n + 2)]}{6n(n + 1)(n + 2) \, (r_m/r_o)^2} \tag{7.119}$$

and degree of consolidation

$$R_T = \frac{2 \, [(\lambda r_m/r_o) - 1] \, [(n + 1) + (\lambda r_m/r_o)]}{(n + 1)(n + 2) \, [(r_m/r_o)^2 - 1]} \tag{7.120}$$

During the second stage, the solution is given with u_i/σ_v as the linking variable:

$$\frac{u_i}{\sigma_v} = \exp\left\{-\frac{2(n + 1)(n + 2)T_r}{[1 - (r_o/r_m)]^2 \, [(n + 1) + (n + 3)(r_m/r_o)]}\right\} \tag{7.121}$$

$$R_T = 1 - \frac{u_i}{\sigma_v} \frac{n\left[(n+1) + (n+3)(r_m/r_o)\right]}{(n+1)(n+2)\left[1 + (r_m/r_o)\right]} \tag{7.122}$$

The complete solution is shown in Fig 7.48 for different values of the exponent n for $r_m/r_o = 10.5$—this value being taken as relevant to typical field applications with drain spacing of 1 m on a triangular grid (Leroueil *et al.*, 1985).

Whereas the use of parabolic isochrones ($n = 2$) has been found to be satisfactory for the one-dimensional situation, for the radial flow situation it might be expected that it would fail to acknowledge the variation of flow area with radius. It is expected that the concentration of flow towards the drain will lead to higher radial pore pressure gradients close to the drain, implying that the ideal value of n will be greater than 2.

To choose a value of n, a full analysis of consolidation with radial flow is in principle required. However, formulae exist from which degree of consolidation around vertical drains can be calculated. Barron's formula (Barron (1947) quoted by Leroueil *et al.* (1985)) indicates that:

$$R_T = 1 - e^{-2T_r/F} \tag{7.123}$$

where F is a function of drain geometry given by:

$$F = \frac{\ln(r_m/r_o)}{1 - (r_o/r_m)^2} - \frac{1}{4}\left[3 - (r_o/r_m)^2\right] \tag{7.124}$$

This expression for degree of consolidation R_T can be used to optimise the selection of the exponent n for our power law expression for radial pore pressure variation. It is found that, for $r_m/r_o = 10.5$ the optimum value of n is about 6.5 (Fig 7.48). Over the range of values of r_m/r_o between 10 and 30 a reasonable estimate of the value of n to make expressions (7.120) and (7.122) conform to expression (7.123) is given by:

$$n \approx 4 + \frac{r_m}{4r_o} \tag{7.125}$$

Having chosen a value of the exponent n to give a good fit to the theoretical variation of degree of consolidation under a constant applied load we can repeat the analysis for varying applied load in just the same way as for the one-dimensional situation. The solution—assuming that the pore pressure at the impermeable distant boundary is always less than the applied vertical total stress σ_c and that the embankment loading σ_c has been applied in time t_c—is given by

$$\frac{u_c}{\sigma_c} = \frac{1}{\beta T_{rc}}\left[1 - e^{-\alpha T_{rc}}\right] \tag{7.126}$$

where the dimensionless construction time is

$$T_{rc} = \frac{c_v t_c}{r_m^2} \tag{7.127}$$

and

$$\alpha = \frac{2(n+1)(n+2)}{\left[1 - (r_o/r_m)\right]^2\left[(n+1) + (n+3)(r_m/r_o)\right]} \tag{7.128}$$

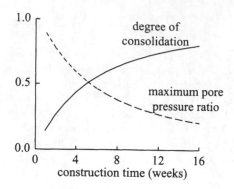

Figure 7.49: One-dimensional consolidation with radial flow and increasing load

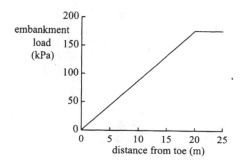

Figure 7.50: Variation of embankment loading with distance from toe

$$\beta = \frac{2n}{[1 - (r_o/r_m)]^2 \, [(r_m/r_o) + 1]} \tag{7.129}$$

The degree of consolidation at the end of construction is given by (7.122) with u_i and σ_v replaced by u_c and σ_c respectively.

Results are shown in Fig 7.49 for the embankment loading considered previously, assuming that the coefficient of consolidation for radial flow is the same as for vertical flow (this will usually be conservative) and assuming drain radius $r_o = 0.05$ m and radius of influence r_m given by $r_m/r_o = 10.5$. Degrees of consolidation between 70 and 80% can now be obtained for construction times between 10 and 16 weeks.

Because of our assumption about local action of the drains we can now go further and calculate spatial variation of excess pore pressures under the sloping edge of the embankment. These pore pressures will be constant with depth but will vary with horizontal position. The local applied embankment load is assumed to vary linearly with distance from the toe of the embankment, with a maximum embankment height of 8 m and side slopes 1:2.5 (Fig 7.50).

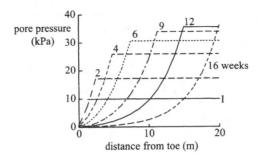

Figure 7.51: Variation of pore pressure with position at different times during construction

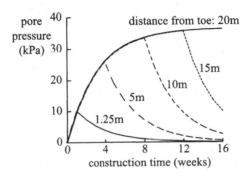

Figure 7.52: Variation of pore pressure with time, during construction, at different distances from toe of embankment

For a total construction time of 16 weeks the time taken for the embankment load to reach its maximum value will also vary linearly with position. For each vertical section, the pore pressure u_c when the local maximum load is reached can be calculated from (7.126). As construction of the embankment continues, no further load is applied to this section and the pore pressure decays according to an adaptation of (7.121) which can be written:

$$\frac{u_i}{u_c} = e^{-\alpha(T_r - T_{rc})} \tag{7.130}$$

where the dimensionless time T_{rc} is the time at which the local maximum load was applied.

Fig 7.51 shows the variation of pore pressure with position for a total construction time of 16 weeks. The effect of the drains in dissipating the pore pressure towards the edge of the embankment is apparent. Clearly this will have a beneficial effect on the stability of the embankment and it is these excess pore pressures that could be used as input to effective stress stability analyses.

The variation with time of the pore pressure at different locations under the edge of the embankment is shown in Fig 7.52: the pore pressure generation curve is the same at all locations but the pore pressures in the clay nearer the edge of the embankment start to dissipate as soon as no further load is being applied.

An analysis has been presented of the generation and dissipation of pore pressures during the construction of an embankment on soft clay. A power law relationship for spatial variation of pore pressures leads to simple expressions for the pore pressures which could be incorporated into stability analyses to provide charts for site control of rates of construction. The value of the exponent which must be used in the power law for analysis of consolidation with radial flow is higher than that required for one-dimensional flow where parabolic isochrones have been found to be adequate.

This analysis pulls in simplifying mathematical statements, that have been found to be useful for other consolidation problems, and provides an example of the adaptation of an existing modelling tool for a new application. More detail is given by Muir Wood (1999a).

7.5   Macroelement models

Numerical modelling using a full finite element or finite difference analysis may be ultimately necessary but may be a heavy-handed way of seeking insight into some aspects of a problem of geotechnical behaviour. Theoretical modelling may only be possible for rather restricted problems. Macroelement modelling may be a helpful intermediate way of introducing some realistic geotechnical non-linearity in order, for example, to compare different constitutive possibilities or perhaps just to provide a rapid 'order-of-magnitude' estimate of response against which the results of more extensive numerical modelling—or physical modelling—can be compared. Equally, physical or numerical modelling may itself provide clues concerning mechanisms of system response which may suggest ways in which simple macroelement models might be devised. It will be seen that this has indeed been the route for the development of some of the macroelement models outlined here.

7.5.1   Box model

In section §3.9 it was shown that, in selecting values of soil parameters for different soil models, there is usually more than one way of describing the same set of experimental data. Here we will show consequences of that lack of uniqueness when applied to a simple boundary value problem, especially when the stress and strain paths followed in the boundary value problem diverge from the stress and strain path followed in the laboratory test from which the soil parameters were determined.

The boundary value problem that we will use will be a very simple 'box model' (see also Nordal, 1983). This model consists of two square elements A (active) and P (passive) separated by a smooth vertical interface and contained within a rigid smooth box (Fig 7.53). The variables are therefore:

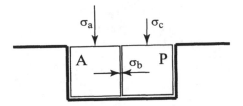

Figure 7.53: Two element 'box model'

- the vertical stress on element A: σ_a;

- the vertical strain for element A: ϵ_a;

- the vertical stress on element P: a surcharge σ_c;

- the vertical strain for element P: ϵ_c;

- the horizontal strains for element A and element P (these are equal and opposite): ϵ_b;

- and the horizontal stress across the interface between the two elements A and P: σ_b.

The imposed boundary conditions are these:

- plane strain: $\delta\epsilon_y = 0$, where y is the out of plane direction;

- constant surcharge on element P: $\delta\sigma_c = 0$; and

- increasing vertical compression strain imposed on element A: $\delta\epsilon_a > 0$.

Given these boundary conditions the changes in all other variables are dependent on $\delta\epsilon_a$ through the constitutive response of the soil elements.

Several constitutive models were presented in Chapter 3 in terms of stiffness relationships linking strain changes (the assumed independent variables) with resulting stress changes (the assumed dependent variables). For the two elements we can therefore define stiffness matrices D_A and D_P and then write down expressions linking changes in strains and changes in stresses for our box model:

$$\delta\sigma_a = D_{A_{11}}\delta\epsilon_a + D_{A_{12}}\delta\epsilon_b \qquad (7.131)$$

$$\delta\sigma_b = D_{A_{21}}\delta\epsilon_a + D_{A_{22}}\delta\epsilon_b \qquad (7.132)$$

$$\delta\sigma_c = D_{P_{11}}\delta\epsilon_c - D_{P_{12}}\delta\epsilon_b = 0 \qquad (7.133)$$

$$\delta\sigma_b = D_{P_{21}}\delta\epsilon_c - D_{P_{22}}\delta\epsilon_b \qquad (7.134)$$

These equations can be solved to give explicit expressions for all the strain and stress increments in terms of $\delta\epsilon_a$ (the independent variable) and the components of the two stiffness matrices:

$$\frac{\delta\sigma_a}{\delta\epsilon_a} = \frac{D_{A_{11}}D_{P_{12}}D_{P_{21}} + D_{P_{11}}D_{A_{12}}D_{A_{21}} - D_{A_{11}}D_{P_{11}}\left(D_{A_{22}} + D_{P_{22}}\right)}{D_{P_{12}}D_{P_{21}} - D_{P_{11}}\left(D_{A_{22}} + D_{P_{22}}\right)} \quad (7.135)$$

$$\frac{\delta\epsilon_b}{\delta\epsilon_a} = \frac{D_{P_{11}}D_{A_{21}}}{D_{P_{12}}D_{P_{21}} - D_{P_{11}}\left(D_{A_{22}} + D_{P_{22}}\right)} \quad (7.136)$$

$$\frac{\delta\sigma_b}{\delta\epsilon_a} = \frac{D_{A_{21}}\left(D_{P_{12}}D_{P_{21}} - D_{P_{11}}D_{P_{22}}\right)}{D_{P_{12}}D_{P_{21}} - D_{P_{11}}\left(D_{A_{22}} + D_{P_{22}}\right)} \quad (7.137)$$

$$\frac{\delta\epsilon_c}{\delta\epsilon_a} = \frac{D_{P_{12}}D_{A_{21}}}{D_{P_{12}}D_{P_{21}} - D_{P_{11}}\left(D_{A_{22}} + D_{P_{22}}\right)} \quad (7.138)$$

For an elastic material, the values of the components of the stiffness matrix, $D_{A_{11}}$, etc., are constant and the stress and strain increments are at all times directly proportional. Once we have an elastic-plastic description of soil behaviour we know that the stiffness depends on the direction of the strain increment (which may or may not be causing yielding) and will, for a hardening model, vary nonlinearly with stress state.

For the present box model a simple explicit solution procedure has been adopted because this can be readily implemented. The loading is divided into increments of strain $\delta\epsilon_a$. For each step the strain, and hence stress, increments that would occur if the strain increments were imposed purely elastically (using the current elastic stiffness properties) are calculated. If these new stresses lie within the present yield locus for either element then only the elastic parts of the components of the corresponding stiffness matrix are used in subsequent calculations. If these new stresses lie outwith the present yield locus for either element then the corresponding stiffness matrix contains both elastic and plastic components. Appropriate elastic or elastic-plastic stiffness matrices are used to calculate new stresses and strains. The process is repeated for the next step.

Such an explicit procedure works well provided the constitutive response is not too nonlinear. Each time the stiffness is calculated using the *current* stresses and used to predict future stresses the stiffnesses may be very different and the stiffnesses may actually vary significantly over the increment. A smaller step size can help to overcome nonlinear effects to some extent but a more subtle numerical scheme would in some way introduce an average stiffness for the increment.

We concentrated, in our presentation of soil models in Chapter 3, on the axisymmetric stress and strain conditions which can be reached in a conventional triaxial apparatus. We are now working in plane strain and need to modify our expressions slightly. For triaxial states of stress we worked in terms of mean stress p' and deviator or distortional stress q. Under plane strain conditions we can define somewhat similar stress variables which include only the stresses in the plane of shearing: a mean stress $s' = (\sigma_1' + \sigma_3')/2 = (\sigma_v' + \sigma_h')/2$ and a shear stress $t = (\sigma_1' - \sigma_3')/2 = (\sigma_v' - \sigma_h')/2$ where σ_1' and σ_3' are major and minor principal stresses. For the special case of our elements A and P, where the principal axes do not rotate, it is convenient to assign them to vertical and horizontal stresses σ'_v and σ'_h respectively, allowing a richer plotting of positive and negative values of shear stress t. The corresponding work conjugate strain increment quantities are volumetric strain $\delta\epsilon_p = \delta\epsilon_v + \delta\epsilon_h$ and shear strain $\delta\gamma = \delta\epsilon_v - \delta\epsilon_h$.

Given stress:strain relationships expressed in terms of mean stress, shear stress, volumetric strain and shear strain we can convert them to relationships expressed in terms of normal stresses and strains—as required for our box model—using the transformations

$$\begin{pmatrix} \delta\epsilon_p \\ \delta\gamma \end{pmatrix} = \begin{pmatrix} 1 & 1 \\ 1 & -1 \end{pmatrix} \begin{pmatrix} \delta\epsilon_v \\ \delta\epsilon_h \end{pmatrix} \tag{7.139}$$

$$\begin{pmatrix} \delta\sigma'_v \\ \delta\sigma'_h \end{pmatrix} = \begin{pmatrix} 1 & 1 \\ 1 & -1 \end{pmatrix} \begin{pmatrix} \delta s' \\ \delta t \end{pmatrix} \tag{7.140}$$

Then, if we have a relationship

$$\begin{pmatrix} \delta s' \\ \delta t \end{pmatrix} = \begin{pmatrix} \Delta_{11} & \Delta_{12} \\ \Delta_{21} & \Delta_{22} \end{pmatrix} \begin{pmatrix} \delta\epsilon_p \\ \delta\gamma \end{pmatrix} \tag{7.141}$$

this becomes

$$\begin{pmatrix} \delta\sigma'_v \\ \delta\sigma'_h \end{pmatrix} =$$

$$\begin{pmatrix} [\Delta_{11} + \Delta_{12} + \Delta_{21} + \Delta_{22}] & [\Delta_{11} - \Delta_{12} + \Delta_{21} - \Delta_{22}] \\ [\Delta_{11} + \Delta_{12} - \Delta_{21} - \Delta_{22}] & [\Delta_{11} - \Delta_{12} - \Delta_{21} + \Delta_{22}] \end{pmatrix} \begin{pmatrix} \delta\epsilon_v \\ \delta\epsilon_h \end{pmatrix} \tag{7.142}$$

Elastic soil model

To generate the elastic stiffness matrix—which will of course be required anyway in order to calculate the elastic-plastic stiffness matrices for elastic-plastic soil models—we need to introduce the plane strain constraint in order to calculate the out of plane stress increment $\delta\sigma'_y$. Writing Hooke's law for the zero strain increment in the out-of-plane direction gives

$$\delta\sigma'_y = 2\nu\delta s' \tag{7.143}$$

In terms of normal stresses and strains, the elastic stiffness relationship is then

$$\begin{pmatrix} \delta\sigma'_v \\ \delta\sigma'_h \end{pmatrix} = \frac{E}{(1-2\nu)(1+\nu)} \begin{pmatrix} 1-\nu & \nu \\ \nu & 1-\nu \end{pmatrix} \begin{pmatrix} \delta\epsilon_v \\ \delta\epsilon_h \end{pmatrix} = \begin{pmatrix} K+4G/3 & K-2G/3 \\ K-2G/3 & K+4G/3 \end{pmatrix} \begin{pmatrix} \delta\epsilon_v \\ \delta\epsilon_h \end{pmatrix} \tag{7.144}$$

so that if the soil behaviour is elastic

$$D_{11} = D_{22} = K + 4G/3 \tag{7.145}$$

and

$$D_{12} = D_{21} = K - 2G/3 \tag{7.146}$$

for both active and passive elements.

In terms of mean stress and shear stress and corresponding strains the elastic stiffness relationship is

$$
\begin{pmatrix} \delta s' \\ \delta t \end{pmatrix} = \begin{pmatrix} K + G/3 & 0 \\ 0 & G \end{pmatrix} \begin{pmatrix} \delta \epsilon_p \\ \delta \gamma \end{pmatrix} \tag{7.147}
$$

demonstrating clearly the decoupling of volumetric and distortional effects. Comparing this with the equivalent expression for axisymmetric stress and strain quantities

$$
\begin{pmatrix} \delta p' \\ \delta q \end{pmatrix} = \begin{pmatrix} K & 0 \\ 0 & 3G \end{pmatrix} \begin{pmatrix} \delta \epsilon_p \\ \delta \epsilon_q \end{pmatrix} \tag{7.148}
$$

we can propose that transformation from axial symmetry to plane strain will require the use of a pseudo-bulk modulus $K^* = K + G/3$ in place of bulk modulus K and the use of G in place of $3G$.

The relationships for the box model can then be deduced

$$
\frac{\delta \sigma_a}{\delta \epsilon_a} = \frac{2E(1 - \nu)}{(1 + \nu)(2 - 4\nu + \nu^2)} \tag{7.149}
$$

$$
\frac{\delta \epsilon_b}{\delta \epsilon_a} = \frac{-\nu(1 - \nu)}{(2 - 4\nu + \nu^2)} \tag{7.150}
$$

$$
\frac{\delta \sigma_b}{\delta \epsilon_a} = \frac{\nu E}{(1 + \nu)(2 - 4\nu + \nu^2)} \tag{7.151}
$$

$$
\frac{\delta \epsilon_c}{\delta \epsilon_a} = \frac{-\nu^2}{(2 - 4\nu + \nu^2)} \tag{7.152}
$$

Elastic-perfectly plastic Mohr-Coulomb soil model

For a soil with frictional strength given by an angle of friction ϕ and with yielding (failure) accompanied by plastic dilation controlled by an angle of dilation ψ it can be shown that the limiting stress ratios in terms of our plane strain stress variables are m for active failure and $-m$ for passive failure

$$
\frac{t}{s'} = \pm m = \pm \sin \phi \tag{7.153}
$$

and the corresponding plastic strain increment ratio, which again changes sign for active and passive conditions, is

$$
-\frac{\delta \epsilon_p^p}{\delta \gamma^p} = \pm m^* = \pm \sin \psi \tag{7.154}
$$

Through these angles we can readily rewrite the elastic-plastic stiffness relationship, previously deduced for axisymmetric conditions (§3.3.4), for our plane strain variables

$$
\begin{pmatrix} \delta s' \\ \delta t \end{pmatrix} =
$$

$$
\left[\begin{pmatrix} K^* & 0 \\ 0 & G \end{pmatrix} - \frac{1}{(mm^*K^* + G)} \begin{pmatrix} mm^*K^{*2} & \mp m^*GK^* \\ \mp mGK^* & G^2 \end{pmatrix} \right] \begin{pmatrix} \delta \epsilon_p \\ \delta \gamma \end{pmatrix} \tag{7.155}
$$

Elastic-hardening plastic Mohr-Coulomb soil model

Once again we can move between the equations for the axisymmetric model (§3.4.1) and the equations for the plane strain model using angles of friction and using expressions similar to (7.153) to link plane strain stress ratios with angles of friction. The plasticity part of the hardening model was defined in terms of: peak angle of friction ϕ_p; critical state angle of friction ϕ_{cv}; strain parameter a.

Let us introduce the symbol $\zeta = t/s'$ to indicate the plane strain stress ratio. Then the peak stress ratio becomes $\zeta_p = \sin\phi_p$ and the critical state stress ratio is $m = \sin\phi_{cv}$.

Writing ζ_y for the yield value of ζ, and assuming that the value of a deduced from triaxial tests remains valid for plane strain conditions, the elastic-plastic stiffness relationship becomes

$$\begin{pmatrix} \delta s' \\ \delta t \end{pmatrix} =$$

$$\left[\begin{pmatrix} K^* & 0 \\ 0 & G \end{pmatrix} - \frac{\begin{pmatrix} -K^{*2}\zeta_y(m-\zeta_y) & \pm GK^*(m-\zeta_y) \\ \mp GK^*\zeta_y & G^2 \end{pmatrix}}{G - K^*\zeta_y(m-\zeta_y) + s(\zeta_p - \zeta_y)^2/(a\zeta_p)} \right] \begin{pmatrix} \delta\epsilon_p \\ \delta\gamma \end{pmatrix} \quad (7.156)$$

Cam clay

For Cam clay (§3.4.2), under conditions of axial symmetry, the bulk modulus is found to be dependent on mean stress and specific volume according to the relationship (3.141)

$$K = \frac{vp'}{\kappa} \quad (7.157)$$

For our plane strain analyses, we can write

$$K^* = \frac{vs'}{\kappa} + \frac{G}{3} \quad (7.158)$$

If we define the model through an elliptical yield locus in the $s' : t$ effective stress plane with geometry defined by a soil parameter m given by $m = \sin\phi_{cv}$, and assume associated plastic flow, then the form of the elastic-plastic stiffness matrix can be deduced directly from the corresponding matrix for axial symmetry (3.164), replacing mean stress p' by plane strain mean stress s' and deviator stress q by shear stress t. For compactness it is convenient to introduce combined variables

$$\mu_1 = 2s' - s'_o \quad (7.159)$$

$$\mu_2 = \frac{t}{m^2} \quad (7.160)$$

Then the stiffness relationship becomes

$$
\begin{pmatrix} \delta s' \\ \delta t \end{pmatrix} = \left[\begin{pmatrix} K^* & 0 \\ 0 & G \end{pmatrix} - \frac{\begin{pmatrix} (\mu_1 K^*)^2 & 2\mu_1\mu_2 GK^* \\ 2\mu_1\mu_2 GK^* & (2\mu_2 G)^2 \end{pmatrix}}{K^*\mu_1^2 + 4G\mu_2^2 + vs's_o'\mu_1/(\lambda - \kappa)} \right] \begin{pmatrix} \delta\epsilon_p \\ \delta\gamma \end{pmatrix}
$$

(7.161)

The use of a very small strain increment becomes especially important with Cam clay. The critical state condition acts as an attractor to which the state of the soil tends at very large strain. However, if the strain increment is too large it is possible (with the very simple explicit integration scheme adopted here) for the state apparently to hop inadmissibly across the critical state. This becomes very obvious when the results are inspected.

Model calibration

We proceed to compare the box model responses of the three elastic-plastic models. Calibration in this exercise becomes a synthetic process of matching the three models for a notional plane strain compression test on normally consolidated soil. This matching is achieved by trial and error in terms of the relationships between shear stress t and shear strain γ and between volumetric strain ϵ_p and shear strain γ. The results are shown in Fig 7.54.

Obviously the elastic-perfectly plastic model (MC) can only provide a very approximate match to the steady nonlinearity of the response—matching the final stress state and a general average initial stiffness. This model is given a small negative angle of dilation so that the continuing plastic compression of the soil can be somewhat incorporated. The hardening Mohr-Coulomb model (MCH) and the Cam clay model (CC) are both able to present rather equivalent degrees of nonlinearity. The Mohr-Coulomb model has a hyperbolic hardening relationship. The Cam clay model has a more elaborate internal structure. In order to give a reasonable match in the early stages of the test it proves necessary to give the Mohr-Coulomb model a slightly higher angle of friction. This will have subsequent consequences for the response of the box model. The soil parameters chosen for the three models are listed in Table 7.4.

Box model performance

The comparison of the response of the three models for the simple two element box model is shown in Fig 7.55 for an initial value of $K_o = 1$ and in Fig 7.56 for an initial value of $K_o = 0.6$. In each figure the response is shown at two scales in order to show both the overall path to ultimate failure (Figs 7.55a, b and 7.56a, b) and the initial stages of loading (Figs 7.55c, d and 7.56c, d).

For all three soil models the ultimate footing stress can be calculated (Figs 7.55a and 7.56a): each of the models has a frictional strength so that the limiting stress ratio in each element is given by the passive pressure coefficient. The theoretical value of the ultimate footing stress should be

$$
\frac{\sigma_a}{\sigma_c} = K_p^2 = \left(\frac{1 + \sin\phi}{1 - \sin\phi} \right)^2
$$

(7.162)

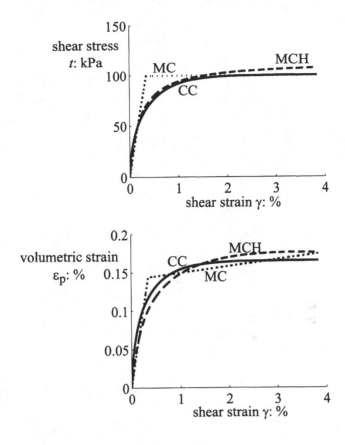

Figure 7.54: Comparison of constitutive models for synthetic plane strain drained compression test (MC: elastic-perfectly plastic Mohr-Coulomb; MCH: elastic-hardening plastic Mohr-Coulomb; CC: Cam clay)

Table 7.4: Soil parameters

Mohr-Coulomb

bulk modulus	K	600 kPa
Poisson's ratio	ν	0.3
angle of shearing resistance	ϕ	30°
angle of dilation	ψ	−0.5°

Hardening Mohr-Coulomb

bulk modulus	K	1500 kPa
Poisson's ratio	ν	0.1
hyperbolic hardening parameter	a	0.09
peak friction angle	ϕ_p	32°
critical state friction angle	ϕ_{cv}	31°

Cam clay

elastic volumetric response	κ	0.05
Poisson's ratio	ν	0.15
plastic volumetric response	λ	0.25
critical state friction angle	ϕ_{cv}	30°
initial specific volume	v_o	2.0

It is now clear that the penalty for choosing a slightly high value of friction angle for the hardening Mohr-Coulomb model is that the ultimate footing stress is correspondingly larger. For $\phi = 30°$ the ratio of footing stress to surcharge stress is 9; for $\phi = 32°$ the ratio is 10.6. Cam clay tends more rapidly than the hyperbolic Mohr-Coulomb model towards the limiting stress. There is of course nothing sacrosanct about the hyperbolic hardening law.

For the isotropically normally consolidated initial conditions, both hardening models generate plastic strains in both elements immediately footing loading begins. Cam clay is slightly less stiff on average over the first stages of loading but it is perhaps surprising that the difference between the two models is not greater (Figs 7.55a, c).

For the soil with initial $K_o = 0.6$, the Mohr-Coulomb model has an extensive region of elastic response for the passive element—it is assumed that yielding does not occur until the same angle of friction is mobilised (or same stress ratio reached) under passive conditions. The higher stiffness for this model is much more apparent (7.56a, c). The stress paths for the active element diverge somewhat initially before converging again at failure (Figs 7.56b, d). The stress paths for the passive element are all identical: the vertical stress is constant and the horizontal stress has to maintain equilibrium with the active element.

So far as the overall load settlement response of the footing is concerned even the perfectly plastic Mohr-Coulomb model produces quite a reasonable match with the nonlinear models in terms of average stiffness and ultimate load (but this is of course built into the model). The two nonlinear models give very similar internal responses (inter-element stress and element strain paths).

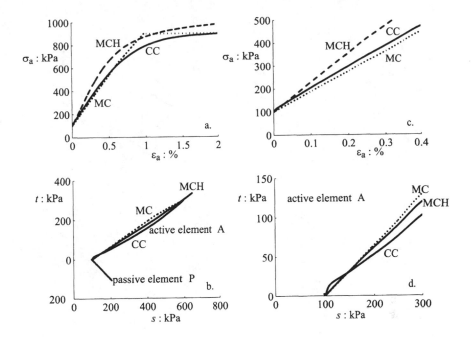

Figure 7.55: Comparison of responses of box model for different constitutive models: $K_o = 1$: (a) and (c) load:settlement response for element A; (b) and (d) stress paths for elements A and P

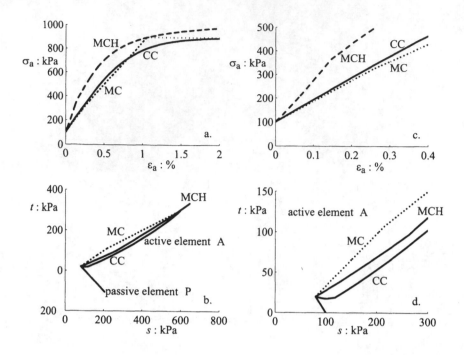

Figure 7.56: Comparison of responses of box model for different constitutive models: $K_o = 0.6$: (a) and (c) load:settlement response for element A; (b) and (d) stress paths for elements A and P

This particular box model provides a simple example of the use of a readily constructed numerical tool which can be used to compare different constitutive models. It is particularly useful as a pedagogic tool to explore and demonstrate differences in constitutive response. For the models considered here the frictional strength appears to dominate the response and may distract us from being aware of the major differences in initial stiffness.

Responses have only been shown for a couple of consolidation histories and initial stress conditions. Evidently if the models were asked to perform in a heavily overconsolidated initial state then they would behave very differently—but of course it would then be appropriate to calibrate the soil parameters for this alternative initial history. Models should be calibrated over as wide a range as possible of the actual stress paths which will influence the overall system response (§3.8).

7.5.2 Wall model

Inspired by displacement observations reported by Bransby and Milligan (1975) from model retaining wall tests at single gravity in sand and from their own observations in centrifuge tests on model diaphragm walls in clay, Bolton and Powrie (1988) have proposed a simple macroelement that can describe the mobilisation of shearing resistance adjacent to a displacing wall.

For a wall rotating about its toe, Bransby and Milligan show that the strains behind the wall roughly follow the pattern shown in Fig 7.57a. Uniform strains occur within the triangle delimited by the wall AB and by the line BC at slope $\pi/4 + \psi/2$ to the horizontal, where ψ is the angle of dilation. All displacements within the triangle ABC occur in a direction at an angle ψ to BC as shown, with the magnitude of displacements increasing linearly with distance from BC. Line BC is a line of zero extension along which the direct strain increment is zero. The principal strain increments are vertical and horizontal: the Mohr circle of strain increment is shown in Fig 7.57b. For a wall rotation of $\delta\theta$ the uniform maximum shear strain increment within the triangle ABC is:

$$\delta\gamma = 2\delta\theta \sec\psi \qquad (7.163)$$

which, for typical angles of dilation below about 25°, Bolton and Powrie suggest can be generally approximated to

$$\delta\gamma \approx 2\delta\theta \qquad (7.164)$$

which of course is now independent of angle of dilation.

Bolton and Powrie then suggest that, since all the actual value of ψ does is to define the limit of the deforming triangle ABC, we can concentrate, in developing displacement mechanisms around moving walls, on 45° triangles of uniform strain, with corresponding constant volume Mohr circle of strain increment and equal and opposite principal direct strains. For a wall rotating about its toe (the Bransby and Milligan problem) the resulting displacement mechanism is shown in Fig 7.58a and the corresponding Mohr circle of strain increment is shown in Fig 7.58b. For a wall rotating about its top Bolton and Powrie suggest the

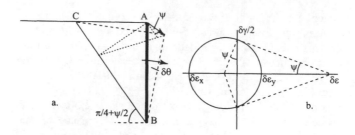

Figure 7.57: Typical schematic strain field in sand behind stiff wall rotating about its toe (after Bransby and Milligan, 1975; Bolton and Powrie, 1988)

combination of a rigidly rotating triangle and a shearing triangle as shown in Fig 7.58c. These two mechanisms can be combined for a translating wall (Fig 7.58d) and kinematic compatibility can be used to deduce the mechanism of displacements around an embedded wall rotating about its top (Fig 7.58e) or about any other point O on its length (Fig 7.58f), where a little circulatory mechanism of deforming triangles is required around the toe of the wall. Where the mechanism includes triangles with two different strain levels at a particular depth (for example, the embedded wall of Fig 7.58e), Bolton and Powrie suggest that the larger of the two strains should be used in subsequent calculation.

Though the patterns may look complicated, all they are really doing is defining a strain at each level and on each side of the wall which will allow us to estimate the mobilised shearing resistance and hence the coefficient of earth pressure and hence the earth pressure itself at that level. This is essentially the free earth support interpretation of equilibrium of the wall. Evidently such a procedure will not work well for flexible walls in which some arching of stresses may take place (§8.7): a procedure that works purely with earth pressure coefficients as a multiplier on directly calculated vertical stresses makes no allowance for interaction of the soil at different levels. Equally this procedure makes no allowance for the fact that below some depth a flexible wall may actually not deflect at all.

As an illustration of the application of this macroelement model we will take the simple case of an embedded rigid wall propped at its top, with excavation proceeding progressively in front of the wall (Fig 7.59) for which the displacement mechanism is shown in Fig 7.58e: for an increment of wall rotation $\delta\theta$ the appropriate strain increments behind the wall (the active region) are equal to $2\delta\theta$ and in front of the wall (the passive region) equal to $2\delta\theta/(1-\alpha)$ where αH is the depth of excavation in front of the wall of overall height H. We imagine that we have a hyperbolic relationship between shear strain and mobilised angle of shearing resistance (Fig 7.60)

$$\phi_{mob} = \phi_i + (\phi_{max} - \phi_i)\frac{\gamma}{b+\gamma} \qquad (7.165)$$

where ϕ_{max} is the maximum value of angle of shearing resistance reached at large strain and b is a constitutive parameter controlling the hyperbolic relationship.

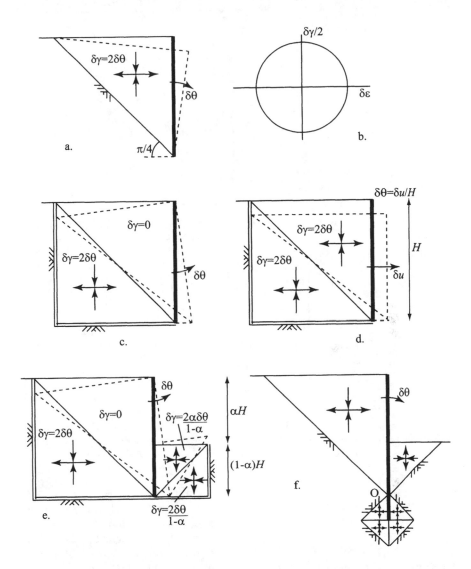

Figure 7.58: Macroelements for soil deformations around stiff walls: (a) rotating about the toe; (c) rotating about the top; (d) translating; and embedded walls: (e) rotating about the top; (f) rotating about some arbitrary point O; (b) corresponding Mohr circle of strain increment (after Bolton and Powrie, 1988)

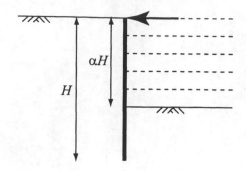

Figure 7.59: Progressive excavation in front of stiff wall of total height H propped at its top

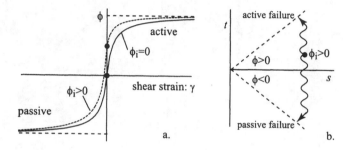

Figure 7.60: Hyperbolic link between mobilised shearing resistance and shear strain for active and passive soil

We give ourselves the possibility of some initial anisotropic stress state ($\phi_i \neq 0$, with different horizontal and vertical stresses in the soil) which will then imply different links between mobilisation of shearing resistance and shear strain on each side of the wall, with negative mobilised angles being assigned to the passive condition $\sigma_h > \sigma_v$ (Fig 7.60b). We will ignore wall friction. (Bolton and Powrie suggest that this can be accommodated by adjusting the limiting value of earth pressure coefficient towards which the soil tends as strain increases, in line with typical design practice.)

Equilibrium of the wall requires, taking moments about the prop (Fig 7.59):

$$\frac{K_a}{K_p} = \frac{1}{2}(1 - \alpha)^2(\alpha + 2) \qquad (7.166)$$

where the current values of active and passive pressure coefficients K_a and K_p depend on the currently mobilised angles of shearing resistance $\phi_{mob,a}$ and $\phi_{mob,p}$ for the active and passive regions in the usual way:

$$K_a = \frac{1 - \sin \phi_{mob,a}}{1 + \sin \phi_{mob,a}}; \qquad K_p = \frac{1 - \sin \phi_{mob,p}}{1 + \sin \phi_{mob,p}} \qquad (7.167)$$

recalling that we are defining *negative* values of ϕ_{mob} as appropriate to the passive region in front of the wall (Fig 7.60).

Tracking of the process of wall rotation as the excavation proceeds then requires the satisfaction of (7.166) with the correct ratio of strains in opposite directions from the initial condition (Fig 7.58e). We know that in the limit, the active and passive pressure coefficients will correspond to those calculated from the available frictional strength of the soil and that the maximum depth of excavation can be thus calculated. The required prop force can be calculated from horizontal equilibrium of the wall in nondimensional form:

$$\frac{2F}{\gamma H^2} = K_a - K_p(1-\alpha)^2 \qquad (7.168)$$

If we assume a maximum angle of shearing resistance of 30°, then the limiting values of K_a and K_p are 1/3 and 3 respectively and, in the limit, $\alpha = 0.714$ and we can evidently not pursue our calculation beyond this depth of excavation in front of the wall. The dimensionless prop force (7.168) is shown as a function of dimensionless excavation depth α and thence of wall rotation in Fig 7.61 for different initial stress conditions ($\phi_i > 0$ implying $K_o < 1$, $\phi_i = 0$ implying $K_o = 1$, and $\phi_i < 0$ implying $K_o > 1$). The units of rotation are arbitrary: rotation is essentially scaled with the constitutive parameter b (7.165). All the curves tend asymptotically to the ultimate state corresponding to full mobilisation of available shearing resistance on both sides of the wall: $2F/\gamma H^2 = 0.0879$. The rotation tends to infinity as the excavation depth approaches the limiting value. There is a more or less gentle peak in the strut force on the way. This is linked with the different rates of mobilisation of friction on each side of the wall and, more particularly, with the progressively diverging moment arms for the active and passive forces. It appears to be harder to prevent movement than to prevent collapse. Of course, it may be regarded as unrealistic to assume uniform mobilisation of earth pressure coefficient over the full height of the wall even for very small depths of excavation (values of α close to 0)—this is an assumption of this particular macroelement model. Allowance for wall flexibility is beyond the scope of this simple procedure (but is discussed within the more general theme of *soil-structure interaction* in section §8.7). Alternative models for the mobilisation of friction, (Fig 7.60), could easily be incorporated.

7.5.3 Footing model

In introducing elastic-plastic constitutive models of soil behaviour (§3.3) an analogy was drawn with the plastic collapse of a steel portal frame in which there is an interaction between the effects of two independent components of load (vertical and horizontal) to cause plastic collapse of the frame. While bearing capacity of footings is primarily focussed on the limiting vertical loads that the footings can sustain, geotechnical engineers are familiar with the introduction of modifying factors to allow for the inclination or eccentricity of the load—in other words to allow for the presence of horizontal and moment loading in addition to the vertical load (Fig 7.62). The need from the 1970s onward to design offshore foundations which were liable to be subjected to horizontal and moment loads which were significant by comparison with the vertical loads led to

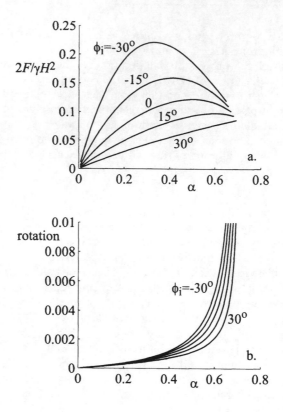

Figure 7.61: Variation of (a) normalised prop force and (b) wall rotation with excavation

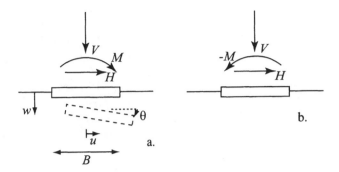

Figure 7.62: Footing under combined vertical V, horizontal H and moment M load: (a) H and M supporting; (b) H and M opposing

extensive experimental study of the response of footings under combined loadings and, from these experimental studies, macroelement models of foundation response have emerged which bear many similarities to the elastic-hardening plastic constitutive models for soil elements of section §3.4.

There have been a number of parallel studies of this problem: some of the key contributions have been made by Nova and Montrasio (1991), Paolucci, (1997), Gottardi, Houlsby and Butterfield (1999), Martin and Houlsby (2001) and Cremer *et al.* (2001). A slightly simplified macroelement model will be described here—it will be clear how this can be modified and extended to reflect more detail of the characteristics of the observed mechanical response.

The four key elements of an elastic-plastic constitutive model are:

- the elastic properties;

- a yield function;

- a flow rule or plastic potential; and

- a plastic hardening rule.

As for an elemental model it is essential that the load components and corresponding displacement components should be properly work-conjugate. For a strip footing of width B there is some advantage in defining the force vector $\boldsymbol{F}$ in a dimensionally consistent form:

$$\boldsymbol{F} = \begin{pmatrix} V \\ H \\ M/B \end{pmatrix} \tag{7.169}$$

with corresponding displacements:

$$\boldsymbol{u} = \begin{pmatrix} w \\ u \\ B\theta \end{pmatrix} \tag{7.170}$$

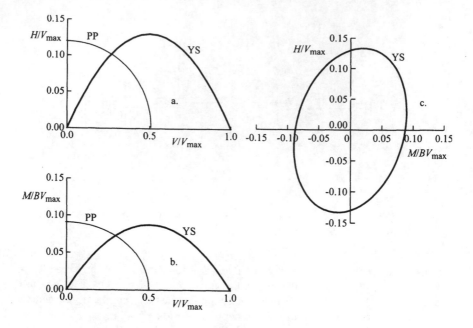

Figure 7.63: Sections through yield surface (YS) and plastic potential (PP) for footing under (a) combined vertical and horizontal load; (b) combined moment and vertical load; (c) combined horizontal and moment load (after Butterfield and Gottardi, 1994 and Cremer *et al.*, 2001)

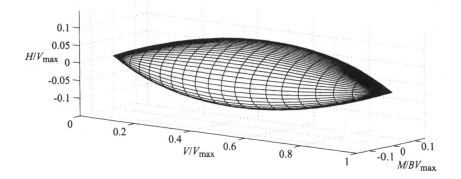

Figure 7.64: Yield surface for footing under combined vertical, horizontal and moment load

Systematic probing of the interaction between V, H and M for strip and circular (spud-can) footings on sand has defined the shape of the yield surface (Butterfield and Gottardi, 1994). To first order, the H/V_{max}, V/V_{max} and M/BV_{max}, V/V_{max} sections are parabolic and the H/V_{max}, M/BV_{max} section is elliptical (Fig 7.63). However, consideration of the nature of the loading should leave us unsurprised that this elliptical section is not symmetric about either the H or the M/B axes: the limiting foundation capacity will depend on whether the horizontal load and moment support or oppose each other (Fig 7.62a, b). The overall yield surface is shown in Fig 7.64 and is described by the equation:

$$f(V, H, M) = \left(\frac{H}{t_h}\right)^2 + \left(\frac{M}{Bt_m}\right)^2 - \frac{2CMH}{Bt_h t_m} - \left[\frac{V}{V_{max}}(V_{max} - V)\right]^2 = 0 \quad (7.171)$$

where t_h, t_m and C are material parameters which, for Butterfield and Gottardi's configuration have the values 0.52, 0.35 and 0.22 respectively.

The inclusion of V_{max} in (7.171) immediately suggests a route by which 'plastic' hardening of the foundation can be introduced. Limiting values of each load component (V, H and M) are expected to be proportional to the maximum vertical load that the foundation has experienced and this load can play the role of a hardening parameter in developing the elastic-plastic model of the foundation. Nova and Montrasio (1991) propose a simple exponential link between V_{max} and the plastic component of foundation settlement w^p:

$$\frac{V_{max}}{V_{ult}} = 1 - e^{-\lambda w^p / V_{ult}} \quad (7.172)$$

where λ controls the initial vertical plastic stiffness of the foundation and V_{ult} represents an ultimate limiting vertical load. There is a very obvious similarity between the roles played by maximum vertical load and plastic settlement in this macroelement model and the size of the yield surface p'_o and plastic volumetric strain in the Cam clay model (§3.4.2).

In order to discover the nature of the flow rule from experimental observations it is necessary to have some model for the elastic displacements so that elastic and plastic displacements can be properly separated:

$$F = Ku \qquad (7.173)$$

$$K = \begin{pmatrix} K_{vv} & 0 & 0 \\ 0 & K_{hh} & 0 \\ 0 & 0 & K_{\theta\theta} \end{pmatrix} \qquad (7.174)$$

where the off-diagonal terms are probably close to zero for surface or shallow foundations but alternative assumptions could be readily incorporated.

While there is then general agreement that an assumption of associated flow is not satisfactory for describing the plastic displacement increments—the vertical displacements are certainly incorrect—there seems to be less agreement about the shape of the plastic potential surfaces from which the flow rule might be deduced. The simple parabolic sections and resulting three-dimensional surface shown in Figs 7.63a, b and 7.64 imply a vertex where the yield surface crosses the V axis whereas we expect that for vertical loads close to V_{max} the displacement will be more or less a pure settlement even in the presence of small horizontal loads or moments. For very low values of V/V_{max} normality to the yield surface seems to imply excessive vertical *heave* of the footing. Cremer *et al.* (2001) propose an elliptical plastic potential centred on the origin: sections through this are included in Fig 7.63a, b.

With all these ingredients in place we can write the overall elastic-plastic stiffness relationship for the foundation macroelement in *exactly* the same way as for the elastic-plastic single element (§3.4, (3.109)):

$$\delta F = \left[K - \frac{K \frac{\partial g}{\partial F} \frac{\partial f}{\partial F}^T K}{\frac{\partial f}{\partial F}^T K \frac{\partial g}{\partial F} + H} \right] \delta u = K^{ep} \delta u \qquad (7.175)$$

where the hardening component H is given by:

$$H = -\frac{\partial f}{\partial V_{max}} \frac{\partial V_{max}}{\partial u}^T \frac{\partial g}{\partial F} \qquad (7.176)$$

and this macroelement can then be used to compute foundation response under static or dynamic loads.

Once the direct similarity with constitutive models for single soil elements is appreciated then it becomes straightforward to extend the simple model presented here to follow more closely the behaviour seen in physical models of foundations or in numerical modelling performed using accepted soil models. For example, it can be expected that footings will show hysteretic response in cycles of unloading and reloading. Elements of kinematic hardening and bounding surface plasticity can be introduced using exactly the same mathematical structures and geometrical procedures (Cremer *et al.*, 2001). Cremer *et al.* also introduce the possibility of 'uplift' into the foundation response so that the foundation is able to lose contact with the soil over part of its width. This requires certain additional modelling assumptions but still forms a theoretically compact macroelement model.

This is a macroelement approach linked very strongly with the physical modelling which has inspired the various ingredients of the elastic-plastic model. *'It is well known that an alternative model of the foundation behaviour obtained by the finite element method, with suitable nonlinear constitutive laws and special contact elements, requires a high degree of modelling competence and is time consuming. The macroelement provides a practical and efficient tool, which may replace efficiently, in a first approach, a costly finite element soil model, and which ensures the accurate integration of the effect of soil-structure interaction.'* (Cremer *et al.*, 2001) It lends itself naturally to application in dynamic analysis where the numerical intricacies of full finite element analysis become greater. A dynamic macroelement provides the possibility of rapidly exploring effects of different classes of input motion in addition to variation of the more obviously geotechnical parameters (see, for example, Paolucci, 1997).

7.5.4 Extended Newmark sliding block model

Given a need to design a geotechnical system to withstand earthquake loading one route which could be followed would be to assess the maximum possible acceleration that could ever occur at the location of the geotechnical structure and then ensure that the failure mode corresponding to this acceleration could not occur. This is a sort of ultimate limit state approach to seismic design which takes no account either of the fact that the maximum acceleration only occurs for a very brief time within an overall earthquake acceleration time history or of the fact that in general it is uneconomic to design any structure to show *no* movement and that some movement can usually be tolerated. The amount of movement that is to be tolerated might well be linked with an assessment of the likely return period of the earthquake. A macroelement approach such as that described in the previous section is ideally suited to estimation of foundation displacements occurring as a result of the application of a real or synthetic time history of acceleration. Newmark's (1965) sliding block approach represents a slightly rougher way of achieving the same goal. We will ignore vertical ground acceleration for simplicity though the procedure can fairly easily be extended to include vertical as well as horizontal ground input motions.

Let us suppose that we can calculate for our geotechnical system—an embankment dam, slope, retaining structure, or foundation—the horizontal base acceleration level $k_h g$ whose *permanent* application will cause the system to fail along a certain mechanism of failure (Fig 7.65a). The application of such an acceleration is equivalent to the rotation and scaling of the gravitational acceleration (Fig 7.65b, c): rotation by an angle $\tan^{-1} k_h$, scaling by a factor $\sqrt{1 + k_h^2}$. If an actual acceleration history falls entirely below $k_h g$ then it is assumed that there will be no permanent displacement because the critical mechanism is never activated. If the acceleration history contains episodes of

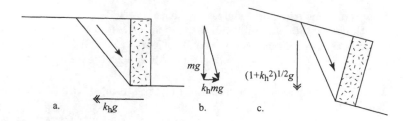

Figure 7.65: (a) Geotechnical system subjected to horizontal acceleration; (c) equivalent system with rotated and scaled gravity as computed in (b)

acceleration greater than $k_h g$ then Newmark's procedure allows us to estimate
the permanent relative displacement of the identified failure mechanism that
will ensue.

The acceleration $k_h g$ represents the largest horizontal acceleration that can
be transmitted from the ground to the geotechnical structure. Consider the
acceleration time history shown in Fig 7.66a. As soon as the input acceleration
exceeds $k_h g$ the structure and the ground must separate: the structure moves
with this limiting value of acceleration and a corresponding steadily increasing
velocity (Fig 7.66b). Evidently the difference in velocities of the ground and
the structure leads to the build up of a permanent relative displacement or
slip (Fig 7.66c, d). This slippage continues until the structure and ground are
able to reattach which occurs when the velocities match once more (Fig 7.66b):
until then the structure continues to slip relative to the ground even though
the ground acceleration is below the critical value $k_h g$. There are thus two
conditions for slip to occur: if $a_g > k_h g$ or $v_s > v_g$ then $a_s = k_h g$ where a_g, a_s
and v_g, v_s are accelerations and velocities of the ground (the input motion) and
the structure respectively. It is then a simple numerical matter of integrating
the velocities to discover the different displacements of ground and structure
and hence the permanent relative movement. Note that inertia does not enter
this calculation: all necessary considerations of strength and mass have already
been used to calculate the limiting acceleration $k_h g$.

An example for a more or less realistic time history is shown in Fig 7.67
where, for sake of illustration, the geotechnical system being analysed is assumed
to be a gravity wall (Fig 7.65). The capping of the wall acceleration at $k_h g$ is
evident (Fig 7.67c). The difference between the displacements of the ground
and the wall (Figs 7.67b, d) leads to relative slip (Fig 7.70).

It is assumed that this permanent movement is occurring because of relative
displacement within the soil on some definite failure mechanism. For many soils
we expect that such relative shear displacement will be linked with a falling
interface strength (Fig 7.68). If we can estimate the link between interface
strength and displacement then we can build this into the calculation of slip.
The simplest assumption will be that the failure mechanism does not change and
that the loss of strength associated with relative displacement merely reduces
the limiting acceleration, so that k_h falls with increasing slip. The calculation

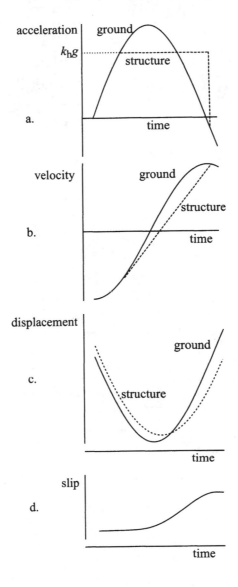

Figure 7.66: Motion of sliding geotechnical structure with critical acceleration $k_h g$

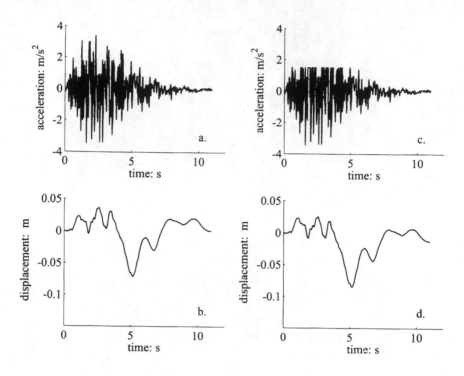

Figure 7.67: Time histories of (a), (c) acceleration and (b), (d) displacement for ground (input) (a), (b) and wall (c), (d)

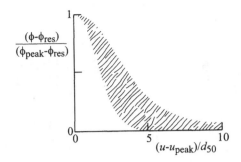

Figure 7.68: Degradation of available strength of granular soils with relative displacement (inspired by Koseki *et al.*, 1998)

is hardly more complicated if this strength variation is included. Figs 7.69 and 7.70 show the result of using an exponential decay of strength and the increased slip that would result.

If we are concerned about the performance of our geotechnical structure against some allowable permanent displacement then this loss of strength with relative displacement should in no way concern us provided that the overall displacement is still acceptable. Zanganeh and Popescu (2002) describe the analysis of the movement of a long submarine slope where the loss of strength is linked with build up of pore pressure. On the other hand, Koseki *et al.* (1998) show how such a procedure can be combined with an assessment of the likelihood of development of new mechanisms, recognising that as the soil softens the initially chosen mechanism may no longer be the most critical.

7.6 Closure

We have gathered together in this chapter a number of examples of problems for which theoretical analysis and essentially exact response can be obtained. If a real geotechnical problem is stripped to its essentials then it too may be capable of similarly exact analysis: this is the key to geotechnical modelling. Evidently we need to retain sufficient geotechnical—and boundary—realism for our theoretical model to appear plausible but even in a severely simplified state such an analytical result may be helpful in confirming at least the order of magnitude of response seen in more elaborate—and more time consuming— physical or numerical modelling.

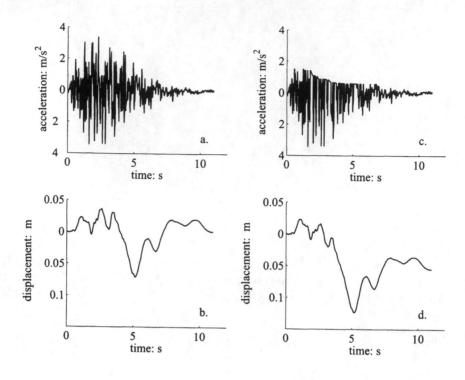

Figure 7.69: Time histories of (a), (c) acceleration and (b), (d) displacement for ground (input) (a), (b) and wall (c), (d): effect of degrading strength on dynamic response

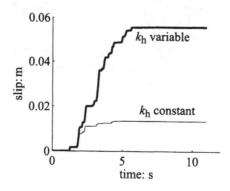

Figure 7.70: Relative displacement (slip) of ground and wall with and without degradation of shear strength

8

Soil-structure interaction

8.1 Introduction

It was argued in section §5.4, based on considerations of dimensional analysis, that physical modelling of geotechnical problems that included structural elements had to take account of and attempt to preserve relative stiffnesses between prototype and model. In this chapter we will investigate the analysis of a number of geotechnical problems which involve interaction of soil with structural elements.

Where the structural elements are infinitely stiff (rigid) or infinitely flexible then analysis is simplified. The interesting problems of soil-structure interaction arise for intermediate stiffnesses and analysis quickly shows, as expected, that it is *relative* stiffness of structural material and the ground that is important rather than the absolute stiffnesses of either component. Analysis of soil-structure interaction is most straightforward when the soil and the structure can both be considered as elastic: the first examples considered here fall into this category. Where soil nonlinearity or plasticity—or nonlinearity of the structural material or of the interface between the soil and the structure—cannot be ignored it is usually necessary to adopt some powerful numerical modelling tool. However, some examples which are capable of theoretical analysis are also included.

8.2 Elastic analyses

8.2.1 Beam on elastic foundation

A first example of soil-structure interaction which lends itself to exact analysis is the beam on an elastic foundation (Fig 8.1) under a central point load P (Winterkorn and Fang, 1975). The foundation is assumed to consist of an infinite number of independent linear Winkler springs so that at each point on the beam the pressure resisting settlement y is directly proportional to that settlement through a spring constant k and for a beam of width B the force per unit length of beam resulting from the settlement is kB. The beam has flexural rigidity EI and the origin of the position coordinate x is taken at the centre of

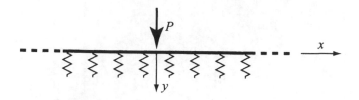

Figure 8.1: Beam on elastic foundation

the beam. The equation of flexural equilibrium of the beam is then:

$$EI\frac{\mathrm{d}^4y}{\mathrm{d}x^4} = -kBy \tag{8.1}$$

for which the general solution is

$$y = \mathrm{e}^{\mu x}\left(A_1\cos\mu x + A_2\sin\mu x\right) + \mathrm{e}^{-\mu x}\left(A_3\cos\mu x + A_4\sin\mu x\right) \tag{8.2}$$

or alternatively

$$y = \left(B_1\cosh\mu x + B_2\sinh\mu x\right)\left(B_3\cos\mu x + B_4\sin\mu x\right) \tag{8.3}$$

where

$$\mu^4 = \frac{kB}{4EI} \tag{8.4}$$

and where A_1, A_2, A_3, A_4 or B_1, B_2, B_3, B_4 must be determined from the boundary conditions of a particular problem.

For an extremely 'long' beam—we will need to assess what 'long' actually means—the boundary conditions are:

- slope $\mathrm{d}y/\mathrm{d}x = 0$ at $x = 0$ from symmetry
- moment $M = EI\mathrm{d}^2y/\mathrm{d}x^2 \to 0$ as $x \to \infty$
- shear force $F = EI\mathrm{d}^3y/\mathrm{d}x^3 = P/2$ at $x = 0$
- shear force $F = EI\mathrm{d}^3y/\mathrm{d}x^3 \to 0$ as $x \to \infty$

These give the solution in the section of the beam for positive x, the boundary conditions for negative x are the same but with the sign of the shear force at the origin reversed, from symmetry. The first form of the solution (8.2) is the more convenient with the final result:

$$y = \frac{P\mu}{2kB}\mathrm{e}^{-\mu x}\left(\cos\mu x + \sin\mu x\right) \tag{8.5}$$

and this profile of deflections, together with the corresponding variations of slope (normalised with $P\mu^2/kB$), bending moment (normalised with $P/4\mu$) and shear force (normalised with $P/2$), is plotted in Fig 8.2 as a function of the normalised position on the beam μx.

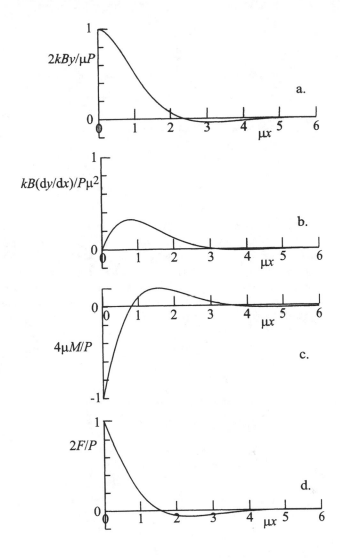

Figure 8.2: Infinite beam on elastic foundation: (a) normalised displacement; (b) normalised slope; (c) normalised moment; and (d) normalised shear force

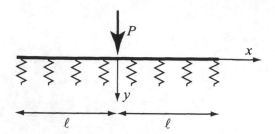

Figure 8.3: Beam of length 2ℓ on elastic foundation

We note various features of this solution. The beam lifts up over part of its length—our simple Winkler spring foundation has to be able to take tension as well as compression. Elastic materials can sense the effects of perturbations over great distances—the solution continues to ripple gently even for large values of μx. However, we discover that once μx is greater than about 6 then the moment and shear force remain below 0.5% of their peak values. We might take this as the definition of a 'long' beam and observe that this definition of 'long' is a function both of the stiffness properties of the beam and the stiffness properties of the ground: relative and not absolute values are important.

For a short beam of length 2ℓ (Fig 8.3) the boundary conditions for $0 < x < \ell$ are:

- slope $\mathrm{d}y/\mathrm{d}x = 0$ at $x = 0$ from symmetry
- moment $M = EI\mathrm{d}^2y/\mathrm{d}x^2 = 0$ at $x = \ell$
- shear force $F = EI\mathrm{d}^3y/\mathrm{d}x^3 = P/2$ at $x = 0$
- shear force $F = EI\mathrm{d}^3y/\mathrm{d}x^3 = 0$ at $x = \ell$

(Again, equivalent conditions can be produced for $-\ell < x < 0$.) The second form of the solution (8.3) is now the more convenient with the final result:

$$\frac{2kB\ell y}{P} = \mu\ell\,[(\cosh\mu x \sin\mu x -$$

$$\sinh\mu x \cos\mu x) + (\alpha\cosh\mu x \cos\mu x - \beta\sinh\mu x \sin\mu x)]$$

where

$$\alpha = \frac{\cosh^2\mu\ell + \cos^2\mu\ell}{\cosh\mu\ell\sinh\mu\ell + \cos\mu\ell\sin\mu\ell} \tag{8.6}$$

and

$$\beta = \frac{\sinh^2\mu\ell + \sin^2\mu\ell}{\cosh\mu\ell\sinh\mu\ell + \cos\mu\ell\sin\mu\ell} \tag{8.7}$$

The quantity $P/2kB\ell$ is the settlement of a load P uniformly distributed over the total length 2ℓ of the beam of width B and resisted by the Winkler springs of stiffness k: $2kB\ell y/P$ thus normalises the profile of settlement of the beam

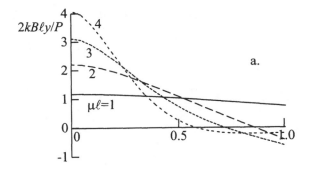

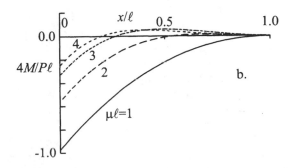

Figure 8.4: Beam on elastic foundation: influence of relative pile stiffness $\mu\ell$: (a) normalised displacements and (b) normalised moments

with this reference settlement. If we normalise the position coordinate x using a dimensionless variable $\xi = x/\ell$ then we see that μ only enters the solution in the dimensionless combination $\mu\ell$. This (or perhaps more conveniently $\mu^4\ell^4 = kB\ell^4/4EI$) is the controlling scaling parameter for this problem. Typical distributions of deflection and moment are shown in Fig 8.4 for different values of $\mu\ell$: low values correspond to stiff beams and soft soil; high values correspond to soft beams and stiff soil. For an infinitely stiff beam ($\mu\ell = 0$) the load P is resisted by a uniform reaction along the length of the beam and the central moment is then $P\ell/4$. Moments are therefore shown in normalised form as $4M/P\ell$.

The range of responses of the beam can be presented in terms of the maximum (central) bending moment on the one hand and the differential settlement between the centre of the beam (under the load) and the free end on the other (Fig 8.5). The central moment is given by:

$$\frac{M_{max}}{P\ell/4} = \frac{1}{\mu\ell} \frac{\sinh^2 \mu\ell + \sin^2 \mu\ell}{\cos \mu\ell \sin \mu\ell + \sinh \mu\ell \cosh \mu\ell} \qquad (8.8)$$

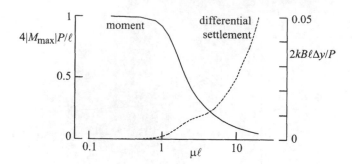

Figure 8.5: Central moment and differential settlement for beam on elastic foundation as function of relative beam stiffness $\mu\ell$

For stiff beams (low values of $\mu\ell$) the differential settlement is evidently negligible. However, as the beam becomes relatively more flexible the central moment reduces—in fact it tends to zero—and the differential settlement continues to increase[1]. We begin to see a key result of soil-structure interaction: there is a trade-off between structural stiffness (which usually implies cost) and deformations. If we can, realistically, accommodate displacement of our structure then we may be able to economise on stiffness.

The concept of *coefficient of subgrade reaction* is intuitively attractive but theoretically questionable because it assumes that the foundation can behave as a series of quite independent springs and that the behaviour of one spring has no influence on the stiffness of an adjacent spring. We know the soil to be continuous so that this independence of local stiffness is clearly an idealisation. The problem confronts us directly when we try to estimate values of coefficient of subgrade reaction from more fundamental constitutive quantities such as shear modulus or bulk modulus. Though we might try an appeal to standard elasticity solutions for foundations on the surface of an elastic half-space we find that those stiffnesses are dependent on the dimensions of the foundation (for example, for a rigid circular footing of radius a the vertical stiffness is $4Ga/(1-\nu)$ where ν is Poisson's ratio) (§3.2.3, §8.2.4) (Terzaghi, 1955). However, for the special case in which the soil is undrained (so that Poisson's ratio $\nu = 0.5$) and the shear stiffness G varies linearly with depth z below the ground surface:

$$G = \lambda z \tag{8.9}$$

Gibson (1967) shows that the soil behaves *exactly* like a bed of Winkler springs with coefficient of subgrade reaction $k = 2\lambda$.[2]

---

[1]Because the beam is supported on a set of completely independent springs, as the stiffness of the beam tends to zero the differential settlement tends to infinity. The Winkler spring foundation breaks down as a representation of real soil at this extreme.

[2]A beam on a Winkler foundation subjected to a uniformly distributed load settles uniformly independent of the relative stiffness of soil and beam.

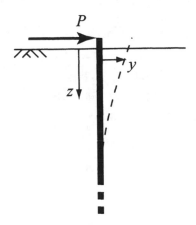

Figure 8.6: Laterally loaded pile in elastic soil

8.2.2 Pile under lateral loading

A pile of width B in an elastic soil loaded by a lateral force P at ground surface can be analysed in exactly the same way as the beam on an elastic foundation if the lateral pressure is assumed to be directly proportional to the relative movement of the pile and the soil according to a coefficient of horizontal subgrade reaction k (Fig 8.6). The governing equation (8.1) is exactly the same, as are the general forms of the solution (8.2), (8.3): the horizontal x coordinate for the beam is replaced by a z coordinate measured down the pile from the ground surface (Fig 8.6). The boundary conditions are slightly different.

For the infinitely long pile the boundary conditions are:

- moment $M = EI\mathrm{d}^2y/\mathrm{d}z^2 = 0$ at $z = 0$

- moment $M = EI\mathrm{d}^2y/\mathrm{d}z^2 \to 0$ as $z \to \infty$

- shear force $F = EI\mathrm{d}^3y/\mathrm{d}z^3 = P$ at $z = 0$

- shear force $F = EI\mathrm{d}^3y/\mathrm{d}z^3 \to 0$ as $z \to \infty$

and the solution is

$$y = \frac{2P\mu}{kB}\mathrm{e}^{-\mu z}\cos\mu z \qquad (8.10)$$

The variation of deflection, slope, moment and shear force down the pile are shown in Fig 8.7. The pile develops an undulating deflected form as the load is transferred down the pile: not much of significance happens below about $z \approx 6/\mu$.

For the short pile the boundary conditions are:

- moment $EI\mathrm{d}^2y/\mathrm{d}z^2 = 0$ at $z = 0$

- moment $EI\mathrm{d}^2y/\mathrm{d}z^2 = 0$ at $z = \ell$

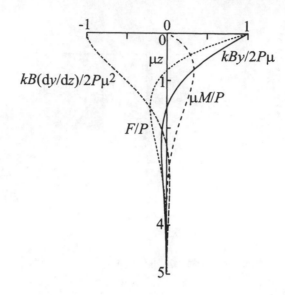

Figure 8.7: Infinite laterally loaded pile in elastic soil: normalised displacement y, slope $\mathrm{d}y/\mathrm{d}x$, moment M and shear force F

- shear force $EI\mathrm{d}^3y/\mathrm{d}z^3 = P$ at $z = 0$
- shear force $EI\mathrm{d}^3y/\mathrm{d}z^3 = 0$ at $z = \ell$

and the solution is

$$\frac{kB\ell y}{2P} = \frac{\mu\ell}{\sin^2 \mu\ell - \sinh^2 \mu\ell} \times$$
$$\left[\sinh^2 \mu\ell \sinh \mu z \cos \mu z + \sin^2 \mu\ell \cosh \mu z \sin \mu z + \right.$$
$$\left. (\sin \mu\ell \cos \mu\ell - \sinh \mu\ell \cosh \mu\ell) \cosh \mu z \cos \mu z\right] \quad (8.11)$$

We deduce that the dimensionless parameter that controls the behaviour of the pile is $\mu\ell$ or $\mu^4\ell^4 = kB\ell^4/4EI$ which is clearly a function of *relative* stiffness of pile and soil (compare §5.4.2). Typical normalised displacements and moments for values of $\mu\ell$ between 1 and 4 are shown in Fig 8.8: for a stiff pile (low values of $\mu\ell$) the pile hardly bends at all but kicks backwards in order to generate the moment to resist the applied load. For more flexible piles the lateral deflection of the top of the pile increases and the flexure of the pile also increases.

8.2.3   Pile under axial loading

If an elastic pile is being loaded axially in an elastic soil then an exact closed form solution for the load distribution within the pile and the settlement distribution down the pile can be obtained, with the aid of one simple assumption. The load

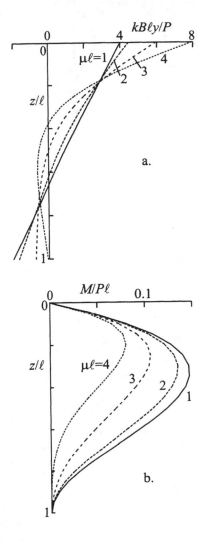

Figure 8.8: Laterally loaded pile of length ℓ in elastic soil: influence of relative pile stiffness $\mu\ell$: (a) normalised displacements and (b) normalised moments

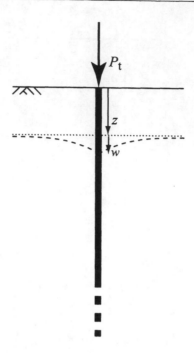

Figure 8.9: Axially loaded pile in elastic soil

in the pile is transmitted to the ground both by shaft friction down the length of the pile and by end bearing at the toe of the pile. We suppose the pile to be circular in cross-section: this is not a necessary requirement but we will have to have some knowledge of the sectional shape (perimeter and cross-sectional area).

At a depth z down the pile (Fig 8.9), the pile has a settlement w relative to the soil at a great distance from the pile. The shear stress τ that develops at the interface between the pile and the soil will be dependent on the stiffness of the soil and on the strains that develop within the soil around the pile. Dimensional analysis suggests that τ should depend on the shear stiffness of the soil G, w, and the radius r_o of the pile through an expression of the form:

$$\frac{\tau}{G} = k \frac{w}{r_o} \tag{8.12}$$

A semi-analytical approach (Fleming *et al.*, 1985) suggests that $k \approx 1/4$: making this assumption unlocks the analytical possibilities for this problem. This assumption is equivalent to assuming that the pile can only be 'felt' to a distance that is some fixed multiple of the radius of the pile. Strictly, in an elastic material, the pile will be 'felt' at infinite radii. However, we are making a gross simplification in treating the soil as a linear elastic material with a constant shear modulus G. In fact we know that the stiffness of soils falls with strain amplitude (eg Fig 2.38) and we can expect the actual shear stiffness to be much

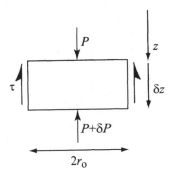

Figure 8.10: Equilibrium of section of axially loaded pile in elastic soil

lower near to the pile than it will be at great distance from the pile. The assumption that the pile load transfer is only felt to a limited radius is thus a convenient but pragmatic compromise.

We can now write down an equation for the equilibrium of a section of the pile at which the local axial load is P (Fig 8.10):

$$\frac{\mathrm{d}P}{\mathrm{d}z} = -2\pi r_o \tau = -\frac{\pi}{2} G w \tag{8.13}$$

The pile is assumed to be behaving elastically so that we can write down an equation governing the elastic axial compression of the pile, introducing the Young's modulus E_p of the pile:

$$\frac{\mathrm{d}w}{\mathrm{d}z} = -\frac{P}{\pi r_o^2 E_p} \tag{8.14}$$

Hence, combining (8.13) and (8.14) we find the general governing equation for the axially loaded pile:

$$\frac{\mathrm{d}^2 w}{\mathrm{d}z^2} = \frac{G w}{2 E_p r_o^2} \tag{8.15}$$

which has the general solution:

$$w = A_1 e^{\eta z} + A_2 e^{-\eta z} \tag{8.16}$$

where

$$\eta = \frac{1}{r_o} \sqrt{\frac{G}{2 E_p}} \tag{8.17}$$

The boundary condition which must be satisfied for all piles is:

$$P_{z=0} = \pi r_o^2 E_p \left(\frac{\mathrm{d}w}{\mathrm{d}z} \right)_{z=0} = P_t \tag{8.18}$$

where P_t is the load applied at the top of the pile.

If the pile is very long (and we will once again discover that 'long' is a relative term) then as $z \to \infty$, $w \to 0$. Consequently, in (8.16), $A_1 = 0$ and $A_2 = (P_t/\pi r_o)\sqrt{(2/E_p G)}$ so that the variation of load down the pile is:

$$\frac{P}{P_t} = e^{-\eta z} \tag{8.19}$$

If we were to introduce an effective length ℓ_{100} at which the axial load in the pile had dropped to 1% of its top value, $P/P_t = 0.01$, then this would imply $\eta \ell_{100} = \ln 100$, $\ell_{100}/r_o = \ln 100 \sqrt{2E_p/G}$. For slender piles in stiff soil very little load reaches the base of the pile.

For *example*, take $E_p = 25$ GPa for concrete, and assume that the shear modulus of the soil is $G = 200c_u$, where the undrained strength $c_u = 200$ kPa corresponding to a firm clay, and $r_o = 0.25$ m. Then $E_p/G = 625$ and for the load to fall to 1% of the surface value $\ell_{100}/r_o = 115$ corresponding to an effective length $\ell_{100} \approx 29$ m.

However, if the pile is short, with finite length ℓ then we expect there to be some load remaining at the base of the pile and we have to impose a boundary condition which ensures compatibility of the base load and the base settlement of the pile. For example, for $z = \ell$, the load P_b and settlement w_b might be related through the settlement characteristics of an elastic pile base (§3.2.3, §8.2.4):

$$w_b = \frac{P_b (1 - \nu_b)}{4r_b G_b} \tag{8.20}$$

where ν_b and G_b are the elastic properties of the ground beneath the pile base of radius r_b (which may be different from r_o for an under-reamed pile). This is the standard expression for the settlement of a rigid circular plate on the surface of an elastic material.

The complete solution of the differential equation for the pile settlement is more cumbersome but still tractable:

$$A_1 = \frac{P_t}{2\pi r_o^2 E_p \eta} \frac{(\xi - 1)(1 - \tanh \eta \ell)}{1 + \xi \tanh \eta \ell} \tag{8.21}$$

$$A_2 = \frac{P_t}{2\pi r_o^2 E_p \eta} \frac{(\xi + 1)(1 + \tanh \eta \ell)}{1 + \xi \tanh \eta \ell} \tag{8.22}$$

where

$$\xi = \frac{(1 - \nu_b) \pi r_o^2 E_p \eta}{4r_b G_b} \tag{8.23}$$

The settlement w_t at the top of pile, at $z = 0$, which is required to estimate overall pile stiffness, is:

$$w_t = \frac{P_t}{\pi r_o^2 E_p \eta} \frac{(\xi + \tanh \eta \ell)}{(1 + \xi \tanh \eta \ell)} \tag{8.24}$$

and the load variation down the pile is given by:

$$\frac{P}{P_t} = \frac{(\xi + 1)(1 + \tanh \eta \ell) e^{-\eta z} - (\xi - 1)(1 - \tanh \eta \ell) e^{\eta z}}{2(1 + \xi \tanh \eta \ell)} \tag{8.25}$$

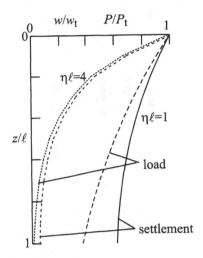

Figure 8.11: Axially loaded pile in elastic soil: distributions of axial load and axial displacement

(Fleming *et al.* present a slightly modified version of this result which allows for a linear variation of shear stiffness of the soil down the length of the pile.)

We note that the key parameter in the solution, η (8.17), is a function of the ratio of the stiffnesses of the soil and the pile. In fact η always appears in the solution in combination with a distance down the pile so that in fact $\eta\ell = (\ell/r_o)\sqrt{G/2E_p}$ is the appropriate dimensionless group which controls the overall response of a pile of length ℓ (compare §5.4.5). In addition we have managed to characterise the entire performance of the pile (8.25) in terms of dimensionless groups for load and position P/P_t and z/ℓ, and for appropriate combinations of material and geometric properties $\eta\ell$ and ξ. The result is thus of completely general applicability. Some typical results are shown in Fig 8.11. For a short pile (low values of $\eta\ell$) the base of the pile takes a significant load—and develops a correspondingly significant settlement. For a longer pile little load reaches the base of the pile.

As an *example* consider a pile of length ℓ =25 m, and radius r_o =0.3 m made of concrete with Young's modulus E_p =25 GPa in soil with shear stiffness G =25MPa and Poisson's ratio ν =0.2. The pile is not under-reamed and the soil beneath the toe of the pile has the same elastic properties. We calculate $\xi = 14.05$ and $\eta\ell = 1.862$, $\tanh\eta\ell = 0.953$ and then, from (8.24), that the axial stiffness of the pile is $P_t/w_t = 505$ MN/m.

8.2.4 Piled raft

Piled rafts (Fig 8.12) allow the load to be spread over a relatively stiff raft but then transferred to the ground partly by the raft contact and partly through transfer to a group of vertical piles. They provide an example of

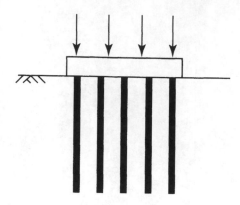

Figure 8.12: Piled raft

structure-soil-structure interaction. The response of the piles and the inter-action of the piles within the group may be influenced by the presence of the raft. Exact analysis requires three-dimensional consideration of interaction of the raft with the ground and hence with the individual piles. However, some simplification of the analysis is possible if we adopt a macroelement approach treating the raft as a single unit, and replacing the piles by an elastic spring with incorporation of *elastic* interaction effects deduced from numerical analysis Randolph, 1986).

Simplification is then possible by separating and superposing the behaviour of the raft and of the pile group (Fleming *et al.*, 1985). Such superposition requires that the response of all elements of the foundation system should be behaving elastically. Let us suppose that we can define the stiffness of the pile group $k_p = P_p/w_p$, where P_p and w_p are the load carried by the pile group and the settlement of the pile group respectively. We define a corresponding raft stiffness $k_r = P_r/w_r$, where P_r and w_r are the load carried by the raft and the settlement of the raft. The combined statement of the interaction between pile group and raft then becomes:

$$\begin{pmatrix} w_p \\ w_r \end{pmatrix} = \begin{pmatrix} 1/k_p & \alpha_{pr}/k_r \\ \alpha_{rp}/k_p & 1/k_r \end{pmatrix} \begin{pmatrix} P_p \\ P_r \end{pmatrix} \qquad (8.26)$$

where the off-diagonal terms indicate the interaction between raft and pile group. For elastic behaviour of all elements of the system we expect symmetry of the flexibility matrix from energy considerations. The factor α_{rp} can be related to the dimensions and spacing of the piles (the spacing controls the raft area associated with each pile). However, Clancy and Randolph (1993) show that for many combinations of pile group size and spacing and material properties a value $\alpha_{rp} \approx 0.8$ will give a reasonable estimate of interaction. This value will be used here for simplicity. (A more detailed treatment is given by Fleming *et al.* (1985).)

The concept of raft stiffness k_r as a single number may appear mysterious. Although one might expect that the overall stiffness of a raft (P_r/w_r) would

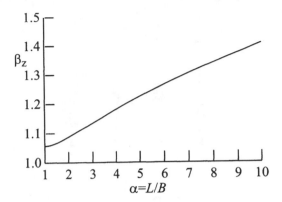

Figure 8.13: Raft stiffness coefficient β_z as function of raft geometry for rectangular rafts

depend on raft flexibility, in fact the flexibility seems to have little influence on the ratio of *total* load to *average* settlement (Barkan, 1962). We can therefore use the results for a rigid raft quoted by Poulos and Davis (1974). For example, for a rectangular raft of width B and length L on the surface of elastic soil with shear modulus G and Poisson's ratio ν;

$$k_r = \frac{P_r}{w_r} = \beta_z \sqrt{BL} \frac{2G}{1-\nu} \tag{8.27}$$

where the dependence of β_z on the proportions of the rectangle, $\alpha = L/B$, is (Barkan, 1962):

$$\beta_z = \frac{\pi\sqrt{\alpha}}{\ln\frac{\gamma+\alpha}{\gamma-\alpha} + \alpha\ln\frac{\gamma+1}{\gamma-1} - \frac{2}{3}\frac{\gamma^3-(1+\alpha^3)}{\alpha}} \tag{8.28}$$

where $\gamma = \sqrt{1+\alpha^2}$. This is shown in Fig 8.13. For a rigid circular raft of radius a:

$$k_r = \frac{P_r}{w_r} = 4a\frac{G}{1-\nu} \tag{8.29}$$

The stiffness of the pile group is lower than the combined stiffness of the individual piles because of the interaction between adjacent piles (Fig 8.14, compare Fig 1.19). Each pile in elastic soil tends to pull down the surrounding soil in which the neighbouring piles are located. For n piles with individual pile stiffness k, the group stiffness k_p is reduced by the pile group efficiency η_w

$$k_p = \frac{P_p}{w_p} = \eta_w n k \tag{8.30}$$

Results of numerical analysis of typical pile groups discussed by Fleming *et al.* show that the pile group efficiency varies with the number of piles in a group:

$$\eta_w = n^{-\alpha} \tag{8.31}$$

and that the exponent $\alpha \approx 0.5$ for typical pile groups, so that

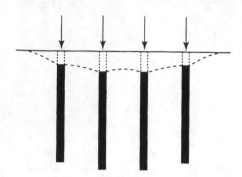

Figure 8.14: Interaction between piles in pile group

$$k_p \approx \sqrt{n}k \tag{8.32}$$

Now we have all the necessary elements to analyse the response of the piled raft. It seems to be reasonable to propose that the settlements of pile group and raft should be equal since the raft provides the cap and connection for the piles. Therefore:

$$w_p = w_r = w \tag{8.33}$$

and the overall stiffness of the piled raft can be obtained:

$$k_{pr} = \frac{P_p + P_r}{w} = \frac{k_p + (1 - 2\alpha_{rp})\,k_r}{1 - \alpha_{rp}^2 \frac{k_r}{k_p}} \tag{8.34}$$

The proportion of load taken by the raft is:

$$\frac{P_r}{P_p + P_r} = \frac{(1 - \alpha_{rp})\,k_r}{k_p + (1 - 2\alpha_{rp})\,k_r} \tag{8.35}$$

As an *example*, consider a raft of dimensions 15×15 m partially supported by a group of piles of length $\ell = 25$ m, radius $r_o = 0.3$ m, $E_p = 25$ GPa, $G = 25$ MPa, $\nu = 0.2$. The group is formed of 10×10 piles at a spacing of 1.5 m.

First we calculate the stiffness of individual piles. The details of the piles are the same as those of the example considered in section 8.2.3 and hence the stiffness is $k = 505$ MN/m.

The pile group contains 100 piles and hence the group stiffness is only 10% of the possible stiffness calculated from the individual piles. Hence: $k_p = \sqrt{n}k = 5051$ MN/m.

The raft stiffness is obtained from (8.27) with $\beta_z = 1.06$ appropriate to $L/B = 1$ and hence $k_r = 1656$ MN/m.

The overall stiffness of the piled raft is then, from (8.34), $k_{pr} = 5135$ MN/m and the proportion of the load taken by the raft is, from (8.35), 8.2%. The

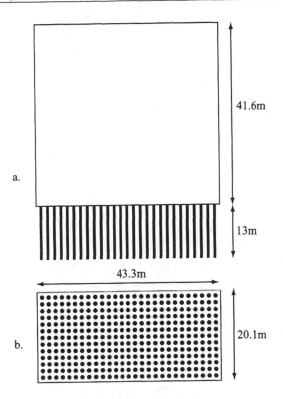

Figure 8.15: Stonebridge Park flats: schematic diagram of foundation arrangement (after Fleming *et al.*, 1985)

stiffness of the raft is about 1/3 of the stiffness of the pile group so it is not surprising that the overall stiffness is close to that of the pile group and the proportion of the load taken by the raft is in fact rather small.

The *example* of the Stonebridge Park flats is presented by Fleming *et al.* (1985) (Fig 8.15). This is a 16 storey building supported on a raft 43.3 m × 20.1 m, founded on London clay on 351 bored piles, 0.45 m diameter, 13 m long at ∼ 1.6 m centres made of concrete with $E_p = 25$ GPa. We analyse it here using the slightly simplified expressions that we have gathered together to illustrate the general performance of this type of foundation.

We first have to assess relevant stiffness properties for London clay. Following Fleming *et al.*, we propose that typically the undrained strength varies with depth according to a relationship $c_u = 100 + 7.2z$ (kPa). We might take $G/c_u \approx 200$ giving $G = 20 + 1.44z$ (MPa). Typically for London clay $\nu = 0.1$. In order to calculate the raft stiffness we might take as typical the shear stiffness at a depth $z = 2B/3$ giving $G = 39.3$ MPa. Then with $L/B = 2.15$, $\beta_z = 1.094$ and $k_r = 2819$ MN/m.

In order to calculate the stiffness of the pile group we might take as the typical governing stiffness of the soil, for estimation of the stiffness of shaft

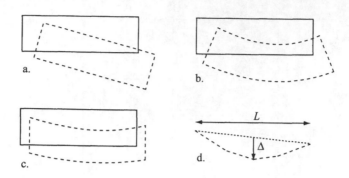

Figure 8.16: (a) Rigid body rotation; (b) bending of foundation; (c) shearing of foundation; (d) definition of deflection ratio Δ/L

load transfer, the value of shear stiffness at a depth $z = 2\ell/3$ giving $G = 32.5$ MPa. For the base of the pile, we should use a value G_b calculated at depth $z = \ell$, giving $G_b = 38.7$ MPa. Then for the calculation of single pile stiffness, $\eta\ell = 1.473$, $\tanh \eta\ell = 0.9$, $\xi = 11.64$, giving $k = 412$ MN/m. The pile group stiffness is calculated from this single pile stiffness: $k_p = k\sqrt{n} = 412\sqrt{351} = 7724$ MN/m. (With a more detailed analysis of pile group behaviour Fleming *et al.* estimate the pile group stiffness as $k_p = 5300$ MN/m.)

Combining these separate stiffnesses, we can estimate the overall stiffness of the piled raft to be $k_{pr} = 7871$ MN/m and for an overall building load of 156 MN the settlement would be 20 mm and the raft would be predicted to take 9.3% of the total building load. In fact measurements at Stonebridge Park showed that the settlement was actually about 25 mm, and the raft transmitted some 23% of the total load. The analysis is simplistic (there is also uncertainty about the choice of average shear stiffnesses for the soil) and does assume that the piles in the group are all behaving linearly and elastically. Back calculation using these actual observed figures suggests that the raft and pile group stiffnesses were nearer $k_r = 4051$ MN/m and $k_p = 5953$ MN/m respectively.

8.3   Serviceability calculations

Geotechnical design based on ultimate limit states provides some confidence that a geotechnical system will not actually fail. The calculation procedures are typically based around plasticity theories and have formed the backbone of geotechnical design for many years. Experience has then shown that satisfactory performance under working loads for standard geotechnical systems can be guaranteed provided the safety factor or load factor (whatever these terms may mean) is set at a sufficiently high level. However, this does not obviously provide a route by which the performance of geotechnical systems under working conditions can actually be computed. Many of the modelling techniques that have been presented in this book have been directed towards this end. Serviceability limits are usually limits on allowable displacements and the challenges relate to the calculation of these displacements. However, whereas the operation of a complete system of structures may be limited by total displacements, damage to individual structures is more controlled by *differential* movements—most frequently differential settlements. As noted by Burland and Wroth (1975) progress towards defining tolerable levels of differential settlement has been slow for various reasons:

- Serviceability is very subjective and must include both the function of the building and the potential reaction of the users.

- Buildings vary so much from one to another that it is difficult to produce general guidelines for allowable movements.

- Buildings, including foundations, seldom perform exactly as designed because of the modelling idealisations adopted during the design process.

- There is a link between development of differential settlement and the stage of construction: much of the eventual structural movement will have occurred before the cladding (which is often particularly susceptible to visible damage) is added.

There are obviously major differences between levels of displacement that lead to concern for overall structural performance and those which produce very visible damage to structural finishes—many of which tend to be quite brittle.

To be able to propose allowable levels of differential settlement it is necessary to have some accepted terms for its characterisation. If a structure merely rotates as a rigid body (Fig 8.16a) then, while it may cause other serviceability difficulties, it should not lead to internal cracking. However, any profile of foundation movements which leads to a 'shearing' of the structure (Fig 8.16b, c) will certainly tend to exacerbate such damage. There are various ways in which relative foundation rotation can be defined (and some will be better suited for different applications): for the present purposes a *deflection ratio*, Δ/L (Fig 8.16d) (Burland and Wroth, 1975) will suffice. Then one can interpret the limiting values quoted by IStructE *et al.* (1989) to give the values in Table 8.1.

Burland and Wroth suggest that a more useful indicator of likely damage from differential settlement can be obtained by estimating the maximum tensile strain that can develop within the structure and then comparing this against critical tensile strains for different configurations. A deep beam of length L and height H (Fig 8.17a) can provide an analogue of a structure. The beam is assumed to be made of elastic material with Young's modulus E and shear modulus G. Tensile strains can be generated through direct tensile strain in sagging or hogging (Fig 8.17b, c) and through diagonal tensile strain resulting from shear (Fig 8.17d).

For such a deep beam loaded with a single central load P, analysed as a plane stress problem with negligible out-of-plane stresses, Timoshenko and Goodier (1970) show that the deflection ratio is:

$$\frac{\Delta}{L} = \frac{PL^2}{48EI}\left(1 + \frac{18EI}{L^2HG}\right) \tag{8.36}$$

Table 8.1: Limiting values of deflection ratio Δ/L (interpreted from values quoted by IStructE *et al.* (1989))

structural type	damage	Δ/L
framed buildings	structural damage	1/500-1/300
framed buildings	cracking in walls	1/1000
reinforced load-bearing walls	structural damage	1/300
reinforced load-bearing walls	cracking	1/1000
unreinforced load-bearing walls	visible cracking (sagging)	1/2500-1/1250
unreinforced load-bearing walls	visible cracking (hogging)	1/5000-1/2500

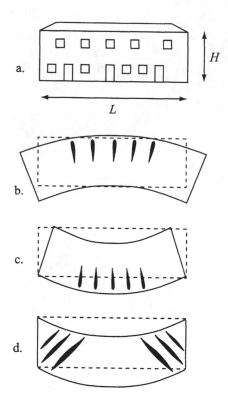

Figure 8.17: (a) Building as equivalent deep beam; (b) 'hogging' tensile cracking; (c) 'sagging' tensile cracking; (d) diagonal shear tensile cracking (after Burland and Wroth, 1975)

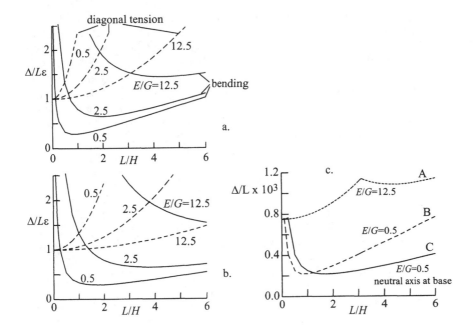

Figure 8.18: (a) Normalised deflection ratio for deep beams with neutral axis at centre; (b) normalised deflection ratio for deep beams with neutral axis at base; (c) tolerable differential settlements (limiting strain 0.075%) A: framed structure; B: load-bearing walls in sagging; C: load-bearing walls in hogging (inspired by Burland and Wroth, 1975)

where the second moment of area $I = H^3 B/12$ for a thin beam of width $B \ll H$. The second term shows the contribution of shearing to the differential settlement of the beam.

Using the expressions that Timoshenko and Goodier give for the field of displacements within the beam, the strain components can be deduced at any location. The maximum direct tensile bending strain occurs at the extreme edge of the beam under the load and is

$$\epsilon_t = \frac{PLH}{8EI} \tag{8.37}$$

The maximum diagonal tensile strain occurs at the centre of the beam at the supports and is

$$\epsilon_d = \frac{PH^2}{32IG} \tag{8.38}$$

If we have some idea about the limiting tensile strain, ϵ_{lim}, that the materials of our structure can withstand then we can use these expressions to deduce the corresponding limiting values of deflection ratio Δ/L. The ratios $\Delta/L\epsilon_t$ (solid lines) and $\Delta/L\epsilon_d$ (dashed lines) are shown as a function of the proportions of the structure L/H in Fig 8.18a for $E/G = 2.5$ corresponding to Poisson's ratio $\nu = 0.25$. It is suggested that load-bearing walls correspond to a modulus ratio of this order whereas a framed structure will be relatively much less stiff in shear so that a modulus ratio $E/G = 12.5$ (say) might be more appropriate (we temporarily suspend our incredulity at the implied value of Poisson's ratio and the consequent implications for the applicability of Timoshenko and Goodier's analysis)—ratios $\Delta/L\epsilon_t$ and $\Delta/L\epsilon_d$ for this value are also shown in Fig 8.18a.

On the other hand Burland and Wroth suggest that load-bearing walls which are subjected to hogging deformations as a result of foundation movement are essentially restrained at their contact with the ground so that they are somewhat equivalent to one half of a deep beam of depth $2H$. All the expressions are modified correspondingly, with H replaced by $2H$ and noting that the value of I will also change. They suggest that such structures are very stiff in shear and that a value $E/G = 0.5$ (say) will be appropriate. The corresponding ratios $\Delta/L\epsilon_t$ and $\Delta/L\epsilon_d$ are shown in Fig 8.18b. As would be expected, load bearing masonry is much more susceptible to damage resulting from differential settlement than framed structures.

Typical values of the limiting strain ϵ_{lim} might be 0.05-0.1% for brickwork and blockwork set in cement mortar; 0.03-0.05% for reinforced concrete (IStructE et al., 1989). Fig 8.18c is plotted using a critical strain of 0.075% to scale the appropriate curves from Figs 8.18a, b for three cases:

- A framed structures, with $E/G = 12.5$;

- B load bearing walls in sagging $E/G = 0.5$; and

- C load bearing walls in hogging $E/G = 0.5$ with the neutral axis at the base.

From comparisons with records of building damage for many different classes of building, Burland and Wroth conclude that a damage criterion based on limiting tensile strain works quite satisfactorily. Further advantages of this approach are noted by IStructE et al. (1989):

- It helps the engineer to decide whether a structure is vulnerable to cracking and where vulnerability to cracking will be greatest.

- It can be combined with complex structural analysis techniques.

- It makes it clear that damage can be controlled by concentrating attention on the modes of deformation of the structure.

- The value of ϵ_{lim} can be chosen to reflect the actual cracking proclivity of the materials that are being used—use of soft bricks and lean mortar (increasing ϵ_{lim}) reduces the likelihood of cracking.

Evidently the onset of cracking does not necessarily indicate the limit of serviceability of the structure or building.

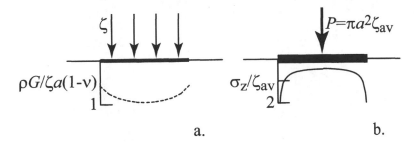

Figure 8.19: (a) Profile of settlement beneath flexible uniformly loaded circular foundation; (b) contact stress distribution beneath rigid circular foundation

8.4   Relative foundation stiffness

If a uniform pressure ζ is applied over a circular area of radius a at the surface of an elastic half-space of shear modulus G and Poisson's ratio ν, then the surface settlement ρ under the footing, normalised as $\rho G/(1-\nu)\zeta a$ varies with radius as shown in Fig 8.19a (adapted from Poulos and Davis, 1974). If a rigid circular foundation of radius a is pushed into the surface of an elastic half-space the settlement under the foundation is of course uniform but the vertical contact stress σ_z varies with radius r:

$$\frac{\sigma_z}{\zeta_{av}} = \frac{1}{2\sqrt{1-(r/a)^2}} \tag{8.39}$$

where ζ_{av} is the average load applied to the foundation. This contact stress variation is shown in Fig 8.19b. Evidently there is a singularity at the edge of the footing where the contact stress is theoretically infinite. The normalised settlement $\rho G/\zeta_{av} a$ of the rigid circular foundation is $(1-\nu)\pi/4$ (§3.2.3, §8.2.4).

In presenting the results of the analysis of the beam on an elastic foundation consisting of a bed of independent springs (section §8.2.1) we observed (Fig 8.5) that *relative* stiffness of beam and foundation had a dramatic effect on the maximum bending moment in the beam and on the differential settlement. Very stiff beams do not develop differential settlement but do develop large bending moments. This beam on an elastic foundation was a simple analogue which was capable of explicit theoretical analysis. We need now to generalise the concept of *relative stiffness* to more realistic foundation forms and to show some of the implications for foundation response (compare §5.2.5).

For circular foundation rafts of radius a, Brown (1969) defines a relative stiffness:

$$K_B = \frac{E_r}{E}\left(1-\nu^2\right)\left(\frac{t}{a}\right)^3 \tag{8.40}$$

where t and E_r are foundation thickness and Young's modulus and E and ν are Young's modulus and Poisson's ratio for the soil.

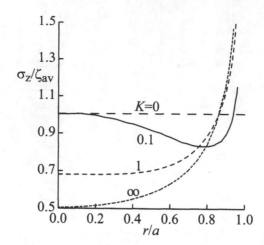

Figure 8.20: Contact stresses beneath circular foundations of different relative stiffnesses (adapted from Brown, 1969)

For other shapes of raft foundation a number of different though somewhat similar expressions have been proposed for relative stiffness. For example, for rectangular rafts of width B, Fraser and Wardle (1976) suggest

$$K_{FW} = \frac{4}{3} \frac{E_r}{E} \frac{(1 - \nu^2)}{(1 - \nu_r^2)} \left(\frac{t}{B}\right)^3 \tag{8.41}$$

and there is an obvious logic to the inclusion of the Poisson's ratio ν_r of the raft to give a combination $E_r t_r^3 / (1 - \nu_r^2)$ which relates (within a constant of proportionality) directly to the flexural rigidity of a plate. (For concrete, if Poisson's ratio $\nu_r \approx 0.2$, then the modifying term $(1 - \nu_r^2) = 0.96$ and its effect may be seen as practically negligible.) However, Horikoshi and Randolph (1997) propose that some rationalisation of results can be obtained if a relative stiffness is defined in such a way that results for rectangular rafts of different proportions (different values of L/B) and for circular rafts can be somewhat unified. Their relative stiffness is:

$$K = \pi^{3/2} \frac{E_r}{E} \frac{(1 - \nu^2)}{(1 - \nu_r^2)} \left(\frac{B}{L}\right)^{1/2} \left(\frac{t}{L}\right)^3 \tag{8.42}$$

where the factor $\pi^{3/2}$ is chosen to give consistency between square and circular rafts of the same area. The exponent on the term $(B/L)^{1/2}$ was chosen to give the best match to numerical results obtained for a wide range of foundation proportions and material properties (for $L/B = 2$, $K = 0.369 K_{FW}$). For a circular foundation equivalence of areas implies that $L = B = a\sqrt{\pi}$ and K differs from K_B (8.40) only by the factor $(1 - \nu_r^2)$.

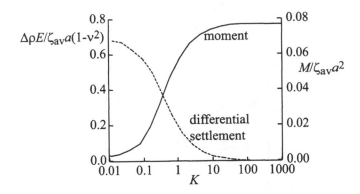

Figure 8.21: Dependence of maximum moment and maximum differential settlement on relative foundation stiffness for circular foundations (adapted from Brown, 1969)

Contact stress distributions for different values of this relative stiffness $K_B \approx K$ are shown in Fig 8.20[3]. The edge singularity remains but becomes less significant as the relative stiffness falls (the curve for $K = \infty$ is the same as the curve for the rigid foundation shown in Fig 8.19b).

More usefully from a structural point of view, Fig 8.21 (following on from Fig 8.5[4]) shows how the maximum moment in the foundation and the differential settlement vary with relative stiffness—the results turn out to be rather independent of the value of ν. For $K > 5$ a circular raft can be classified as rigid, for $K < 0.08$ it can be classified as flexible. (As the flexibility of the raft increases, K reduces, the location of the points of maximum moment moves from the centre towards the edge of the raft.)

Detail of settlements and differential settlements of points on rectangular rafts with aspect ratio $L/B = 2$ is shown in Fig 8.22 (adapted from Fraser and Wardle (1976) to show the effect in terms of relative stiffness K). The settlements or differential settlements ρ_i can be calculated from the influence factors I_i plotted: $\rho_i = I_i \zeta_{av} B(1 - \nu)/2G$, where ζ_{av} is the average foundation pressure ($\zeta_{av} = P/BL$ for total foundation load P). Plausible weighting of the several settlements could be used to test the assertion that the average settlement is rather independent of the relative foundation stiffness (§8.2.4).

Horikoshi and Randolph suggest, from numerical analysis, that, using (8.42) to characterise relative foundation stiffness, the differential settlement between the centre of a rectangular raft and the mid-point of the short edge is rather independent of the aspect ratio L/B, provided this differential settlement is normalised with the average foundation settlement for the rectangular raft calculated using (8.27) (Fig 8.23). Differential settlement is obviously slightly

---

[3]Fig 8.20 is adapted from numerical results presented by Brown (1969) computed for $\nu_r = 0.3$ but the effect of changing ν_r to 0.15 is negligible—and merely changes the ratio K/K_B from 0.91 to 0.98.

[4]$\mu\ell$ in Fig 8.5 is a ratio of ground stiffness to structural stiffness; K in (8.42) is a ratio of structural stiffness to ground stiffness.

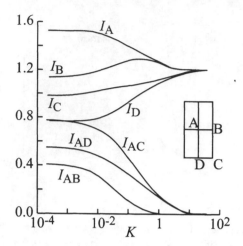

Figure 8.22: Settlement and differential settlement coefficients for rectangular foundations with $L/B = 2$ as function of relative foundation stiffness (adapted from Fraser and Wardle, 1976)

greater between a corner and the centre, but Horikoshi and Randolph suggest that the corners of rafts will in practice often be stiffened by the presence of walls. The result for circular rafts in Fig 8.23 falls between the edge and corner values of differential settlement.

Further unification of results is proposed by Horikoshi and Randolph for estimation of maximum moments in rectangular rafts. Normalised bending moments $M_{yy}/\zeta_{av}L^2$ for bending of rectangular rafts supporting average foundation pressure ζ_{av} along their length (with the y axis parallel to the shorter sides) are shown in Fig 8.24a for rigid rafts (Horikoshi and Randolph, 1997). The variation of maximum moment in *flexible* rafts normalised with the value for rigid rafts is then found to be essentially independent of raft aspect ratio (Fig 8.24b).

There are obvious parallels in the complementary links between differential settlement and maximum bending moment and relative stiffness shown in Figs 8.5, 8.21, 8.23 and 8.24b[5].

8.5   Downdrag on pile

All the examples of soil-structure interaction that we have considered so far have treated all aspects of the problem as entirely elastic. The next few examples allow in different ways for some nonlinearity of material or interface response to enter the analysis.

---

[5]Noting again the essentially reciprocal versions of relative stiffness implied by the definitions of $\mu\ell$ (Fig 8.5) and K (Figs 8.21, 8.23 and 8.24b).

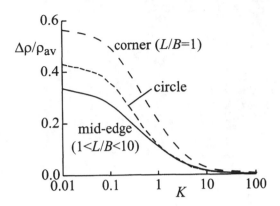

Figure 8.23: Differential settlements as function of relative foundation stiffness (after Horikoshi and Randolph, 1997)

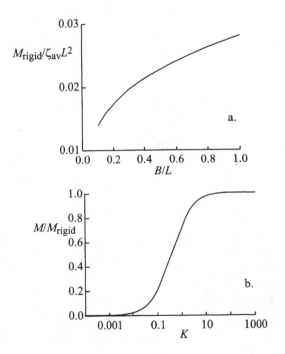

Figure 8.24: (a) Maximum central moment in rigid rectangular rafts; (b) maximum central moment in flexible rafts (after Horikoshi and Randolph, 1997)

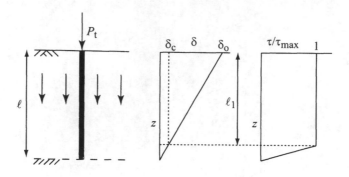

Figure 8.25: Downdrag of end-bearing pile

Table 8.2: Pile shaft friction: key properties (Lancellotta, 1987; Fleming *et al.*, 1985)

soil	$\beta = K \tan \phi'$	δ_c/r_o
clay	0.2-0.25	1-4%
silt	0.25-0.35	
sand	0.35-0.50	0.25%

Piles are often rather stiff structural members transferring load *through* soft soils to firm underlying ground (Fig 8.25). We might naïvely assume that the load that is imposed on the pile at its top from the above-ground structure is always the largest load that the pile will experience. However, if the soft soils around the pile deform then this deformation imposes additional loads on the piles which the piles must be designed to resist. We concentrate here on axial load effects but transverse load effects, leading to unintended bending moments in the piles are often more serious: some aspects of these are considered in section 8.9.

Skin friction τ develops with relative movement between pile and soil. Typically, the maximum shaft friction that can be developed can be related to *in-situ* vertical effective stress σ'_v through an earth pressure coefficient K and the angle of internal friction ϕ' (where the value of K incorporates allowance for any reduction in ϕ' occurring at the soil-pile interface):

$$\tau_{max} = K\sigma'_v \tan \phi' \tag{8.43}$$

with typical values of $\beta = K \tan \phi'$ given in Table 8.2:

The relative movement δ_c required to mobilise the maximum shaft friction is proportional to pile radius r_o. Typical values are also shown in Table 8.2. We assume for simplicity that the shaft resistance is mobilised linearly with relative movement (Fig 8.26).

As an *example*, consider an end bearing pile of length ℓ and radius r_o in consolidating clay (Fig 8.25). The ultimate load that can be carried by the pile

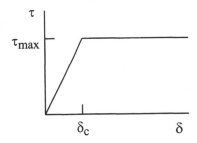

Figure 8.26: Assumed mobilisation of shaft friction with relative pile-soil movement

in shaft friction—when the pile is moving *down* relative to the soil—is $P_{ult} = \pi\beta\gamma'r_o\ell^2$ assuming that vertical effective stress is being generated uniformly with depth in a soil with buoyant unit weight γ'.

Suppose that the settlement of the clay is δ_o at the ground surface, decreasing linearly with depth (Fig 8.25). Then the depth to which shaft friction is fully mobilised is ℓ_1 given by, for $\delta_o > \delta_c$,

$$\frac{\ell_1}{\ell} = 1 - \frac{\delta_c}{\delta_o} \tag{8.44}$$

The extra load at the toe of pile is ΔP given by

$$\Delta P = \int_0^\ell 2\pi r_o \tau \mathrm{d}z = 2\pi r_o \left[\int_0^{\ell_1} \beta\sigma'_v \mathrm{d}z + \int_{\ell_1}^\ell \frac{z - \ell_1}{\ell - \ell_1} \beta\sigma'_v \mathrm{d}z \right] \tag{8.45}$$

giving

$$\frac{\Delta P}{P_{ult}} = \frac{2\ell^2 - \ell\ell_1 + 2\ell_1^2}{3\ell^2} = 1 - \frac{\delta_c}{\delta_o} + \frac{2}{3} \left(\frac{\delta_c}{\delta_o} \right)^2 \tag{8.46}$$

Fig 8.27 shows how the load builds up as the surface displacement of the settling soil increases relative to the critical interface displacement δ_c. For $\delta_o \gg \delta_c$, $\Delta P/P_{ult} = 1$ and the axial load arising from the settlement of the surrounding ground is in addition to any axial load the pile may already be carrying. For *example*, with $\gamma' = 5$ kN/m^3 (for soft clay), $r_o = 0.25$ m, $\ell = 15$ m, $\beta = 0.25$, $\Delta P = 221$ kN.

The ultimate load capacity of the pile is not affected by downdrag: failure will imply sufficient *downward* movement of the pile to generate full *positive* shaft friction. However, if the pile design has essentially neglected the contribution to ultimate load that the shaft friction in the soft soils may provide then there may, even under loads close to failure, be insufficient downward movement of the pile relative to the settling ground to ensure that negative shear stresses are not mobilised throughout its length.

In a closely spaced pile group, equilibrium shows that the skin friction on the piles must in fact reduce the vertical effective stress in the soil. For *example*,

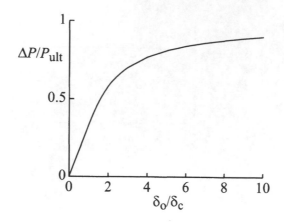

Figure 8.27: Development of additional axial downdrag load in end-bearing pile with increasing ground settlement

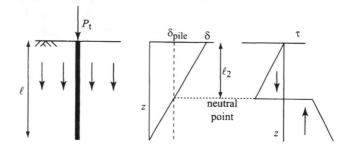

Figure 8.28: Downdrag of 'floating' pile

with piles of length ℓ at spacing D on a square grid the effective weight of the unit cell block of soil around each pile is $\gamma' D^2 \ell$ and the load shed into the pile from the settling soil is potentially $\pi \beta \gamma' r_o \ell^2$. For the piles of the previous example at $D = 2$ m spacing the ratio of shaft friction to block weight $\gamma' D^2 \ell$ appears to be 74% which would obviously be too high to be neglected (the reduction in vertical effective stress would actually affect the calculation of skin friction). With a spacing of 5 m this ratio falls to 12%.

Floating friction piles (Fig 8.28) are slightly more complicated. The skin friction contributions need to balance the applied load P_t to give overall equilibrium. There is a neutral point at depth ℓ_2 at which there is no relative displacement of pile and soil and at which the direction of the shaft shear stress reverses. For simplicity we will assume that $\delta_c \approx 0$ so that shaft friction—either upwards or downwards—is always fully mobilised. We also assume that the pile is rigid. Then equilibrium requires that:

$$P_t + \int_0^{\ell_2} \beta\gamma'2\pi r_o zdz = \int_{\ell_2}^{\ell} \beta\gamma'2\pi r_o zdz \qquad (8.47)$$

giving

$$\left(\frac{\ell_2}{\ell}\right)^2 = \frac{1}{2}\left(1 - \frac{P}{\pi\beta\gamma'r_o\ell^2}\right) = \frac{1}{2}\left(1 - \frac{P_t}{P_{ult}}\right) \qquad (8.48)$$

where $P_{ult} = \pi\beta\gamma'r_o\ell^2$ is the overall capacity of this friction pile. The variation of axial load in the pile is then, for $0 < z < \ell_2$:

$$\frac{P}{P_{ult}} = \frac{P_t}{P_{ult}} + \left(\frac{z}{\ell}\right)^2 \qquad (8.49)$$

and for $\ell_2 < z < \ell$:

$$\frac{P}{P_{ult}} = \frac{P_t}{P_{ult}} + 2\left(\frac{\ell_2}{\ell}\right)^2 - \left(\frac{z}{\ell}\right)^2 = 1 - \left(\frac{z}{\ell}\right)^2 \qquad (8.50)$$

The variation of axial load is shown in Fig 8.29a for different values of P_t/P_{ult}.

Again we note that the ultimate capacity of the pile is not affected. If the pile is moving downwards sufficiently far to generate *positive* shaft friction, the limiting curve in Fig 8.29a corresponds to $P_t/P_{ult} = 1$ and then the axial load falls parabolically from the top of the pile. However, with the applied load at the top of the pile P_t using only a fraction of the available shaft resistance the maximum load in the pile P_{max}, which will occur at the neutral point $z = \ell_2$, may be greatly magnified. The variation of this magnification ratio P_{max}/P_t with apparent pile load factor P_t/P_{ult} is shown in Fig 8.29b. The input load at the ground surface P_t may provide only a very unsafe estimate of the maximum axial load in the pile.

8.6 Settlement reducing piles

In analysing the interaction of pile and raft in piled raft foundations in section 8.2.4 we assumed that both the raft and the piles were behaving elastically. The movement necessary to fully mobilise shaft friction may, as just seen, be quite small. Inelastic shaft load transfer will vitiate the elastic analysis but it is possible to exploit this characteristic of pile response.

The cost of a foundation raft will vary with its stiffness. We have seen that rafts need to be stiff (relative to the soil) in order to maintain differential settlements at an acceptable level (which may be necessary to reduce the likelihood of development of cracking in the structure being supported (§8.3) and we have observed a 'complementarity' between variation of contact pressure and variation of foundation settlement across the foundation as the relative stiffness is changed from 0 to ∞ (Fig 8.19). However, the flexible foundations that we have studied only develop large differential settlements because they have been

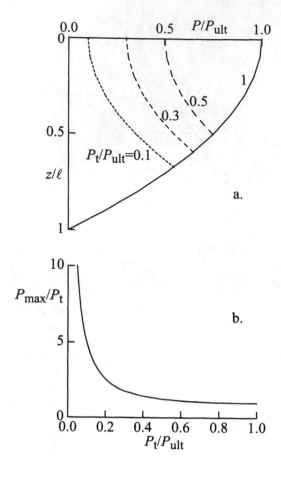

Figure 8.29: (a) Axial load distribution in floating piles; (b) magnification of axial load as function of load applied at top of pile

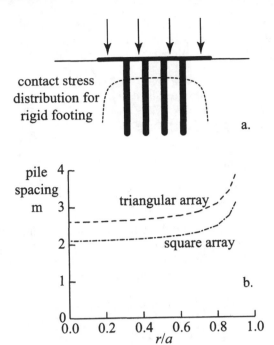

Figure 8.30: (a) Central piles used to relieve pressure under flexible foundation (contact stresses for rigid circular foundation shown); (b) spacing of settlement-reducing piles under flexible circular tank foundation

subjected to uniform loads. If we subjected our flexible foundation to a distribution of pressures which exactly matched the pressure distribution for a rigid foundation of the same shape then it would settle uniformly with no differential settlement. That is an ideal which we cannot realise but we can use friction piles, judiciously located, and operating more or less at their ultimate capacity to achieve something tending towards this ideal result (Fig 8.30a) (Burland *et al.*, 1977).

As an *example*, consider a tank 17 m diameter founded on soft clay applying an average design loading of 160 kPa. The object is to find an arrangement of individual piles which could be combined with a flexible raft to produce greatly reduced differential settlements. Let us assume that the piles are 250 mm square concrete friction piles, 10 m long generating a shaft capacity of 500 kN each in soil with undrained strength of the order of 50 kPa. We will assume—but must subsequently check—that this shaft capacity will be fully mobilised by the actual tank settlements.

An iterative approach is required in order to choose the number of piles. The tank supplies a loading 160 kPa over a circle of 8.5 m radius; use n piles to reduce average loading to $\zeta_{av} = 160 - (500 \times n)/(\pi \times 8.5^2)$. Try $n = 30$: then $\zeta_{av} = 94$ kPa. Now compare the required loading (160 kPa) with the contact stress for a rigid raft and use piles, at an appropriate spacing, to take the difference. The contact stress for a rigid circular raft depends on radius (8.39), and consequently the required distribution of piles also depends on radius.

- At the centre, $r/a = 0$, $\sigma_z/\zeta_{av} = 0.5$, $\sigma_z = 47$ kPa. The piles have to take $160 - 47 = 113$ kPa which requires $113/500 = 0.226$ piles/m^2 or 4.4 m^2/pile. This can be achieved with a square grid with spacing 2.1 m or a triangular grid with spacing 2.6 m.

- At $r/a = 0.5$, $\sigma_z/\zeta_{av} = 0.577$, $\sigma_z = 54$ kPa, and the pile spacing becomes 2.2 m (square) or 2.7 m (triangular).

- At $r/a = \sqrt{3/2}$, $\sigma_z/\zeta_{av} = 1$, $\sigma_z = 94$ kPa, and the pile spacing becomes 2.8 m (square) or 3.4 m (triangular).

The variation of pile spacing with radius is shown in Fig 8.30b. In principle the contact stress for the rigid raft is only greater than the average applied stress for radii greater than $\sqrt{3}/2 = 0.87$ times the foundation radius so that the load needs to be shed into the piles until we get quite close to the edge of the foundation.

To estimate the raft settlement we need to estimate the shear stiffness using perhaps $G = 200c_u = 10$ MPa; we might take Poisson's ratio for the soil $\nu = 0.3$. For a foundation of radius 8.5 m carrying an effective load of 94 kPa the settlement, from (8.29) is 44 mm which is about 18% of the pile size and should certainly be adequate to mobilise full shaft resistance.

We might note that the 351 piles supporting the raft for Stonebridge Park (§8.2.4) are not being used very efficiently. Let us explore an alternative design using just 40 piles of the same size (0.45 m diameter, 13 m long). A similar treatment can be used as for the circular tank foundation looking at the contact pressure distribution for a rigid rectangular raft. The raft at Stonebridge Park has $L/B = 43.3/20.1 = 2.15$ so some modest interpolation is required to produce Fig 8.31 from the distributions presented by Poulos and Davis (1974) for rafts of proportions $L/B = 1.5$ and 4.

First we need to assess the shaft capacity of the piles. For the heavily overconsolidated London clay we might choose to use a total stress approach in which the shaft friction is linked with undrained strength $\tau_{max} = \alpha c_u$. In our previous analysis (§8.2.4) we assumed that the undrained strength variation with depth was given by $c_u = 100 + 7.2z$ kPa. The strength at the mid-height of the piles is thus 146.8 kPa. With a shaft friction factor $\alpha = 0.4$ for an overconsolidated clay, the shaft friction is then $\pi \times 0.45 \times 13 \times 0.4 \times 146.8 = 1079$ kPa and 40 piles, working close to their ultimate load, will absorb 43.2 MN of the total load, leaving $156 - 43.2 = 112.8$ MN to be transferred by the raft. (Base resistance will be a stress of the order of $9c_u$ giving a contribution to capacity of 277 kN per pile but this will be mobilised much more slowly.) The average raft stress is then $112.8/(43.3 \times 20.1) = 130$ kPa. The overall raft stiffness will be independent of its flexural rigidity and remains 2819 MN/m. The settlement will now be 40 mm, adequate to mobilise the shaft friction of our piles which are being assumed to respond essentially perfectly plastically and therefore contribute no stiffness to the combined piled raft.

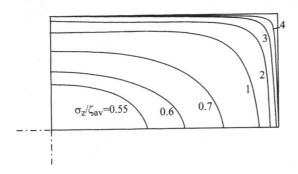

Figure 8.31: Contact stresses under rigid rectangular foundation with $L/B = 2$ (interpolated from results presented by Poulos and Davis, 1974)

At the centre of the raft the contact pressure is about half the average pressure, 65 kPa. The average contact stress without the piles is $156/(43.3 \times 20.1) = 179$ kPa so that the centre piles must support $179 - 65 = 114$ kPa requiring 9.46 m^2/pile or spacing 3.1 m. Where the contact stress is reduced to $0.7\zeta_{av}$ the spacing can be increased to 3.5 m. The need for iteration becomes clear—we need to ensure that we have enough piles to provide the necessary reduction in contact stress over a sufficient proportion of the raft foundation.

Horikoshi and Randolph (1995) perform a more elaborate analysis of the Stonebridge Park raft and emerge with a redesign using just 18 piles, slightly larger and longer, placed over the central part of the raft. The design is supported by numerical modelling using a specialist piled raft analysis program. The computed settlement is 42 mm but the differential settlements are very small (these are harder to estimate using the simplistic approach outlined here).

Love (2003) describes the recent design of a foundation in London making use of settlement reducing piles but notes both that this type of foundation has not been widely used and that, for it to be successful, the pile capacity must be rather accurately assessed so that the necessary perfect plasticity under working conditions of the structure can be guaranteed. The guiding principle is evidently not 'how many piles are needed to carry the entire weight of this building?' but 'how many piles are needed to reduce differential settlements to an acceptable level?' (Fleming *et al.*, 1985).

8.7 Flexible retaining wall

Commonly adopted calculation procedures used for the design of retaining structures assume that the earth pressure varies linearly with depth according to some coefficient of earth pressure computed from a Rankine limiting stress field or a Coulomb critical wedge. Even when the design approach attempts to allow for a less than complete mobilisation of the available frictional strength of the soil—either because of guidance provided by a code of practice (for example, EC7, 1995 or BS8002, 1994) or as a result of theoretical insight (§7.5.2)—the

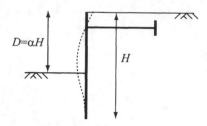

Figure 8.32: Anchored flexible wall

calculation procedure still usually uses the language of coefficients of earth pressure. There are three reasons why this approach may be inadequate:

- The soil behaviour is not rigid perfectly plastic but usually steadily non-linear. This nonlinearity can to some extent be accommodated through the use of a nonlinear mobilisation model.

- The adoption of coefficients of earth pressure assumes that the interaction of the wall and the soil is effectively through a series of independent springs. The response of one spring is in no way influenced by the response of adjacent springs—whereas we might anticipate that there would be variations in degree of mobilisation as the soil manages to arch across regions of reduced stiffness[6].

- The flexibility of the structure has not entered the calculation but it will certainly influence the way in which the wall deforms and the earth pressures are generated.

Allowance for these effects can be achieved through full numerical analysis of the soil and wall together. However, some empirical allowance for the flexibility of the structure can be made using as inspiration results of experiments performed by Rowe (1952) and associated numerical analyses (Rowe, 1955). We will concentrate on anchored flexible walls (Fig 8.32). Further discussion is given by Powrie (1997).

Rowe (1952) tested model walls of heights up to 0.91 m, anchored at various points near the top, carrying various levels of surcharge behind the wall, and retaining various different dry granular materials. The soil was first filled uniformly on both sides of the wall and then excavated in stages in front of the wall, with measurements of prop loads and bending strains in the wall being taken during this excavation. Typical effects of wall flexibility and soil density on the wall displacements and distributions of stress on the wall in the passive zone are shown in Fig 8.33. The simplistic calculation assumes that the wall rotates rigidly about the prop. A flexible wall bows out towards the excavation

---

[6]Compare Fig 8.5, where the differential settlement for an extremely flexible beam on a bed of springs increases indefinitely with reduction in beam stiffness (increase in $\mu\ell$), with Fig 8.21 where the continuity of the elastic material manages to restrain the differential settlement even when the relative stiffness K becomes very small.

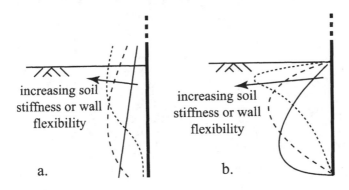

Figure 8.33: Schematic effects of soil density and wall stiffness on (a) deflected profile and (b) horizontal stresses in passive region (inspired by Rowe, 1952 and Clayton and Mililitsky, 1986)

but experiences hardly any outward movement towards its embedded toe. Consequently very little passive pressure is generated towards the toe and, further, the centre of action of the passive pressure moves up towards the dredge level. The maximum moment that is generated in the wall will of course be strongly influenced by the lines of action of the active and passive forces: if the passive force moves up then the maximum moment will tend to fall.

Rowe presented his results in terms of the reduction of moment as a function of some measure of wall flexibility expressed as a 'flexibility number' $\rho = H^4/EI$ for a wall of *total* height H and flexural rigidity EI per unit width (Fig 8.34). The results of many tests show modest variation of moment reduction with embedment ratio $\alpha = D/H$ where D is the depth of excavation (Fig 8.32), and more significant variation with soil density.

Unfortunately the flexibility number ρ is not dimensionless. Rowe measured E in pounds per square inch, I in inches[4] per foot and H in feet. The conversion to SI units requires that $\rho_{SI} = 916.5\rho_{Rowe}$. With E in kPa, I in m^4/m and H in metres the value of ρ has units m^3/kN. Fig 8.34 is plotted with these units for ρ.

Rowe notes that there is a critical value of ρ below which there is no reduction of moment, reached when the deflection of the wall at the level of the base of the excavation is equal to the deflection of the wall at its toe. He deduces from his model wall tests and from one-dimensional compression tests on the various soils that this critical flexibility number $\rho_{crit} \approx 630/E_{oed}$ m^3/kN, where E_{oed} (measured in kiloPascals) is an estimate of the one-dimensional stiffness of the soil.

Rowe is emphatic that this moment reduction factor is linked with the rise in location of the resultant passive force and 'has nothing whatever to do with arching on the active pressure side of the wall'. This then justifies the use by Rowe (1955) of a numerical analysis of the bending of a wall which treats the soil as a series of independent springs in order to explain his earlier

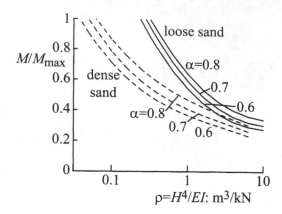

Figure 8.34: Reduction in moment dependent on soil density, wall geometry, and wall flexibility: collected results from model tests (after Rowe, 1952)

experimental findings. He uses a spring stiffness k for the development of passive pressure which varies linearly with depth below the excavation $k = mz$. We could interpret this as indicating a soil stiffness directly dependent on overburden stress. The parameter m has dimensions of stress/length: Rowe's results indicate $m \approx 200$ MPa/m for dense sand and $m \approx 9$ MPa/m for loose sand. The moment reduction factors (and reduction factors for the prop force) are now presented in terms of $m\rho$ which is a dimensionless factor (Fig 8.35). The mean curves from Fig 8.34 are also included in this figure using Rowe's values of m.

Alternative guidance on selection of values of m is provided by Terzaghi (1955) (using a different notation). He proposes values of about 7 MPa/m and 0.9 MPa/m for dense and loose sand respectively on the basis of large model wall tests. The order of magnitude difference presumably reflects the greater range of deflection and greater dimensions in Terzaghi's tests and hence the greater degree of soil nonlinearity and decrease in stiffness from an initial value.

As a simple *example* of the application of Rowe's correction curves let us consider the design of an anchored quay wall with free height 10 m in dense sand with $\phi' = 38°$ and total unit weight $\gamma = 20$ kN/m^3 (Fig 8.36). The water level is at the ground surface and the anchor is also at the ground surface. We need to estimate the required depth of embedment, the anchor force, and the maximum bending moment. From the maximum moment (allowing for moment reduction through pile flexibility) we will be able to choose an appropriate sheet pile section.

The standard route to design uses the free earth support method which we have introduced previously (§7.5.2) and which assumes stress distributions associated with ultimate limit states (though possibly with some modification of soil parameters). Thus we assume active pressure behind the wall, and passive pressure in front. Moment equilibrium and horizontal equilibrium allow us to

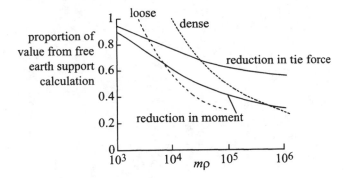

Figure 8.35: Reduction of maximum moment and anchor force from numerical analysis of anchored flexible walls (after Rowe, 1955): dotted curves are taken from experimental curves of Fig 8.34, for $\alpha = 0.7$, with $m = 200$ MPa/m and 9 MPa/m for dense and loose sand respectively

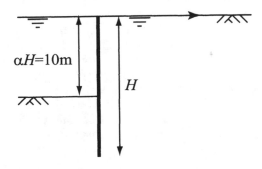

Figure 8.36: Example anchored flexible wall

Table 8.3: Estimate of ϕ' for siliceous sands and gravels (after BS8002, 1994)

contributory factor	value
angularity	$A°$
rounded	0
sub-angular	2
angular	4
grading	$B°$
uniform $(D_{60}/D_{10} < 2)$	0
moderate $(2 < D_{60}/D_{10} < 6)$	2
well graded $(D_{60}/D_{10} > 6)$	4
modified SPT N value	$C°$
$N' \approx 225/(\sigma'_v + 55) - 0.2$ with σ'_v in kPa	
$N' < 10$	0
$10 < N' < 20$	2
$20 < N' < 40$	6
$40 < N' < 60$	9

$$\phi' = 30 + A + B + C°$$
$$\phi'_{crit} = 30 + A + B°$$

calculate the depth of embedment, the anchor force and the moments in the wall.

First of all we have to choose an appropriate route by which to introduce factors of safety in our initial ultimate limit state calculation. There are many different proposals. We have already seen (§7.5.2) the consequences of the requirement for kinematic compatibility of displacements on the two sides of the wall and the consequence that limiting active conditions are usually attained more rapidly than limiting passive conditions (but this will depend on the initial horizontal pressures in the ground). One standard scheme is then to apply a reduction factor of 2 to the passive pressures.

BS8002 (1994) proposes that in the analysis of earth retaining structures the angle of shearing resistance should be computed using a mobilisation factor $M = 1.2$. The design strength is then the *lower* of ϕ'_{peak}/M and the critical state angle ϕ'_{crit} for constant volume shearing. Guidance on estimation of angles of shearing resistance for siliceous sands and gravels is shown in Table 8.3.

EC7 (1995) on the other hand proposes that a partial factor of 1.25 should be applied to $\tan \phi'$ and makes no mention of critical state values. Both BS8002 and EC7 lead to design with the same factored strength on each side of the wall and this single value of strength is used to compute appropriate values of active and passive earth pressure coefficients K_a and K_p.

For a generic anchored wall of total height H and free height (excavation depth) αH (Fig 8.36), and with water table at the same level on both sides of

Table 8.4: Wall geometries for different design assumptions

rule	design value of ϕ'	K_a	K_p	$K_p/2$	α
$K_p/2$	$\phi' = 38°$	0.24	4.20	2.10	0.710
BS8002	$\phi' = \tan^{-1}[(\tan 38)/1.2] = 33.0°$	0.29	3.40		0.751
EC7	$\phi' = \tan^{-1}[(\tan 38)/1.25] = 32.0°$	0.31	3.25		0.737

the wall, moment equilibrium about the anchor tells us that (compare (7.166))

$$\frac{1}{3}\gamma' K_a H^3 = \frac{1}{6}\gamma' K_p H^3 (1 - \alpha)^2 (\alpha + 2) \qquad (8.51)$$

which can be solved to give the values shown in Table 8.4 for the three different design approaches (where wall friction has been ignored in the calculation of values of K_a and K_p). The choice of factors is in some way trying to ensure that serviceability conditions for the wall will be acceptable but the 'deformation calculation' is still based on an ultimate limit state approach to the problem.

We will adopt the EC7 approach here, so that $\alpha - 0.737$ and the total length of our wall is $10/0.737 = 13.56$ m. Then horizontal equilibrium of the wall gives the anchor force T:

$$T = \frac{1}{2}K_a \gamma' H^2 - \frac{1}{2}K_p \gamma' H^2 (1 - \alpha)^2 = 287.8 - 210.4 = 77.4 \text{ kN/m} \qquad (8.52)$$

The maximum moment M_{max} occurs where the shear force is zero, at depth x_m below the top of wall:

$$\frac{1}{2}K_a \gamma' x_m^2 = T; \quad \rightarrow \quad x_m = 7.03 \text{ m} \qquad (8.53)$$

and the value of this moment is:

$$M_{max} = T x_m - \frac{1}{2}K_a \gamma' x_m^2 \frac{x_m}{3} = \frac{2}{3}T x_m = 363 \text{ kN-m/m} \qquad (8.54)$$

We have completed our initial design and can now start to explore the moment reduction associated with flexibility of our sheet pile wall. The wall properties are not initially known. The free earth analysis just performed provides values of H, α and M_{max}. The correction curve appropriate for the soil type can be used to determine the moment reduction factor $\lambda = M_{req}/M_{max}$ as a function of ρ and hence a curve relating required moments M_{req} and second moment of area I of the sheet pile section can be produced (Table 8.5 and Fig 8.37). Candidate sheet pile sections can then be directly compared with this curve and available values of second moment of area and allowable steel moments can be studied to choose an appropriate economical section—recognising that there may be other considerations that will limit the amount of wall deformation that can be allowed in certain circumstances.

Care needs to be taken with the units in generating the numbers in Table 8.5: the units of I and M_{req} have been chosen to match those of tabulated sheet pile

Table 8.5: Application of moment reduction chart: $\alpha = 0.737$, dense sand

$\log_{10}\rho$	$\rho = H^4/EI$ m^3/kN	I cm^4/m	M_{req}/M_{max}	M_{req} kN-m/m
-0.5	0.316	50939	0.64	232
-0.3	0.501	32140	0.57	207
-0.1	0.794	20279	0.49	178
0.1	1.259	12795	0.43	156
0.3	1.995	8073	0.37	134
0.5	3.162	5094	0.33	120
0.7	5.012	3214	0.30	109
0.9	7.943	2028	0.27	98

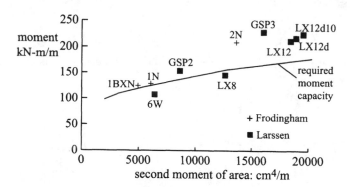

Figure 8.37: Required link between maximum moment and second moment of area of wall compared with available sheet pile sections

Table 8.6: Sheet pile section data

pile type	section	I cm^4/m	$M_{allowable}$ kN-m/m
Frodingham	1BXN	4947	125
	1N	6072	129
	2N	13641	209
Larssen	LX8	12863	149
	LX12	18727	217
	LX12d	19217	223
	LX12d10	19866	231
	6W	6508	110
	GSP2	8756	158
	GSP3	16316	235

(from http://www.corusconstruction.co.uk/indexes/idxph002.htm)

section data (Table 8.6). We take Young's modulus of steel $E = 210$ GPa and then use sheet pile section data to calculate allowable maximum moments using a maximum steel stress $\sigma_{max} = 180$ MPa corresponding to medium tensile steel (EN10248) and permanent works (British Steel, 1997). The resulting sheet pile data have been plotted in Fig 8.37. This figure shows very clearly the way in which increased pile stiffness leads to increased moments. The range of values of I that has been explored is much greater than that required for the purposes of design of this particular wall. From inspection Frodingham 1BXN or 1N and Larssen GSP2 sections will be satisfactory, having adequate moment capacity for their flexural rigidity.

8.8 Tunnel lining

A simple analysis of the ground-structure interaction implied by the construction of a lined tunnel can be achieved by assuming that the tunnel is of circular section, radius r_o, and sufficiently deep in uniform ground for the ground surface to be assumed infinitely distant and the initial stress state to be isotropic and homogeneous (Fig 8.38)[7]. The lining is of thickness $t \ll r_o$. We neglect the difficulty that would be caused by the three-dimensional nature of a tunnel heading and treat the tunnel as a plane strain problem. This is obviously an idealised situation but one that is capable of closed form analysis[8].

From symmetry stresses can only vary with radius. We take a sign convention that assumes that *compressive* stresses and strains are *positive*. The first statement to be made concerns equilibrium

---

[7]I am grateful to Alan Muir Wood for initial discussion on this section.

[8]The axisymmetric *collapse* of a tunnel as ground support is removed is directly analogous to the problem of *expansion* of a cylindrical cavity such as the pressuremeter.

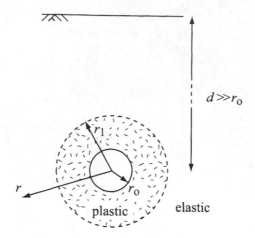

Figure 8.38: Circular tunnel in elastic-plastic ground

$$\frac{\mathrm{d}}{\mathrm{d}r}(r\sigma_r) = \sigma_\theta \tag{8.55}$$

In the far field the radial and circumferential stresses are assumed equal $\sigma_r = \sigma_\theta = \sigma_\infty$. (Some assumption about *stresses* and not just *stress increments* is necessary because we are going to introduce the possibility of plasticity in the soil.) Since both radial and circumferential stresses are assumed to be changing from the same initial value the equilibrium equation then applies equally to the *changes* in stresses $\Delta\sigma_r$ and $\Delta\sigma_\theta$.

The second statement to be made concerns kinematic compatibility. Radial and circumferential strains are linked to *inward* radial displacements u and radial gradients of radial displacement.

$$\Delta\epsilon_\theta = \frac{u}{r} \tag{8.56}$$

$$\Delta\epsilon_r = \frac{\mathrm{d}u}{\mathrm{d}r} \tag{8.57}$$

The third statements that are made concern the constitutive response of the ground. It is assumed that there is a region of plastic yielding which extends from the tunnel boundary as far as a radius r_1 (Fig 8.38). Within the plastic zone the circumferential and radial stresses are related through some Mohr-Coulomb criterion with cohesive and frictional terms (which do not need to exist simultaneously)

$$\sigma_\theta = A + B\sigma_r \tag{8.58}$$

where

$$A = \frac{2c\cos\phi}{1 - \sin\phi} \tag{8.59}$$

and

$$B = \frac{1 + \sin \phi}{1 - \sin \phi} = K_p \qquad (8.60)$$

for a soil with cohesion c and angle of shearing resistance ϕ.

For the tunnel construction, where the excavation of the cavity results in a reduction of radial stress, it is the circumferential stress that will be the major compressive principal stress (whereas for the pressuremeter being expanded it is the radial stress which is the major principal stress).

In the plastic zone a statement is required to describe the deformation condition: a flow rule needs to be introduced. In general, there will be a link between volumetric strain and radial strain which implies a link between circumferential strain and radial strain (expressed here, for simplicity, as a link between *total strains* rather than *plastic strain increments*):

$$\frac{\Delta \epsilon_\theta}{\Delta \epsilon_r} = -\frac{1 - \sin \psi}{1 + \sin \psi} = -K_\epsilon \qquad (8.61)$$

where ψ is the angle of dilation. For constant volume shearing, $\psi = 0$ and radial and circumferential strain increments are equal and opposite. For tunnel convergence with $\psi > 0$ the magnitude of the radial extension exceeds the magnitude of the circumferential compression[9].

In the plastic zone, combination of the flow rule (8.61) and the kinematic compatibility conditions ((8.56), (8.57)) leads to

$$\frac{u}{r} = -K_\epsilon \frac{du}{dr} \qquad (8.62)$$

and hence

$$u = \frac{C}{r^{1/K_\epsilon}} \qquad (8.63)$$

The constant C is given by the need for compatibility at the boundary between the plastic and elastic zones at radius r_1 where the radial displacement is u_1.

In the elastic zone, for radii $r > r_1$, from Hooke's law (§3.2) for plane strain, strain increments and stress increments are related by

$$\Delta \sigma_\theta = \frac{E}{(1 + \nu)(1 - 2\nu)} [(1 - \nu) \Delta \epsilon_\theta + \nu \epsilon_r] \qquad (8.64)$$

$$\Delta \sigma_r = \frac{E}{(1 + \nu)(1 - 2\nu)} [(1 - \nu) \Delta \epsilon_r + \nu \epsilon_\theta] \qquad (8.65)$$

Using the compatibility equations ((8.56), (8.57)), and substituting into the equilibrium equation (8.55), we obtain

$$r^2 \frac{d^2 u}{dr^2} + r \frac{du}{dr} - u = 0 \qquad (8.66)$$

---

[9]For radial cavity expansion with $\psi > 0$, the magnitude of the circumferential extension exceeds the magnitude of the radial compression. This can be treated by changing the sign of the angle of dilatancy in (8.61).

with the solution

$$u = \alpha r + \frac{\beta}{r} \tag{8.67}$$

where $\alpha = 0$ because $u \to 0$ as $r \to \infty$. The integration constant β will be found from the need for radial continuity of stresses and displacements at the boundary between the elastic and plastic regions at $r = r_1$.

Hence

$$\Delta \epsilon_r = -\Delta \epsilon_\theta = -\frac{\beta}{r^2} \tag{8.68}$$

The elastic deformation occurs at constant volume and is thus a pure shearing process irrespective of the value of Poisson's ratio of the soil. The changes in stresses are then

$$\Delta \sigma_r = -\Delta \sigma_\theta = -\frac{2G\beta}{r^2} \tag{8.69}$$

At the boundary between the elastic and plastic zones at radius r_1 the elastic stresses must just satisfy the yield criterion

$$\sigma_\infty + \frac{2G\beta}{r_1^2} = A + B \left(\sigma_\infty - \frac{2G\beta}{r_1^2} \right) \tag{8.70}$$

and hence

$$\beta = \frac{r_1^2 \left[A + (B-1)\, \sigma_\infty \right]}{2G\,(B+1)} \tag{8.71}$$

and at the boundary between elastic and plastic zones the radial stress is given by

$$\sigma_{r_1} = \frac{2\sigma_\infty - A}{B+1} \tag{8.72}$$

The radial displacement u must be continuous at the boundary $r = r_1$ between the plastic zone (8.63) and the elastic zone (8.67). Hence

$$u_1 = \frac{C}{r_1^{1/K_\epsilon}} = \frac{\beta}{r_1} \tag{8.73}$$

so that the radial movement at the boundary of the tunnel at radius r_o is u_o:

$$u_o = \frac{\beta}{r_1} \left(\frac{r_1}{r_o} \right)^{1/K_\epsilon} \tag{8.74}$$

Within the plastic zone the combination of the yield criterion (8.58) with the equilibrium equation (8.55) gives

$$\frac{d}{dr}\,(r\sigma_r) = A + B\sigma_r \tag{8.75}$$

and hence

$$\sigma_r = \frac{1}{B-1} \left(\lambda r^{B-1} - A \right) \tag{8.76}$$

The integration constant λ must be chosen to give stress compatibility with the radial stress at the boundary with the elastic zone $r = r_1$ (8.72):

$$\lambda = \frac{1}{r_1^{B-1}} \frac{2\left[A + \sigma_\infty(B-1)\right]}{B+1} \qquad (8.77)$$

and sufficient information is now available to establish the relation between radial stress at the tunnel boundary and radial convergence at this radius (Fig 8.39).

Some special cases can be considered. For purely elastic ground (Fig 8.39a) (8.67), (8.69):

$$\sigma_{r_o} = \sigma_\infty - 2G\frac{u_o}{r_o} \qquad (8.78)$$

For purely cohesive ground ($\phi = \psi = 0$, $c = c_u$, $A = 2c_u$, $B = 1$) (Fig 8.39b):

$$\sigma_{r_1} = \sigma_\infty - c_u \qquad (8.79)$$

$$\sigma_{r_o} = (\sigma_\infty - c_u) + 2c_u \ln\left(\frac{r_o}{r_1}\right) \qquad (8.80)$$

$$\beta = \frac{c_u}{2G} r_1^2 \qquad (8.81)$$

$$u_o = \frac{c_u}{2G}\frac{r_1^2}{r_o} \qquad (8.82)$$

For purely frictional ground ($c = A = 0$) (Fig 8.39c):

$$\sigma_{r_o} = \frac{2\sigma_\infty}{B+1}\left(\frac{r_o}{r_1}\right)^{B-1} \qquad (8.83)$$

$$\beta = r_1^2 \left(\frac{B-1}{B+1}\right)\frac{\sigma_\infty}{2G} \qquad (8.84)$$

$$u_o = r_1 \left(\frac{B-1}{B+1}\right)\frac{\sigma_\infty}{2G}\left(\frac{r_1}{r_o}\right)^{1/K_\epsilon} \qquad (8.85)$$

The variation of radial tunnel stress with inward movement at the tunnel boundary is shown in Fig 8.40 for the cohesive and frictional soils. For the elastic soil the tunnel pressure varies linearly with inward movement.

The tunnel lining is also assumed to be elastic with Young's modulus E_s and Poisson's ratio ν_s. Under an external radial pressure σ_{r_o} the circumferential stress will be

$$\sigma_{\ell\theta} = \sigma_{r_o}\frac{r_o}{t} \qquad (8.86)$$

The axial stress in the lining will adopt the value necessary to allow the lining to deform in plane strain. If $t \ll r_o$, the average radial stress in the lining will be negligible by comparison with the circumferential stress. The resulting change in radius of the lining can be deduced from the circumferential strain:

$$\frac{u_o}{r_o} = \frac{1 - \nu_s^2}{E_s}\frac{r_o}{t}\sigma_{r_o} \qquad (8.87)$$

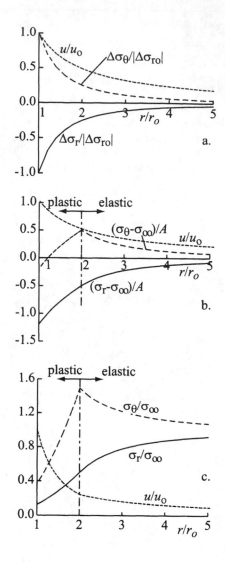

Figure 8.39: Stresses and displacements around tunnel in (a) elastic ground; (b) elastic-plastic cohesive ground ($r_1/r_o = 2$); (c) elastic-plastic frictional ground ($r_1/r_o = 2$, $\phi = 30°$, $K_p = 3$, $\psi = 20°$, $K_\epsilon = 0.49$)

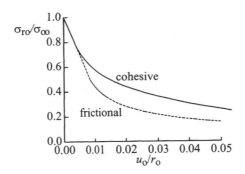

Figure 8.40: Variation of tunnel stress with inward tunnel movement (cohesive ground: $G/c_u = 150$, $G/\sigma_\infty = 30$; frictional ground: $\phi = 30°$, $\psi = 20°$, $G/\sigma_\infty = 30$)

The equilibrium tunnel lining deflection is then the consequence of combining the elastic loading and displacement of the lining with the elastic-plastic unloading of the ground (Fig 8.41). A soft lining will allow the ground to deflect and a low lining stress will be generated (Fig 8.41a). A stiff lining on the other hand will attract load and in the limit would be required to support a radial stress equal to the *in-situ* stress in the ground (Fig 8.41a). Delay in installing the lining will also reduce the stress that the lining is called upon to support (Fig 8.41b). We deduce again that there is an economic benefit to be gained (through the use of a thinner lining) if we can tolerate some displacement.

8.9   Pile in displacing ground

The use of correction factors for calculation of stress resultants in flexible retaining walls based on Rowe's (1952) experimental observations (§8.7) represents an empirical approach to the conversion of an ultimate limit state geotechnical calculation to an assessment of soil-structure interaction. The analyses that Rowe (1955) subsequently performed provide some justification for the nature of the correction procedure and were based on an analysis of the bending of the flexible wall experiencing the presence of the soil through a set of linear springs with stiffnesses varying with depth. An analysis will be presented here in which the interaction of a pile with the soil is modelled as a series of *nonlinear* springs—thus introducing a slightly greater degree of geomechanical realism.

This analysis is of interest because it shows how a problem which initially appears somewhat intractable can be reduced to a set of simple controlling variables whose effect on the character of the response can be readily explored. The analysis is inspired by a real problem in which landfill was known to be moving slowly down a gentle slope past a piled structure. The structural question is: what are the moments induced in the piles as a result of the ground movement? A subsidiary question is: if we can measure the lateral movements of the piles

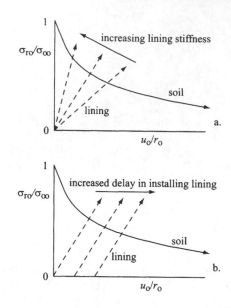

Figure 8.41: Interaction of deforming tunnel lining with contracting tunnel: (a) effect
of lining stiffness; (b) effect of delay in installing lining

at the ground surface, does this provide sufficient information to enable us to
estimate the maximum moments in the piles?

Modelling the structure-soil interaction with a series of independent nonlinear springs may be more appropriate for a pile, where one can imagine the
dominant soil movements being somewhat confined to sub-horizontal planes,
flowing round the pile, than for a long wall where vertical soil movement is
certainly likely to occur.

The analysis of active or passive lateral loading of piles using a set of subgrade
reaction springs is not novel (Reese and Matlock, 1956; Poulos, 1973) but such
analyses have usually assumed that the soil is elastic so far as generation of
displacements and stresses is concerned, and exact solutions can be invoked for
the magnitudes of these stresses (though Poulos introduces the possibility of a
limiting soil resistance). The problem to be analysed here is approached in a
different way.

There are a number of situations where structures are supported on piles
through weak strata which are expected to undergo large movements—for example, offshore structures on continental margins where recent slope deposits
are often in a metastable state; or structures piled through landfill where the
effects of lateral movement of the fill can be less easily avoided than those of
settlement. For such materials the mechanical properties are often rather poorly
defined and there are some advantages in exploring the possibilities of rather
simple analyses and in ascertaining the dimensionless combinations of soil and
structural parameters which control the problem.

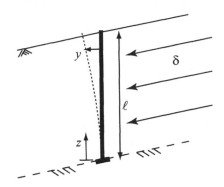

Figure 8.42: Pile loaded by translating ground

The generic problem to be analysed is therefore described as follows. A volume of ground—soil or fill—is known to be moving sideways as a result of internal redistribution of stresses, as a result of movement on a gently inclined basal layer, or as a result of a changing pore pressure regime. Observations suggest that the movements are likely to continue for some time and although estimates can be made (from direct observation) of the magnitude of horizontal movement at the ground surface the distribution of horizontal movement with depth through the ground is unknown. A structure supported on piles, of length ℓ, and uniform flexural rigidity EI, and built in at their base, is founded through the ground on the underlying competent material (Fig 8.42). (Other degrees of end fixity could be studied by changing the boundary conditions of the analysis.) A dimensionless coordinate $\eta = z/\ell$ defines position on the pile with $\eta = 0$ at the base and $\eta = 1$ at the ground surface. The pile is assumed to have no restraint at the top, $\eta = 1$.

The ground is moving with a profile of horizontal movements

$$\delta = \delta_o \eta^\alpha \qquad (8.88)$$

where δ_o is the movement at the ground surface, $\eta = 1$, and α is a parameter which characterises the profile of movement (Fig 8.43). A value of $\alpha = 1$ implies linear variation of movement with depth; $\alpha > 1$ implies that the movement is more concentrated towards the surface (in principle $\alpha = \infty$ implies that movement is everywhere zero except for $\eta = 1$); $\alpha < 1$ implies that the movement is more concentrated towards the base of the ground ($\alpha = 0$ implies that the ground is moving as a block with $\delta = \delta_o$ at all depths). It is assumed that individual piles are sufficiently far apart that their influence on the overall ground movement and their interaction with each other can be ignored.

The translating ground imposes a load on the pile dependent on the relative movement of ground and pile. It is assumed that the ground has strength characteristics such that the lateral stress σ_h exerted on the pile depends on the overburden pressure or depth

$$\sigma_h = K\gamma\ell(1 - \eta) \qquad (8.89)$$

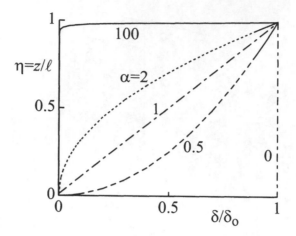

Figure 8.43: Profiles of ground displacement

where γ is an appropriate unit weight for the soil and the lateral stress coefficient K is a function of relative movement $\Delta = \delta - y$ where y is the lateral movement of the pile.

It seems realistic to assume that the lateral stress coefficient cannot exceed some limiting value K^* when the relative movement becomes large, and an appropriate expression for $K = f(\Delta)$ could be (Fig 8.44).

$$\frac{K}{K^*} = \tanh\left[\lambda\left(\frac{\Delta}{d}\right)\right] = \tanh\left[\beta\left(\eta^\alpha - \zeta\right)\right] \tag{8.90}$$

where $\zeta = y/\delta_o$ and $\beta = \lambda\delta_o/d$; λ and K^* are subgrade reaction parameters; and d is a typical pile dimension related to pile diameter or pile width. The dimensionless group β combines information about the stiffness of the pile-ground response, λ, with a ratio of length parameters: ground movement at the free surface, δ_o and typical pile dimension, d.

Such a relationship seems reasonable for a material with purely frictional strength for which the asymptote K^* would be linked with the passive pressure coefficient K_p for the soil. Alternatively, for a normally consolidated clay with undrained strength $c_u = \rho\sigma'_v$, where σ'_v is overburden pressure and ρ might typically be ~ 0.2 (Muir Wood, 1990), then K^* would be a simple multiple of ρ. For example, Randolph and Houlsby (1984) suggest that $\sigma_h/c_u \approx 10$ for flow of cohesive soil around a cylindrical pile so that $K^* \approx 2$.

The tanh function has the benefit that it is symmetric and will produce reasonable lateral pressures whatever the sign of the relative movement between ground and pile. The initial stiffness of the interaction between the pile and the ground is

$$\frac{\mathrm{d}K}{\mathrm{d}\Delta} = \lambda\frac{K^*}{d} \tag{8.91}$$

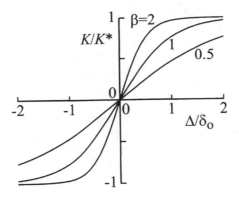

Figure 8.44: Development of pressure on pile as result of relative movement between pile and ground

The actual coefficient of subgrade reaction expressed as stress per unit relative displacement of pile and ground is

$$k_h = \frac{d\sigma_h}{d\Delta} = \lambda K^* \gamma \frac{\ell}{d}(1 - \eta) \qquad (8.92)$$

with the value $k_{ho} = \lambda K^* \gamma \ell/d$ at the base of the pile ($\eta = 0$).

The degree of nonlinearity introduced by the tanh function is not great at low values of Δ/δ_o. The relative movement required to move halfway to the limiting value (corresponding to $K/K^* = 1/2$) is $\Delta_{50} = (d \ln 3)/(2\lambda) = (\delta_o \ln 3)/(2\beta)$ or

$$\beta = \frac{\ln 3}{2} \frac{\delta_o}{\Delta_{50}} \qquad (8.93)$$

which provides a potential link between β and other parameters describing the problem.

In order to complete the equation of elastic deformation of the pile it is convenient to assume that the lateral stress σ_h acts over an equivalent width d which might be assumed to be greater than the diameter of the pile in order to allow for side friction as the ground flows past the pile and for any 'boundary layer' effect of influence of the pile on the ground movements: d is the equivalent pile size in freely moving soil. Deformation of the pile is then governed by the equation

$$EI\frac{d^4 y}{dz^4} = K\gamma(\ell - z)d \qquad (8.94)$$

which can be conveniently rewritten in dimensionless form

$$\frac{d^4\zeta}{d\eta^4} = \frac{1-\eta}{\chi} \frac{K}{K^*} = \frac{1-\eta}{\chi} \tanh[\beta(\eta^\alpha - \zeta)] \qquad (8.95)$$

with

$$\chi = \frac{EI\delta_o}{\ell^2} \frac{1}{K^* \gamma \ell^3 d} \qquad (8.96)$$

This dimensionless parameter χ indicates the ratio of two moments, one a structural property and the other a loading characteristic. For a cantilever of length ℓ and flexural rigidity EI subjected to a tip displacement δ_o with no other loading, the root moment is $M_r = 3EI\delta_o/\ell^2$. For a cantilever of length ℓ subjected to lateral pressures over a width d given by the limiting value of the lateral stress coefficient $K = K^*$ throughout the depth, the root moment is $M_f = K^* \gamma \ell^3 d/6$. Thus $\chi = M_r/18M_f$.

The boundary conditions for the cantilever pile shown in Fig 8.42 are zero deflection and slope at the base of the pile (assuming complete fixity at the base): $\zeta = d\zeta/d\eta = 0$ for $\eta = 0$; and zero moment and shear force at the top of the pile $d^2\zeta/d\eta^2 = d^3\zeta/d\eta^3 = 0$ for $\eta = 1$.

Once the deformation equation (8.95) has been solved to give a profile of normalised displacement ζ with normalised depth η, the variation of moment M within the pile can be presented in various dimensionless ways:

$$\mu = \frac{d^2\zeta}{d\eta^2} = \frac{M}{EI\delta_o/\ell^2} \tag{8.97}$$

normalises the moment with input parameters of the problem. The displacement of the tip of the pile is an output quantity, $y_{max} = \zeta_1 \delta_o$, where $\zeta = \zeta_1$ at $\eta = 1$, which will in general be different from the ground displacement δ_o. A cantilever in air whose tip is moved sideways by a distance y_{max} develops a root moment $M_{max} = 3EIy_{max}/\ell^2$ so that the alternative dimensionless group μ_1

$$\mu_1 = \frac{M}{M_{max}} = \frac{\mu\delta_o}{3y_{max}} = \frac{\mu}{3\zeta_1} \tag{8.98}$$

allows us to compare the moments in a pile whose lateral movement is brought about by the translation of the ground with the maximum (root) moment in a pile in air given the same movement at the top. The dimensionless group μ_2:

$$\mu_2 = \frac{M}{M_f} = \frac{M}{K^* \gamma \ell^3 d/6} = 6\chi\mu \tag{8.99}$$

normalises the moment with the maximum moment that can be generated when the soil is slipping past the pile and fully mobilising the resistance coefficient K^* over the full length of the pile.

Even if we do not have detailed information available, we need to choose reasonable values of the three controlling parameters α, β and χ. For parametric studies, a reference value $\alpha = 1.0$, corresponding to linear variation of ground movement with depth (Fig 8.43) has been taken and the effect of changing this by a factor of 2 (to 0.5 and 2.0) has been explored.

The parameter β controls the stiffness of the pile-ground interaction but its dimensionless definition (8.93) introduces both the relative movement Δ_{50} required to generate half the limiting load (a parameter of the ground-pile interaction), and the magnitude of the ground surface movement δ_o—which is assumed to be quite independent of the pile response. It might be assumed that Δ_{50} is about half the pile diameter and that the ground surface movement is of the same order as the pile diameter. Then, for pile diameter $d = 0.5$ m

$(= 2\Delta_{50})$, $\beta = 1.1$. A reference value $\beta = 1.0$ has been used for parametric studies and the effect of increasing and decreasing this has been explored. Stiffer pile-ground interaction (lower Δ_{50}) will imply increased β. An earlier stage in the overall process of ground movement after installation of the piles will imply lower δ_o and hence a decreased value of β.

Typical values of χ can be deduced for appropriate assumed values for the parameters defining the problem:

- ground surface movement: $\delta_o = 0.5$ m (as before)

- pile diameter: $d = 0.5$ m (implying $I = \pi d^4/64 \approx 0.0031$ m^4)

- Young's modulus for concrete: $E = 30$ GPa

- pile length: $\ell = 10$ m

- ground unit weight: $\gamma = 10$ kN/m^3

- limiting lateral pressure coefficient: $K^* = 2$ (corresponding to Rankine passive pressure ratio K_p for angle of friction $\phi' \approx 20°$ or to the Randolph and Houlsby (1984) treatment of normally consolidated clay noted above).

Then $\chi = 0.0464$. A reference value $\chi - 0.05$ has been used for parametric studies and the effect of increasing and decreasing this has been explored over the range 0.0004 to 0.1. The value of χ is very sensitive to the details of the pile geometry: $\chi \propto d^3$ and ℓ^{-5} (see (8.96)).

The governing equation (8.95) is extremely nonlinear but is capable of numerical solution using standard routines[10]. Results are shown in plots of dimensionless pile displacement $\zeta = y/\delta_o$ (Fig 8.45) and of dimensionless moment μ_1 (Fig 8.46) with dimensionless position on the pile $\eta = z/\ell$.

The effect of varying the profile of ground movement through the parameter α is shown in Figs 8.45a and 8.46a. The greater the average movement of the ground (the lower the value of α) the greater the load on the pile and the greater the pile movement and root moment. Fig 8.46a appears to show little influence of α but recall that μ_1 scales moment with pile tip movement (8.98) and this is certainly influenced by α (Fig 8.45a). The value of α is not something over which the engineer has much control, but any tendency for mass movement of the ground to occur will certainly be very damaging for any structure trying to impede the motion. Most analyses have been performed with α kept constant at 1.0 corresponding to linear variation of ground translation with depth.

The pile displacements and moments are very sensitive to the stiffness of the interaction between the pile and the ground: the higher the value of β the larger the displacements and moments (Figs 8.45b, c and 8.46b, c). However, the effect of changing β is dependent on the pile stiffness, as reflected in the parameter χ. With a lower value of $\chi = 0.002$, the tip deflection of the pile is hardly affected by the value of β for $\beta > 1.0$ (Fig 8.45c) although the greater curvature at the toe of the pile leads to much higher moments.

Increase of pile stiffness through χ has the expected effect of reducing pile deflection (compare Figs 8.45b, c). However, reducing χ below 0.002 has little

---

[10] I am grateful to Jörgen Johansson for programming the solution of (8.95).

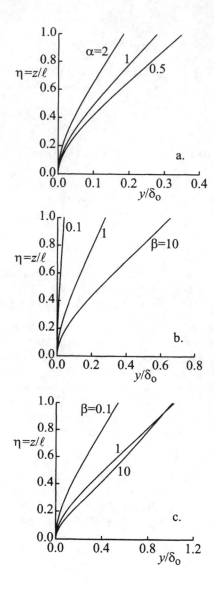

Figure 8.45: Deflected shape of pile for (a) different values of α ($\beta = 1$, $\chi = 0.05$); (b) different values of β ($\alpha = 1$, $\chi = 0.05$); (c) different values of β ($\alpha = 1$, $\chi = 0.002$)

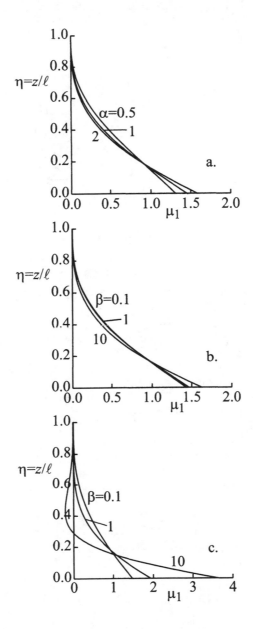

Figure 8.46: Normalised moments in pile for (a) different values of α ($\beta = 1$, $\chi = 0.05$); (b) different values of β ($\alpha = 1$, $\chi = 0.05$); (c) different values of β ($\alpha = 1$, $\chi = 0.002$)

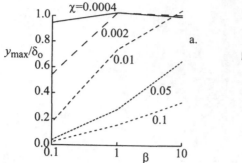 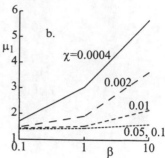

Figure 8.47: Influence of β and χ on (a) maximum deflection and (b) maximum moment developed in pile

additional effect on the pile displacements although the curvature at the toe of the pile is slightly increased. The profile of dimensionless moment, μ_1, remains unchanged for values of pile stiffness $\chi > 0.05$ (Fig 8.46b). For the most flexible pile considered the maximum moment is nearly 6 times the free air value; for the stiffer piles it is still more than 50% higher than this value. It can be concluded that observation of pile tip movement at the ground surface gives a very poor indication of the magnitude of moments in the pile.

The interaction of values of β and χ in influencing the tip movement and root moment is shown in Fig 8.47. It is clear that for a very flexible pile ($\chi = 0.0004$) the tip movement is more or less equal to the ground movement for all values of β, whereas for less flexible piles the proportion of ground movement at the tip increases with β (or, which is equivalent, with δ_o). Similarly, the scaled moment, μ_1, is independent of β for higher values of pile stiffness; these scaled moments are always at least 50% higher than those obtained for the cantilever displaced in free air.

Since the two dimensionless parameters β and χ both contain the surface ground movement δ_o, the process of gradual mobilisation of ground-pile interaction for a given pile can be followed by varying both β and χ in appropriate constant proportion.

$$\frac{\beta}{\chi} = \frac{\ln 3}{2} \frac{\delta_o}{\Delta_{50}} \frac{\ell^2}{EI\delta_o} \left(K^*\gamma\ell^3 d\right) = \frac{\ln 3}{2} \frac{K^*\gamma\ell^5}{EI} \frac{d}{\Delta_{50}} \qquad (8.100)$$

and for the typical values that we have suggested here, $\beta/\chi \approx 20$. We can follow the gradual mobilisation of pile moment by looking at the variation of the normalised moment μ_2 (8.99) with increasing normalised ground displacement $\delta_o/d \approx \beta$ (Fig 8.48). As ground displacement builds up this normalised moment approaches 1: the ground:pile earth pressure coefficient is close to K^* over the whole length of the pile. A lower ratio β/χ indicates a stiffer pile: for stiffer piles the limiting moment is reached more rapidly.

The moments generated in a pile by translating ground have been shown to be dependent on three dimensionless parameters and thus the results are

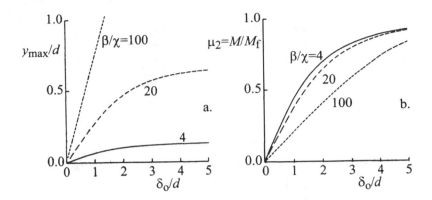

Figure 8.48: Development of (a) pile top movement and (b) pile toe moment with ground movement

of completely general application. Not all of these parameters can be easily controlled by a designer. The profile of ground movement with depth, through parameter α, has a major influence on pile response and deep mass movement of the ground will be particularly damaging. The stiffness of the pile and the stiffness of the interaction between the pile and ground combine in their effect on pile movement and moments. The ground-pile stiffness is not easy to specify. It is linked with some estimate of coefficient of subgrade reaction—but subgrade reaction is a notoriously empirical concept. The link with some assessment of the relative movement required to generate limiting loads on the pile may be more useful. However, this dimensionless form of β that emerges from the analysis includes characteristics of both the pile-ground interaction (Δ_{50}) and the ground deformation (δ_o) which are essentially independent. For a poorly compacted fill β is likely to be towards the lower end of the range of values that has been studied. The parameter χ is easier to assess provided that some estimate of the limiting pile-ground interaction coefficient K^* can be obtained. All the other quantities are directly known.

Faced with an apparently difficult problem the geotechnical engineer is not permitted just to throw up his or her hands in despair. This section has tried to show how such a problem can be logically dissected and made ultimately amenable to analysis.

8.10 Integral bridge abutment

The examples of soil-structure interaction that we have considered so far have either dealt only with elastic materials (soil and structure), or have permitted nonlinear soil behaviour but with a very simple geometry (tunnel), or have treated the nonlinearity in a rather simplistic way (nonlinear subgrade reaction springs). Such analyses are important in providing qualitative or quantitative

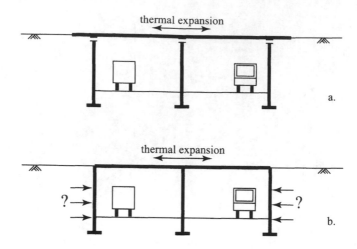

Figure 8.49: (a) Bridge with bearings separating deck and columns; (b) bridge with integral connections between deck and columns

insight into the character of the response but in order to obtain a more complete picture of the consequences of soil-structure interaction a full numerical analysis will usually be unavoidable. We present here the results of such an analysis—albeit using a rather basic soil model—which has assisted in understanding the behaviour of integral bridge abutments. The results are initially slightly counterintuitional. Fuller details of the analyses are provided by Muir Wood (1999b) and Muir Wood and Nash (2000).

Traditionally highway bridges have been constructed with their deck sections structurally separated from their supporting abutments by bearings which permit relative sliding movement to occur as the length of the deck changes as a result of thermal effects (Fig 8.49a). However, these bearings tend to be adversely affected by the de-icing salts that are used on United Kingdom roads and by other sources of corrosion and consequently are in need of regular inspection and maintenance. In order to reduce some of these continuing costs associated with highway bridges it has become popular recently in the United Kingdom to construct so-called integral bridges in which the deck is connected directly to the abutments with no intervening bearings (Fig 8.49b). Bearings do of course make it possible for the daily and seasonal thermal expansion of the bridge deck to be accommodated as movement relative to the abutments. Without this freedom the thermal expansion must be accommodated through interaction of the abutments themselves with the fill that is placed behind them. From a structural point of view, the design question is: what are the magnitudes of the earth pressures that are generated on the abutments and what are the magnitudes of the resulting bending moments?

It appears that this is a simple problem of calculation of earth pressures and, since thermal expansion causes the abutment wall to move towards the fill, it must be passive pressures that will dominate the loading. However, it is known that passive pressures are mobilised rather slowly with increasing wall displacement so it must be appropriate to make some allowance for actual wall movement in estimating the passive pressure coefficient to be used. On the other hand, the process of construction of the abutment will involve compaction of the fill and this will itself lock in certain initial stresses into the wall.

In fact, this is a classic example of soil-structure interaction where the structural consequences of the thermal movements are dependent on the relative stiffness of the fill and the structure. We present some numerical analyses of a typical integral bridge abutment which explore the way in which earth pressures are generated and calculate the consequent abutment bending moments. Parametric studies have been performed to indicate those properties of the backfill or of the structural materials which have the greatest influence on the earth pressures and bending moments. The structural stiffness properties will usually be known quite closely. However, the properties of the backfill may be more uncertain. In particular, since this is a problem of soil-structure interaction, it is important to be able to estimate the stiffness of the fill—this will usually be neither known nor controlled.

So far as earth pressures on integral bridge abutments are concerned, the design guidance in the United Kingdom at the time of the analyses (Highways Agency, 1996) proposed that the earth pressure coefficient should be equal to K^* over the top half of the retained height H of the wall (Fig 8.50)

$$K^* = K_p \left(\frac{\delta}{0.05H} \right)^{0.4} \tag{8.101}$$

where K_p is the passive pressure coefficient (calculated taking appropriate account of the interface friction between the abutment and the backfill) and δ is the horizontal displacement at the top of the abutment wall due to thermal expansion of the deck. The lateral pressure then remains constant with depth as the earth pressure coefficient drops to the earth pressure coefficient at rest K_o and then the earth pressure coefficient remains constant at K_o below this depth (Fig 8.50). There is an additional limit on K^* that it should not be lower than either K_o or $K_p/3$ and it is this latter restriction that tends to be limiting once the angle of shearing resistance of the backfill becomes high so that the deformation (and hence flexibility) of the abutment wall is then not allowed to have any effect on the resulting stresses. The value of the earth pressure coefficient at rest K_o is linked with the angle of shearing resistance ϕ' (§3.8):

$$K_o = 1 - \sin \phi' \tag{8.102}$$

These proposals for estimation of the pressures resulting from the thermal expansion of integral bridges make no allowance for the flexibility of the abutment. The bridge abutment is primarily required to resist the vertical load from the bridge deck so that the main requirement is that it should be stiff vertically. It must be sufficiently strong laterally to contain the fill (and possibly resist loads generated by the traffic on the approach embankment) but lateral stiffness is not essential. An extremely flexible abutment will withstand deck

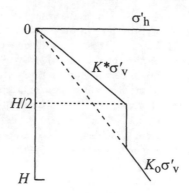

Figure 8.50: Design guidance for horizontal stresses on integral abutment

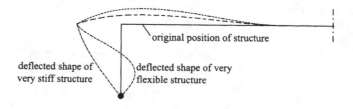

Figure 8.51: Schematic influence of abutment flexibility on abutment deflections (abutment assumed pinned at base)

expansion by flexure near its top and may not move towards the fill lower down its height sufficiently to generate any significant passive pressures: indeed it may retain an outward deflection from the initial construction placement and compaction of the backfill (Fig 8.51). A much stiffer or rigid abutment would tend to move into the backfill more monolithically (Fig 8.51) and would be expected to generate passive pressures over much of its height. Since structural costs will be controlled by the magnitude of the bending moments that arise in the abutment there is some advantage in aiming for a flexible structural element which tries to reduce the mobilisation of high earth pressures. This is yet another example of soil-structure interaction where the structural consequences of the thermal movements are dependent on the relative stiffness of the fill and the structure.

Numerical analyses have been performed using the finite difference program FLAC (Itasca, 2000). The model chosen for analysis is a slightly simplified version of a prototype integral bridge with a continuous deck across two equal spans constructed in a limestone rock cutting with side slopes cut at 1:1.5 (Fig 8.52a). The bridge has two equal spans of 26.3 m. The abutments are 8.8 m high and are pinned at their bases to pad footings founded on the rock. The section through the abutment has been analysed in plane strain: it is assumed

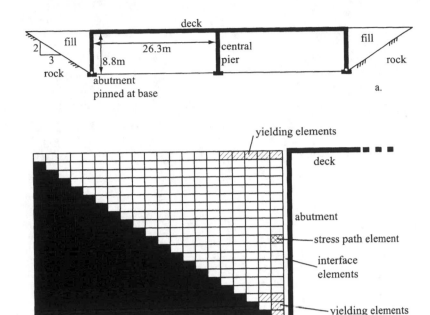

Figure 8.52: Integral abutment (a) prototype dimensions; (b) FLAC model (see also Muir Wood and Nash, 2000)

that the abutment earth wing walls provide sufficient restraint to make this an appropriate assumption. A plane strain analysis will anyway tend to overestimate the earth pressures if there is in reality some deformation out of plane. The grid used for the analyses is shown in Fig 8.52b.

The fill is built up from 20 layers of elements with heights chosen to fit in with the changes in the structural abutment section properties as shown in Table 8.7. The rock cutting beyond the backfill is modelled as a stiff and strong Mohr-Coulomb material (friction angle $30°$, cohesion 10^{10} kPa, dilation angle zero, shear modulus 10^5 kPa, bulk modulus 2×10^5 kPa). The bottom boundary of the backfill and the left hand boundary of the rock are both smooth and rigid.

The abutment and bridge deck are composed of 'beam' elements. The abutment wall, also of height 8.8 m, is composed of 20 elements with lengths chosen to fit in with changes in section properties (taken from the prototype) as shown in Table 8.7. The deck is composed of 20 elements with total length 26.3 m representing half the bridge. Individual lengths have been chosen to fit in with the changes in structural section as shown in Table 8.7.

The abutment elements are separated from the backfill by a layer of interface elements. At its top the abutment has full moment connection to the deck. At the centre of the bridge, where there is a supporting pier, the deck is prevented

Table 8.7: Section properties for structural elements

	number of elements	element length m	cross-sectional area m^2/m	second moment of area m^4/m	note
Abutment					
0 - 6.9 m	16	0.43125	0.8	0.0427	uncracked
			0.2533	0.02	cracked
6.9 - 8.8 m	4	0.475	0.9	0.0607	uncracked
			0.28	0.0273	cracked
Deck					
0 - 2 m	2	1	0.713	0.124	diaphragms cracked
2 - 6.35 m	4	1.0875	0.413	0.093	slab cracked
6.35 - 19.95 m	8	1.7	0.553	0.143	uncracked
19.95 - 24.3 m	4	1.0875	0.413	0.093	slab cracked
24.3 - 26.3 m	2	1	0.713	0.124	diaphragms cracked

from vertical movement and rotation and a plane of symmetry is assumed for the analysis.

In the analysis the backfill is placed layer by layer, and each layer is 'compacted' by adding and removing a vertical stress of 20 kPa to model, at least partially, the real construction operation using a moving vibrating roller moving over the surface of each layer of fill. Once construction of the backfill is complete, no surcharge loading has been applied to either the backfill or the bridge deck during the thermal expansion. The results that are obtained thus relate purely to the expansion process since this is the loading case that is expected to be limiting for structural design of the abutment wall.

Thermal expansion of the bridge deck is represented by a horizontal movement of the centre of the bridge sufficient to produce a movement of 11 mm at the top of the abutment. With a coefficient of thermal expansion of concrete of $12 \times 10^{-6}/°C$ this corresponds to a seasonal temperature range of about 35°C. The actual thermal expansion is of course distributed along the deck but the structural consequences for the abutment will be identical using this more convenient mode of loading of the numerical model.

The abutment and deck beam elements are modelled as elastic concrete. The reinforced concrete was specified to have Young's modulus 34 GPa. The formulation of the beam elements in FLAC assumes that they are plane stress elements. In order to match them to a plane strain analysis it is necessary to modify the Young's modulus by a factor $1/(1-\nu^2)$ (Itasca, 2000). Poisson's ratio ν has been taken as 0.3 for concrete, so that the analyses have been performed using a value of concrete Young's modulus of 37.36 GPa.

As is typical for many structures of this type, the information available from which the properties of the backfill can be deduced is extremely limited. Since the design procedure makes use of a terminology of frictional limiting pressures it seems appropriate to model the backfill using a Mohr-Coulomb model (§3.3.4).

This model assumes that the soil behaves elastically (and isotropically) until a limiting frictional strength is reached. It certainly provides a rather crude representation of the way in which real soils deform before failure (and fails to reproduce the steady fall in soil stiffness that is known to occur in reality prior to failure). Equally the unavailability of information about the fill gives no justification for greater modelling sophistication. The model requires the specification of two elastic stiffness properties (for example, shear modulus and bulk modulus or Young's modulus and Poisson's ratio) together with a frictional strength, and an angle of dilatancy for the material.

Because the design guidance links earth pressure generation to passive pressure coefficients the result is very sensitive to the value that is chosen for the angle of friction of the fill. In practice, there are competing requirements. The fill is required to have a low strength in order to reduce these passive pressures: and that might typically lead to a specification of a uniform rounded granular material. However, the fill is also the material which supports the approach roadway and for this purpose it is required to have high stiffness so that it will not generate excessive settlements during the life of the structure. For high stiffness a well-graded fill may be more appropriate but this is likely to have a higher angle of friction. Since it might be intuitively assumed that angle of friction would be the principal variable affecting the performance of the integral bridge abutment, analyses have been performed using angles of friction $\phi' = 35$ to $55°$ in order to enable calculations to be made across a full potential range of actual fill strengths.

A zero angle of dilation has been used for all the analyses presented here. The angle of dilation describes the volumetric deformations that occur when frictional failure is reached. The present problem has a large unrestrained free surface and it is found that dilatancy has negligible influence on the system response.

There is no available information concerning the stiffness properties of the prototype granular backfill material (crushed limestone). It has been assumed that Young's modulus for the backfill varies according to some power of the confining stress:

$$E = E_o \left(\frac{\sigma}{\sigma_o} \right)^\alpha \tag{8.103}$$

with the reference stress $\sigma_o = 100$ kPa (approximately equal to atmospheric pressure) and the exponent $\alpha = 0.5$. With the range of vertical overburden stresses in the different layers of backfill this implies that the stiffness will vary from $0.216E_o$ in the surface layer to $1.298E_o$ in the bottom layer of the model. The range of possible values of the reference stiffness E_o is large. Muir Wood and Nash (2000) show, from a brief review of the literature, that a minimum value $E_o = 20$ MPa and a maximum value $E_o = 180$ MPa would be reasonable with a value $E_o = 60$ MPa seen as a 'best estimate'. Poisson's ratio for the backfill material has been taken as 0.3 throughout.

The interface between the abutment and the backfill has to be given stiffness properties as well as strength properties. The stiffness relates movements and forces from nodes on opposite sides of the interface but plays no role in analyses of the type reported here: normal and shear stiffnesses of 10^{10} kN/m have been used. The limiting frictional strength on the interface has been set at half the angle of shearing resistance of the backfill. (An analysis with a smooth interface showed little difference in the structural response.)

Detailed results of the numerical analyses are provided by Muir Wood and Nash (2000): here we will concentrate on the structural consequences of the soil-structure interaction at the end of thermal expansion of the bridge deck into the backfill. The analysis with $\phi' = 45°$ and $E_o = 60$ MPa is taken as a reference analysis.

The principal surprise is that backfill strength has negligible effect on the horizontal stresses in the abutment (Fig 8.53a). Backfill stiffness on the other hand has a dramatic effect (Figs 8.53b, 8.54). That it is in fact *relative* stiffness that is important is shown by performing analyses in which the structural section properties are first modestly reduced to allow for cracking of some of the tensile concrete (Table 8.7), and then artificially reduced by a factor of 10 (10% concrete stiffness in Fig 8.53c). The pattern of horizontal stresses depends on the relative stiffness; the absolute values of the horizontal stresses depend on the absolute soil stiffness. It may be noted that the shapes of the horizontal stress profiles are very similar to those observed by Rowe (1952) in his tests on model flexible retaining walls (Fig 8.33b). Except for the highest backfill stiffness—and even then only at the very top of the abutment—the stresses are much lower than the design line assuming an earth pressure coefficient $K_p/3$ shown for $\phi' = 45°$.

Horizontal displacements of the abutment, for the various values of backfill stiffness, are shown in Fig 8.54a. Unless the flexural stiffness of the abutment is made artificially low the abutment does indeed move towards the fill through its entire height. The deflections are always lower than those of the structure subjected to thermal expansion in the absence of the fill: the curvature of the wall is reversed through most of the height as a result of the earth pressures that are generated (Fig 8.54a).

The horizontal stresses, horizontal deflections and bending moments computed in the reference analysis ($\phi' = 45°$, $E_o = 60$ MPa) are compared in Fig 8.55 with the corresponding values that emerge from application of the design guidance with $\phi' = 35°$ and $45°$. The profile of horizontal stress is completely different (Fig 8.55a). The design guidance results in wall deflections *away from* the backfill over much of the height of the abutment (Fig 8.55b) which is of course in contradiction with the basic underpinning hypothesis of the guidance that the earth pressures are generated by passive soil deformation. And the bending moments are greatly overpredicted (Fig 8.55c)—the maximum sagging moment is some 7 times greater and the maximum hogging moment, at the top of the wall, about 3 times greater.

The apparently counterintuitional almost complete independence of the results from the angle of friction of the backfill over the range from 35° to 55° can be understood when the stress path for a typical element of soil at mid-height just behind the wall (indicated in Fig 8.52b) is considered (Fig 8.56a). There is of course some rotation of principal axes (Fig 8.56b) resulting from the generation of shear stresses on vertical planes in the backfill from the shear resistance on the back of the wall so that principal stresses are not actually

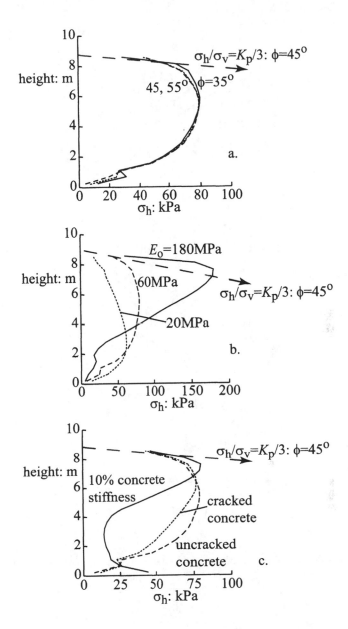

Figure 8.53: Horizontal stresses on abutment (a) effect of angle of shearing resistance of fill; (b) effect of fill stiffness; (c) effect of abutment stiffness

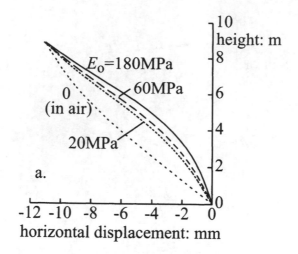

a.

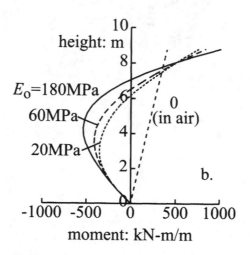

b.

Figure 8.54: Effect of fill stiffness on (a) abutment deflections; (b) abutment moments

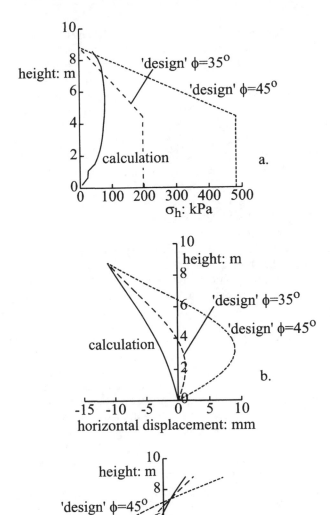

Figure 8.55: Comparison of typical calculation with design values of (a) horizontal stress; (b) deflection; (c) moment

sufficient to describe the stress state completely. However, in terms of principal stress quantities—the major and minor principal stresses σ_1 and σ_3—a plane strain mean stress $s = (\sigma_1 + \sigma_3)/2$ and shear stress $t = (\sigma_1 - \sigma_3)/2$ can be defined (compare §2.4, §3.8, §7.5.1).

Yielding of the backfill—mobilisation of the available shearing resistance— is confined to a small row of elements close to the top of the wall where the stress level is low and the strain forced on the fill by the deck expansion brings the fill to Mohr-Coulomb failure (Fig 8.52b). There is also a group of yielding elements near the bottom of the wall where, although the lateral movements are low (the toe of the abutment is pinned and not allowed to move), the fill is being squeezed between the abutment and the rock. The detail revealed from the numerical analysis is a little crude in this region because of the coarseness of the mesh and the approximate way in which the inclined boundary of the rock cutting has been modelled.

Around the mid-height of the wall, where the most significant stress changes occur, the effect of the movement of the abutment into the backfill is to move the state of the soil *away* from failure (Fig 8.56a). At the end of construction the major principal compressive stress is approximately vertical (Fig 8.56b). The compaction process, as modelled, is a somewhat one-dimensional loading and unloading of the backfill, but with the flexibility of the abutment allowing a little lateral stress relief. The movement of the wall towards the backfill then tends to increase the horizontal stress producing a major change in mean stress s but a rather small change in the shear stress t, thus moving the stress state towards a more isotropic condition with a *lower* mobilised angle of friction. The wall friction is also helping to generate some additional vertical stress and hence provide some additional confinement to the fill. The majority of the backfill is therefore being loaded entirely elastically (as indicated by the absence of yielding in Fig 8.52b) and its frictional strength plays no important role—except just towards the top of the abutment—but the stresses here have negligible effect on the generation of moments in the abutment.

The stiffness of the actual backfill is not known. The Mohr-Coulomb model has been used here in the absence of sufficient data on the behaviour of the backfill materials to warrant any greater sophistication. This model assumes that the soil behaves elastically until the frictional failure state is reached. It is known that all real granular materials show a steady fall in stiffness from any initial stress state towards failure. The incremental stiffness that is mobilised is dependent on the direction of the current stress changes in relation to previous stress changes (§2.5.3): persistent travel along a stress path in one direction leads to reducing stiffness; reversal of movement leads to higher stiffness.

The highest value of reference stiffness for the backfill, $E_o = 180$ MPa, probably tends towards the dynamic stiffness which will only be relevant for very small strain excursions and will not be appropriate for a complete episode of bridge deck expansion. It can be deduced from the analyses that a fall in incremental backfill stiffness will lead to a reduction in the lateral stresses and resulting moments. Any tendency to further compaction of the backfill through the life of the structure, as a result of the cyclic seasonal expansion of the bridge or as a result of steady low level cyclic traffic loading, would tend to lead to an

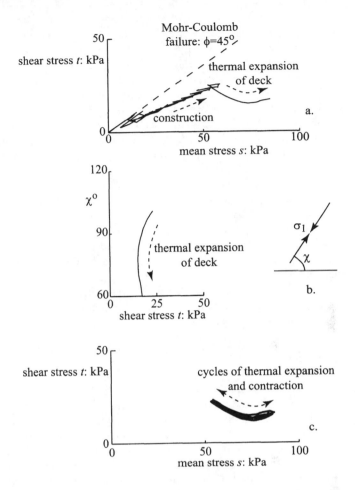

Figure 8.56: (a) Stress path for element at mid-height behind abutment; (b) rotation of principal axes for this element; (c) stress path during cycles of thermal expansion and contraction

increase in the stiffness of the fill and hence a tendency to generate increased stresses towards the top of the backfill and increased moments during any one expansion episode. The standard Mohr-Coulomb model is not sufficiently advanced to be able to make any useful prediction for this situation. However, if, using this simple model, a series of cycles of bridge deck expansion and contraction is applied, the extent of the already small region in which yielding occurs diminishes until, after three complete cycles, there is no further yielding. The stress paths during these cycles more or less reverse and repeat those shown for the first episode of deck expansion (Fig 8.56c)—a further indication that though there are nonlinearities in the system the process is predominantly elastic.

The analyses have deliberately concentrated on a structure which has relatively low abutment flexural stiffness: this seems to be efficient in performing the bridge support role that is required of it. Passive pressures can only be generated if the wall moves in towards the soil; design horizontal pressures based on partial mobilisation of passive pressures will lead to outward movement of the wall when applied to a flexible structural element. Once the stress paths for soil elements have been inspected the computed behaviour of the overall backfill-abutment system is seen to be entirely plausible. There is a message here: do not pause in the assessment of results of numerical modelling until any anomalies have been satisfactorily explained.

8.11   Closure

The numerical analysis of the final case study of soil-structure interaction confirms a message that has been repeated several times in this chapter. Structural resultants—which will be required for structural design purposes—are controlled by relative stiffness of structure and ground and not primarily by soil strength. The 'sad, perpetual strain' is that problems of soil-structure interaction *must* be considered as complete systems. Attempts to consider the soil and the structure separately are doomed to failure since their intricate interaction cannot then be understood.

A corollary of this might be that close control over structural consequences of soil-structure interaction can only be achieved through reliable knowledge or control of soil stiffness. The word 'stiffness' is intended to imply a model of the deformation response of the soil which might (or might not) be an elastic model. Certainly it implies laboratory or *in-situ* testing designed to reveal deformation and not just strength characteristics. For man-made ('designer') soils, such as the crushed rock fill of the final example, stiffness is expected to be correlated in a general way with strength so that although stiffness effects are not regularly reliably determined, a link between compactive effort and density and strength might be used to infer a value of stiffness. However, this is not really a satisfactory substitute for direct stiffness measurement.

9

Envoi

It is said that Gerd Gudehus, visiting Cambridge from Karlsruhe in the 1960s, described with Cartesian conviction the way in which mathematical modelling held the key to understanding and reproducing soil behaviour. In response Andrew Schofield, coming from an environment of Baconian scepticism, opined that 'The soil will defeat you!' and returned to his direct observation of soil behaviour in early British geotechnical centrifuges. I have tried in this book to give a flavour of many of the different aspects of theoretical and physical modelling that might help us to at least obtain a truce in our battles with the soil.

Modelling is personal. This book contains an eclectic, personal collection of tools, techniques and technologies that seem to have some place in geotechnical modelling. However, whatever modelling you undertake, you need to feel comfortable with the approaches and simplifications that have been adopted. It is much easier to defend your own decisions than those which you have had reluctantly thrust upon you[1].

> My modelling choices
> Are founded in logic.
> Can I convince you?

Modelling should attempt to surprise. Modelling to support design may be performed to provide reassurance. Modelling to further geotechnical understanding should deliberately set out to surprise.

> If you look down,
> Are you surprised to see
> Only your feet?

If we only perform routine tests and explore the response of our constitutive and theoretical models only within the context of these tests then, not only is there

---

[1]When I spoke on a related theme in Japan a few years ago I introduced a number of *haiku*-like apophthegms to encapsulate some of the messages that I was attempting to convey (Muir Wood, 1995). I reproduce some similar messages here. Purists will remark that *haiku* should have 5-7-5 syllables and have a seasonal allusion.

the danger that we will fail to discover possible irregularities in our theoretical models, but also we are neglecting potentially wide and fascinating tracts of geotechnical knowledge which are awaiting exploration.

> The train keeps to the tracks
> But the country around
> Remains unexplored.

Modelling should be justifiable. Modelling is concerned with appropriate simplification of reality. The emphasis is obviously on *appropriate*.

Modelling must be auditable. The steps taken along the way and the assumptions and simplifications must be declared so that your findings—whether experimental or numerical—can be reproduced.

Modelling should be adequately complex. I suspect that the more complex the model the more the modelling itself will obscure the underpinning mechanics of soil behaviour and geotechnical system behaviour which we are trying to probe and understand. The art of successful modelling is to include just enough detail for the implied simplification to be reasonable for the particular application. Modelling assumptions will differ in different contexts.

> Show the trees in the wood:
> but must we include
> all the reeds on the muir?

'*Scientific understanding proceeds by way of constructing and analysing models of the segments or aspects of reality under study. The purpose of these models is not to give a mirror image of reality, not to include all its elements in their exact sizes and proportions, but rather to single out and make available for intensive investigation those elements which are decisive. We abstract from non-essentials, we blot out the unimportant to get an unobstructed view of the important, we magnify in order to improve the range and accuracy of our observation. A model is, and must be, unrealistic in the sense in which the word is most commonly used. Nevertheless, and in a sense, paradoxically, if it is a good model it provides the key to understanding reality.*' (Baran and Sweezy, 1968)

Modelling is local. It follows that geotechnical modelling will be 'local' in the sense that it will try to describe well the behaviour under modest perturbations of engineering interest from the current state. This is about extrapolation. The more distant the extrapolation the more chaotically uncertain the reliability of the outcome will be. A detailed distant model may be so complex that it becomes unwieldy.

> The excellent map
> misses no detail, but it is
> as large as the world.

Modelling should be elegant? Dirac is supposed to have written on a blackboard in Moscow: 'It is more important to have beauty in one's equations than to have them fit experiment.' This seems to be a misguided sentiment in the context of modelling for geotechnical engineering. We know that soils are complex materials—I do not believe that we yet understand enough of the richness of their behaviour to be dogmatic about the ways in which we should model them. Beauty and elegance in modelling are satisfying but we should ensure that they do not unduly distract.

In the end there is a need for compromise between several of the competing demands but also a need to exploit the complementary benefits of different modelling possibilities. The realism of physical modelling may provide insights of which numerical modelling is incapable—especially where novel engineering solutions are being proposed. But the observations of experiment can provide the data necessary to refine the numerical models.

Those involved in modelling tend to become more interested in the process than its purpose:
the stimulation of simulation is greater than the pleasurement of measurement: but it makes you go blind.

Bibliography

AIREY, DW, BUDHU, M AND WOOD, DM (1985) Some aspects of the behaviour of soils in simple shear. Chap. 6 in *Developments in soil mechanics and foundation engineering—2* (eds PK Banerjee and R Butterfield) Elsevier Applied Science Publishers, London 185-213.

ALLARD, MA AND SCHENKEVELD, FM (1994) The Delft Geotechnics model pore fluid for centrifuge tests. *Proc. Int. Conf. Centrifuge 94*, Singapore (eds CF Leung, FH Lee and TS Tan) AA Balkema, Rotterdam 133-138.

ALLERSMA, HGB (1998) Development of cheap equipment for small centrifuges. *Proc. Int. Conf. Centrifuge 98*, Tokyo (eds T Kimura, O Kusakabe and J Takemura) AA Balkema, Rotterdam 1 85-90.

ALMEIDA, MSS, BRITTO, AM AND PARRY, RHG (1986) Numerical modelling of a centrifuged embankment on soft clay. *Canadian Geotechnical Journal* **23** 103-114.

ALMEIDA, MSS AND PARRY, RHG (1994) Penetrometer apparatus for use in the centrifuge during flight. *Application of centrifuge modelling to geotechnical design* (ed WH Craig) AA Balkema, Rotterdam 47-65.

AL-TABBAA, A AND MUIR WOOD, D (1989) An experimentally based 'bubble' model for clay. *Numerical Models in Geomechanics NUMOG III* (eds S Pietruszczak and GN Pande) Elsevier Applied Science, London 91-99.

AL-TABBAA, A AND MUIR WOOD, D (1991) Horizontal drainage during consolidation: insights gained from analyses of a simple problem. *Géotechnique* **41** (4) 571-585.

ARROYO, M (2001) *Pulse tests in soil samples.* PhD thesis, University of Bristol.

ARTHUR, JRF, CHUA, KS AND DUNSTAN, T (1977) Induced anisotropy in a sand. *Géotechnique* **22** (1) 13-30.

ARULANANDAN, K AND SCOTT, RF (EDS) (1993/1994) *Verification of numerical procedures for the analysis of soil liquefaction problems* AA Balkema, Rotterdam (2 vols).

ATKINSON, JH AND SÄLLFORS, G (1991) Experimental determination of stress-strain-time characteristics in laboratory and *in-situ* tests. in *Deformation of soils and displacements of structures (Proc X ECSMFE)* (Ed Associazione Geotecnica Italiana) AA Balkema, Rotterdam **3** 915-956.

ATTERBERG, A (1911) Lerornas förhållande till vatten, deras plasticitetsgränser och plasticitetsgrader. *Kungl. Landtbruks-akademiens Handlingar och Tidskrift* **50** (2) 132-158.

BAGUELIN, F, JÉZÉQUEL, J-F AND SHIELDS, DH (1978) *The pressuremeter and foundation engineering.* Trans Tech Publications, Clausthal.

BAKER, JF AND HEYMAN, J (1969) *Plastic design of frames: 1. Fundamentals.* Cambridge University Press, Cambridge.

BAKIR, NE, GARNIER, J AND CANEPA, Y (1994) Loading of shallow foundations: importance of testing procedures. *Proc. Int. Conf. Centrifuge 94*, Singapore (eds CF Leung, FH Lee and TS Tan) AA Balkema, Rotterdam 553-558.

BARAN, PA AND SWEEZY, PM (1968) *Monopoly capital: An essay on the American economic and social order.* Penguin Books, Harmondsworth.

BARKAN, DD (1962) *Dynamics of bases and foundations.* (translated by L Drashenska and GP Tschebotarioff) McGraw-Hill Book Company, New York.

BARRON, RA (1947) Consolidation of fine-grained soils by drain wells. *Proc. ASCE, J. of Soil Mechs and Foundations Div.* **73** (SM6) 811-835.

BASSETT, RH AND CRAIG, WH (1988) The development of geotechnical centrifuges in the United Kingdom. *Centrifuges in soil mechanics* (eds WH Craig, RG James and AN Schofield) AA Balkema, Rotterdam 35-60.

BEEN, K AND JEFFERIES, MG (1985) A state parameter for sands. *Géotechnique* **35** (2) 99-112.

BEEN, K AND JEFFERIES, MG (1986) Discussion: A state parameter for sands. *Géotechnique* **36** (1) 127-132.

BENAHMED, N (2001) *Comportement mécanique d'un sable sous cisaillement monotone et cyclique: application aux phénomènes de liquéfaction et de mobilité cyclique.* Thèse de doctorat, Ecole Nationale des Ponts et Chaussées, Paris.

BISHOP, AW (1959) The principle of effective stress. *Teknisk Ukeblad*, Oslo **39** 859-863.

BISHOP, AW AND HENKEL, DJ (1962) *The measurement of soil properties in the triaxial test.* (2nd edition) Edward Arnold, London.

BJERRUM, L (1972) Embankments on soft ground. *Proc. Specialty Conf. on Performance of earth and earth-supported structures*, Purdue, ASCE **2** 1-54.

BOLTON, MD AND POWRIE, W (1988) Behaviour of diaphragm walls in clay prior to collapse. *Géotechnique* **38** (2) 167-189.

BOLTON, MD AND STEEDMAN, RS (1985) The behaviour of fixed cantilever walls subject to lateral shaking. *Application of centrifuge modelling to geotechnical design* (ed WH Craig) AA Balkema, Rotterdam 301-313.

BOYCE, JR (1980) A nonlinear model for the elastic behaviour of granular materials under repeated loading. *Soils under cyclic and transient loading.* (eds GN Pande and OC Zienkiewicz) AA Balkema, Rotterdam **1** 285-294.

BRANSBY, PL AND MILLIGAN, GWE (1975) Soil deformations near cantilever sheet-pile walls. *Géotechnique* **25** (2) 175-195.

BRITISH STEEL (1997) *Piling handbook* (7th edition) British Steel plc (see also http://www.corusconstruction.co.uk/piling/7th_edition/index.html)

BRITTO, AM AND GUNN, MJ (1987) *Critical state soil mechanics via finite elements.* Ellis Horwood, Chichester.

BROWN, PT (1969) Numerical analyses of uniformly loaded circular rafts on deep elastic foundations. *Géotechnique* **19** (3) 399-404.

BS1377 (1990) *British Standard Methods of test for Soils for civil engineering purposes: Part 2: Classification tests.* British Standards Institution, London.

BS8002 (1994) *Code of practice for earth retaining structures.* British Standards Institution, London.

BURLAND, JB, BROMS, B AND DE MELLO, VFB (1977) Behaviour of foundations and structures (State-of-the-art report) *Proc. 9th Int. Conf. on Soil Mechanics and Foundation Engineering* Japanese Society of Soil Mechanics and Foundation Engineering, Tokyo **2** 495-546.

BURLAND, JB AND WROTH, CP (1975) Settlement of buildings and associated damage: Review paper. *Settlement of structures*, Pentech Press, London 611-654.

BUTTERFIELD, R (1999) Dimensional analysis for geotechnical engineers. *Géotechnique* **49** (3) 357-366.

BUTTERFIELD, R AND GOTTARDI, G (1994) A complete three-dimensional failure envelope for shallow footings on sand. *Géotechnique* **44** (1) 181-184.

CALLADINE, CR (1985) *Plasticity for engineers.* Ellis Horwood, Chichester.

CALLISTO, L (1996) *Studio sperimentale su un'argilla naturale: il comportamento meccanico del'argilla di Pisa.* Thesis, Dottorato di ricerca in ingegneria geotecnica, Università di Roma 'La Sapienza'.

CALLISTO, L, GAJO, A AND MUIR WOOD, D (2002) Simulation of stress probe tests on natural and reconstituted Pisa clay. *Géotechnique* **52** (9) 649-666.

CARSLAW, HS AND JAEGER, JC (1959) *Conduction of heat in solids.* (2nd edition) Clarendon Press, Oxford.

CATES, ME, WITTMER, JP, BOUCHAUD, J-P AND CLAUDIN, P (1998) Development of stresses in cohesionless poured sand. *Phil. Trans. Roy. Soc. London* **A356** 2535-2560.

CHEN, Y-C, ISHIBASHI, I AND JENKINS, JT (1988) Dynamic shear modulus and fabric: Part I: Depositional and induced anisotropy. *Géotechnique* **38** (1) 25-32.

CLANCY, P AND RANDOLPH, MF (1993) An approximate analysis procedure for piled raft foundations. *Int. J. for Numerical and Analytical Methods in Geomechanics* **17** 849-869.

CLAYTON, CRI AND HEYMANN, G (2001) Stiffness of geomaterials at very small strains. *Géotechnique* **51** (3) 245-255.

CLAYTON, CRI AND MILITITSKY, J (1986) *Earth pressure and earth-retaining structures.* Surrey University Press, Glasgow.

COOK, RD (1995) *Finite element modelling for stress analysis.* John Wiley, New York.

COOK, RD, MALKUS, DS AND PLESHA, ME (1989) *Concepts and applications of finite element analysis.* (3rd edition) John Wiley, New York.

CRAIG, WH (1985) Centrifuge modelling for site-specific prototypes. *Application of centrifuge modelling to geotechnical design* (ed WH Craig) AA Balkema, Rotterdam 485-501.

CRAMPIN, S (1981) A review of wave motion in anisotropic and cracked elastic media. *Wave motion* **3** 343-391.

CREMER, C, PECKER, A AND DAVENNE, L (2001) Cyclic macro-element for soil-structure interaction: material and geometrical nonlinearities. *International Journal for Numerical and Analytical Methods in Geomechanics* **25** (13), 1257-1284.

DERKX, F, MERLIOT, E, GARNIER, J AND COTTINEAU, LM (1998) On-board remote-controlled centrifuge robot. *Proc. Int. Conf. Centrifuge 98*, Tokyo (eds T Kimura, O Kusakabe and J Takemura) AA Balkema, Rotterdam **1** 97-102.

DE SOUZA, E (2002) A centrifuge for solving mining problems. *Physical modelling in geotechnics: ICPMG'02*, Newfoundland (eds R Phillips, PJ Guo and R Popescu) AA Balkema Publishers, Lisse 49-54.

DIBENEDETTO, H, TATSUOKA, F AND ISHIHARA, M (2002) Time-dependent shear deformation characteristics of sand and their constitutive modelling. *Soils and Foundations* **42** (2) 1-22.

DRESCHER, A AND DE JOSSELIN DE JONG, G (1972) Photoelastic verification of a mechanical model for the flow of a granular material. *J. Mech. Phys. Solids* **20** 337-351.

DYSON, GJ AND RANDOLPH, MF (1998) Installation effects on lateral load-transfer curves in calcareous sands. *Proc. Int. Conf. Centrifuge 98*, Tokyo (eds T Kimura, O Kusakabe and J Takemura) AA Balkema, Rotterdam **1** 545-550.

ESQUIVEL, ER AND KO, H-Y (1994) Development of a miniature piezocone. *Proc. Int. Conf. Centrifuge 94*, Singapore (eds CF Leung, FH Lee and TS Tan) AA Balkema, Rotterdam 89-94.

EC7 (1995) *Eurocode 7: Geotechnical design - Part 1: General rules, together with the United Kingdom National Application Document.* DD ENV 1997-1:1995 British Standards Institution, London.

FLEMING, WGK, WELTMAN, AJ, RANDOLPH, MF AND ELSON, WK (1985) *Piling engineering.* Surrey University Press, Glasgow and John Wiley, New York.

FORAY, P, BALACHOWSKI, L AND RAULT, G (1998) Scale effect in shaft friction due to the localisation of deformations. *Proc. Int. Conf. Centrifuge 98*, Tokyo (eds T Kimura, O Kusakabe and J Takemura) AA Balkema, Rotterdam 1 211-216.

FRASER, RA AND WARDLE, LJ (1976) Numerical analysis of rectangular rafts on layered foundations. *Géotechnique* 26 (4) 613-630.

GAJO, A, BIGONI, D AND MUIR WOOD, D (2001) Stress induced elastic anisotropy and strain localisation in sand. *Localisation and bifurcation theory in geomechanics* (ed H Mühlhaus). Swets & Zeitlinger, Lisse 37-44.

GAJO, A AND MUIR WOOD, D (1999) Severn-Trent sand: a kinematic hardening constitutive model for sands: the $q - p$ formulation. *Géotechnique* 49 (5) 595-614.

GARNIER, J (2001) *Modéles physiques en géotechnique: état des connaissances et évolutions récentes* (1ère Biennale Coulomb) Comité Français de Mécanique des Sols et de Géotechnique, Paris.

GARNIER, J (2002) Properties of soil samples used in centrifuge models. *Physical modelling in geotechnics: ICPMG'02*, Newfoundland (eds R Phillips, PJ Guo and R Popescu) AA Balkema Publishers, Lisse 5-19.

GAUDIN, C, GARNIER, J, GAUDICHEAU, P AND RAULT, G (2002) Use of a robot for in-flight excavation in front of an embedded wall. *Physical modelling in geotechnics: ICPMG'02*, Newfoundland (eds R Phillips, PJ Guo and R Popescu) AA Balkema Publishers, Lisse 77-82.

GIBSON, RE (1967) Some results concerning displacements and stresses in a non-homogeneous elastic half-space. *Géotechnique* 17 (1) 58-67.

GOTTARDI, G, HOULSBY, GT AND BUTTERFIELD, R (1999) Plastic response of circular footings on sand under general planar loading. *Géotechnique* 49 (4) 453-469.

GRAHAM, J AND HOULSBY, GT (1983) Elastic anisotropy of a natural clay. *Géotechnique* 33 (2) 165-180.

GUDEHUS, G (1979) A comparison of some constitutive laws for soils under radially symmetric loading and unloading. *Proc. 3rd Conf. Numerical methods in geomechanics*, Aachen (ed W Wittke) AA Balkema, Rotterdam 1309-1323.

GUI, MW, BOLTON, MD, GARNIER, J, CORTÉ, J-F, BAGGE, G, LAUE, J AND RENZI, R (1998) Guidelines for cone penetration tests in sand. *Proc. Int. Conf. Centrifuge 98*, Tokyo (eds T Kimura, O Kusakabe and J Takemura) AA Balkema, Rotterdam 1 155-160.

GURUNG, SB, KUSAKABE, O AND KANO, S (1998) Behaviour of reconstituted and natural soil models under pullout force. *Proc. Int. Conf. Centrifuge 98*, Tokyo (eds T Kimura, O Kusakabe and J Takemura) AA Balkema, Rotterdam 1 477-482.

HAMBLY, EC (1969) A new true triaxial apparatus. *Géotechnique* **19** (2) 307-309.

HANSEN, JB (1968) *A revised extended formula for bearing capacity.* Bulletin 28, Danish Geotechnical Institute, Copenhagen.

HARRIS, HG AND SABNIS, GM (1999) *Structural modelling and experimental techniques.* (2nd edition) CRC Press, Boca Raton.

HERLE, I AND GUDEHUS, G (1999) Determination of parameters of a hypoplastic constitutive model from properties of grain assemblies. *Mechanics of cohesive-frictional materials* **4** (5) 461-486.

HIGHWAYS AGENCY (1996) *The design of integral bridges.* Design manual for roads and bridges 1:3:Part 12, BA42/96 Highways Agency, London.

HORIKOSHI, K AND RANDOLPH, MF (1995) *Optimum design of piled rafts.* University of Western Australia, Department of Civil Engineering, Geomechanics Group, Research report G1179.

HORIKOSHI, K AND RANDOLPH, MF (1997) On the definition of raft-soil stiffness ratio for rectangular rafts. *Géotechnique* **47** (5) 1055-1061.

HOULSBY, GT (1985) The use of a variable shear modulus in elastic-plastic models for clays. *Computers and Geotechnics* **1** 3-13.

HVORSLEV, MJ (1937) *Über die Festigkeitseigenschaften gestörter bindiger Böden.* Danmarks Naturvidenskabelige Samfund, København. Ingeniørvidenskabelige Skrifter A45 (English translation (1969) *Physical properties of remoulded cohesive soils.* US Waterways Experimental Station, Vicksburg, Mississippi, report 69-5).

IAI, S (1989) Similitude for shaking table tests on soil-structure-fluid model in 1*g* gravitational field. *Soils and Foundations* **29** (1) 105-118.

IMAMURA, S, HAGIWARA, T, MITO, K, NOMOTO, T AND KUSAKABE, O (1998) Settlement trough above a model shield observed in a centrifuge. *Proc. Int. Conf. Centrifuge 98,* Tokyo (eds T Kimura, O Kusakabe and J Takemura) AA Balkema, Rotterdam **1** 713-719.

INSTITUTION OF STRUCTURAL ENGINEERS, INSTITUTION OF CIVIL ENGINEERS, INTERNATIONAL ASSOCIATION FOR BRIDGE AND STRUCTURAL ENGINEERING (1989) *Soil-structure interaction: The real behaviour of structures.* Institution of Structural Engineers, London.

ISHIBASHI, I, CHEN, Y-C AND JENKINS, JT (1988) Dynamic shear modulus and fabric: Part II: Stress reversal. *Géotechnique* **38** (1) 33-37.

ITASCA (2000) *FLAC: Fast Lagrangian Analysis of Continua.* Itasca Consulting Group, Minneapolis.

JENG, FS, LU, CY AND LEE, CJ (1998) Major scale effects influencing model simulation of neotectonics. *Proc. Int. Conf. Centrifuge 98,* Tokyo (eds T Kimura, O Kusakabe and J Takemura) AA Balkema, Rotterdam **1** 911-916.

JOHNSON, KL (1985) *Contact mechanics.* Cambridge University Press, Cambridge.

KARLSSON, R (1977) *Consistency limits: a manual for the performance and interpretation of laboratory investigations: Part 6.* Statens råd för byggnadsforskning, Stockholm.

KIMURA, T (2000) Development of geotechnical centrifuges in Japan. *Proc. Int. Conf. Centrifuge 98*, Tokyo (eds T Kimura, O Kusakabe and J Takemura) AA Balkema, Rotterdam **2** 945-954.

KIMURA, T, KUSAKABE, O AND TAKEMURA, J (EDS) (1998/2000) *Proc. Int. Conf. Centrifuge 98*, Tokyo AA Balkema, Rotterdam (2 vols).

KO, H-Y AND SCOTT, RF (1967) A new soil testing apparatus. *Géotechnique* **17** (1) 40-57.

KOLBUSZEWSKI, JJ (1948) An experimental study of the maximum and minimum porosities of sands. *Proc. 2nd Int. Conf. on Soil mechanics and foundation engineering*, Rotterdam **1** 158-165.

KÖNIG, D, JESSBERGER, HL, BOLTON, MD, PHILLIPS, R, BAGGE, G, RENZI, R AND GARNIER, J (1994) Pore pressure measurement during centrifuge model tests: experience of five laboratories. *Proc. Int. Conf. Centrifuge 94*, Singapore (eds CF Leung, FH Lee and TS Tan) AA Balkema, Rotterdam 101-108.

KOSEKI, J, TATSUOKA, F, MUNAF, Y, TATEYAMA, M AND KOJIMA, K (1998) Modified procedure to evaluate active earth pressure at high seismic loads. *Soils and Foundations*, Special issue No. 2 on Geotechnical aspects of the January 17 1995 Hyogoken-Nambu Earthquake, 209-216.

KRAWINKLER, H (1979) Possibilities and limitations of scale-model testing in earthquake engineering. *Proc. 2nd US National Conf. on Earthquake Engineering*, Stanford University 283-292

KUHLEMEYER, RL AND LYSMER, J (1973) Finite element method accuracy for wave propagation problems. *Proc ASCE, J. Soil Mechs and Foundations Div.* **99** (SM5) 421-427.

LADE, PV AND DUNCAN, JM (1975) Elasto-plastic stress-strain theory for cohesionless soil. *Proc ASCE, Journal of the Geotechnical Engineering Division* **101** (GT10) 1037-1053.

LAMBE, TW AND WHITMAN, RV (1979) *Soil mechanics, SI version.* John Wiley, New York.

LANCELLOTTA, R (1987) *Geotecnica.* Zanichelli, Bologna.

LANGHAAR, HL (1951) *Dimensional analysis and theory of models.* John Wiley, New York.

LAUE, J, NATER, P, SPRINGMAN, SM AND GRÄMIGER (2002) Preparation of soil samples in drum centrifuges. *Physical modelling in geotechnics: ICPMG'02*, Newfoundland (eds R Phillips, PJ Guo and R Popescu) AA Balkema Publishers, Lisse 143-148.

LEE, FH (2002) The philosophy of modelling versus testing. *Constitutive and centrifuge modelling: two extremes* (ed SM Springman) AA Balkema Pubilshers, Lisse 113-131.

LEE, FH, JUNEJA, A, WEN, C, DASARI, GR AND TAN, TS (2002) Performance of total stress cells in model experiments in soft clays. *Physical modelling in geotechnics: ICPMG'02*, Newfoundland (eds R Phillips, PJ Guo and R Popescu) AA Balkema Publishers, Lisse 101-106.

LEROUEIL, S, MAGNAN, JP AND TAVENAS, F (1985) *Remblais sur argiles molles.* Technique et Documentation Lavoisier, Paris.

LINGS, ML AND GREENING, PD (2001) A novel bender/extender element for soil testing. *Géotechnnique* **51** (8) 713-717.

LINGS, ML, PENNINGTON, DS AND NASH, DFT (2000) Anisotropic stiffness parameters and their measurement in a stiff natural clay. *Géotechnique* **50** (2) 109-125.

LIVESLEY, RK (1983) *Finite elements: an introduction for engineers.* Cambridge University Press, Cambridge.

LOPRESTI, DCF, JAMIOLKOWSKI, M, PALLARA, O, CAVALLARO, A AND PEDRONI, S (1997) Shear modulus and damping of soils. *Géotechnique* **47** (3) 603-617.

LOVE, JP (2003) Use of settlement reducing piles to support a raft structure. *Proc. ICE, Geotechnical Engineering* **156** (GE4) 177-181.

LUPINI, JF, SKINNER, AE AND VAUGHAN PR (1981) The drained residual strength of cohesive soils. *Géotechnique* **31** (2) 181-213.

MCNAMARA, AM AND TAYLOR, RN (2002) Use of heave reducing piles to control ground movements around excavations. *Physical modelling in geotechnics: ICPMG'02*, Newfoundland (eds R Phillips, PJ Guo and R Popescu) AA Balkema Publishers, Lisse 847-852.

MADABHUSHI, SPG, SCHOFIELD, AN AND LESLEY, S (1998) A new Stored Angular Momentum (SAM) based earthquake actuator. *Proc. Int. Conf. Centrifuge 98*, Tokyo (eds T Kimura, O Kusakabe and J Takemura) AA Balkema, Rotterdam **1** 111-116.

MAIR, RJ (1979) *Centrifugal modelling of tunnel construction in soft clay.* PhD thesis, University of Cambridge.

MAIR, RJ AND WOOD, DM (1987) *Pressuremeter testing—methods and interpretation.* CIRIA Ground Engineering Report: *In-situ* testing, CIRIA and Butterworths, London.

MARTIN, CM AND HOULSBY, GT (2001) Combined loading of spudcan foundations on clay: numerical modelling. *Géotechnique* **51** (8) 687-699.

MATSUOKA, H AND NAKAI, T (1982) A new failure criterion for soils in three-dimensional stresses. *Proc. IUTAM Symp. on Deformation and failure of granular materials* (eds PA Vermeer and HJ Luger) AA Balkema, Rotterdam 253-263.

MATSUSHITA, M, TATSUOKA, F, KOSEKI, J, CAZACLIU, B, DiBENEDETTO, H AND YASIN, SJM (1999) Time effects on the pre-peak deformation properties of sands. *Pre-failure deformation characteristics of geomaterials* (eds M Jamiolkowski, R Lancellotta and DCF LoPresti) AA Balkema, Rotterdam **1** 681-689.

MEIGH, AC (1987) *Cone penetration testing: methods and interpretation.* CIRIA Ground Engineering Report: *In-situ* testing, CIRIA and Butterworths, London.

MEYERHOF, GG (1963) Some recent research on the bearing capacity of foundations. *Canadian Geotechnical Journal* **1** (1) 16-26.

MUIR WOOD, D (1990) *Soil behaviour and critical state soil mechanics.* Cambridge University Press, Cambridge.

MUIR WOOD, D (1995) Discussion summary : Modelling of shear deformation of geomaterials - Modelling of material properties. *Pre-failure deformation characteristics of geomaterials* (eds S Shibuya, T Mitachi and S Miura) AA Balkema, Rotterdam **2** 1229-1230.

MUIR WOOD, D (1999a) Pore pressures for stability analysis of embankment on soft clay. *Numerical models in geomechanics: NUMOG VII*, Graz (eds GN Pande, S Pietruszczak and HF Schweiger) AA Balkema, Rotterdam 521-526.

MUIR WOOD, D (1999b) Numerical analysis of earth pressures on an integral bridge abutment. *FLAC and Numerical modelling in geomechanics* (eds C Detournay and R Hart) AA Balkema, Rotterdam 443-450.

MUIR WOOD, D (2000) The role of models in civil engineering. *Constitutive modelling of granular materials* (ed D Kolymbas) Springer-Verlag, Berlin 37-55.

MUIR WOOD, D (2002) Some observations of volumetric instabilities in soils. *International Journal of Solids and Structures* **39** (13-14) 3429-3449.

MUIR WOOD, D (2004) Experimental inspiration for kinematic hardening models of soil. *Proc ASCE, Journal of Engineering Mechanics* (to appear).

MUIR WOOD, D, BELKHEIR, K AND LIU, D-F (1994) Strain-softening and state parameter for sand modelling. *Géotechnique* **44** (2) 335-339.

MUIR WOOD, D, CREWE, AJ AND TAYLOR, CA (2002) Shaking table testing of geotechnical models. *International Journal of Physical Modelling in Geotechnics* **2** (1) 1-13.

MUIR WOOD, D, CREWE, AJ, TAYLOR, CA AND GAJO, A (1998) Localisation in earthquake simulation experiments. *Localisation and bifurcation theory for soils and rocks* (eds T Adachi, F Oka and A Yashima) AA Balkema, Rotterdam 117-126.

MUIR WOOD, D, HU, W, AND NASH, DFT (2000) Group effects in stone column foundations: model tests. *Géotechnique* **50** (6) 689-698.

MUIR WOOD, D AND KUMAR, GV (2000) Experimental observations of behaviour of heterogeneous soils. *Mechanics of Cohesive-Frictional Materials* **5** (5) 373-398.

MUIR WOOD, D, MACKENZIE, NL AND CHAN, AHC (1993) Selection of parameters for numerical predictions. *Predictive soil mechanics (Proc. Wroth Memorial Symposium)* (eds GT Houlsby and AN Schofield) Thomas Telford, London 496-512.

MUIR WOOD, D AND NASH, DFT (2000) Earth pressures on an integral bridge abutment: a numerical case study. *Soils and Foundations* **40** (6), 23-38.

NASH, DFT (2001) Modelling the effect of surcharge to reduce long term settlement of relamation over soft clays: a numerical case study. *Soils and Foundations* **41** (5) 1-13.

NASH, DFT, LINGS, ML AND PENNINGTON, DS (1999) The dependence of anisotropic G_o shear moduli on void ratio and stress state for reconstituted Gault clay. *Pre-failure deformation characteristics of geomaterials*, (eds M Jamiolkowski, R Lancellotta and DCF LoPresti) AA Balkema, Rotterdam **1** 229-238.

NASH, DFT AND RYDE, SJ (2001) Modelling consolidation accelerated by vertical drains in soils subject to creep. *Géotechnique* **51** (3) 257-273.

NEWMARK, NM (1965) Effects of earthquakes on dams and embankments. (5th Rankine Lecture). *Géotechnique* **15** (2) 139-160.

NG, YW, LEE, FH AND YONG, KY (1998) Development of an in-flight sand compaction piles installer. *Proc. Int. Conf. Centrifuge 98*, Tokyo (eds T Kimura, O Kusakabe and J Takemura) AA Balkema, Rotterdam **1** 837-843.

NICOLAS-FONT, J (1988) Design of geotechnical centrifuges. *Centrifuge 88: Proc. Int. Conf. on Geotechnical centrifuge modelling*, Paris (ed J-F Corté) AA Balkema, Rotterdam 9-15.

NORDAL, S (1983) *Elasto-plastic behaviour of soils analysed by the finite element method*. Doctor of Engineering thesis, Norges Tekniske Högskole, Trondheim.

NOVA, R, CASTELLANZA, R AND TAMAGNINI, C (2002) A constitutive model for bonded geomaterials subject to mechanical and/or chemical degradation. *International Journal for Numerical and Analytical Methods in Geomechanics* **27** (9) 705-732.

NOVA, R AND MONTRASIO, L (1991) Settlements of shallow foundations on sand. *Géotechnique* **41** (2) 243-256.

NOVA, R AND WOOD, DM (1979) A constitutive model for sand. *Int. J. for Numerical and Analytical Methods in Geomechanics* **3** (3) 255-278.

NUÑEZ, I AND RANDOLPH, MF (1984) Tension pile behaviour in clay - centrifuge modelling techniques. *Application of centrifuge modelling to geotechnical design* (ed WH Craig) AA Balkema, Rotterdam 87-102.

ODA, M AND IWASHITA, K (EDS) (1999) *Mechanics of granular materials: an introduction*. AA Balkema, Rotterdam.

PAOLUCCI, R (1997) Simplified evaluation of earthquake-induced permanent displacements of shallow foundations. *Journal of Earthquake Engineering* **1** (3) 563-579.

PENNINGTON, DS, NASH, DFT AND LINGS, ML (1997) Anisotropy of G_o shear stiffness in Gault clay. *Géotechnique* **47** (3) 391-398.

PENNINGTON, DS, NASH, DFT AND LINGS, ML (2001) Horizontally mounted bender elements for measuring anisotropic shear moduli in triaxial clay specimens. *Geotechnical Testing Journal* **24** (2) 133-144.

PHILLIPS, R, GUO, PJ AND POPESCU, R (EDS) (2002) *Physical modelling in geotechnics: ICPMG'02*, Newfoundland, AA Balkema Publishers, Lisse.

POTTS, DM (2003) Numerical analysis: a virtual dream or practical reality? (42nd Rankine Lecture). *Géotechnique* **53** (6) 535-573.

POTTS, DM AND ZDRAVKOVIĆ, L (1999) *Finite element analysis in geotechnical engineering: theory.* Thomas Telford, London.

POULOS, HG (1973) Analysis of piles in soil undergoing lateral movement. *Proc. ASCE, Journal of the Soil Mechanics and Foundations Division* **99** (SM5) 391-406.

POULOS, HG AND DAVIS, EH (1974) *Elastic solutions for soil and rock mechanics.* John Wiley, New York.

POULOS, HG AND DAVIS, EH (1980) *Pile foundation analysis and design.* John Wiley, New York.

POWRIE, W (1997) *Soil mechanics: concepts and applications.* E&FN Spon, London.

RANDOLPH, MF (1986) Design of piled raft foundations. *Recent developments in laboratory and field tests and analysis of geotechnical problems* (eds AS Balasubramaniam, S Chandra and DT Bergado) AA Balkema, Rotterdam 525-537.

RANDOLPH, MF AND HOULSBY, GT (1984) The limiting pressure on a circular pile loaded laterally in cohesive soil. *Géotechnique* **34** (4) 613-623.

RANDOLPH, MF, JEWELL, RJ, STONE, KJL AND BROWN, TA (1991) Establishing a new centrifuge facility. *Proc. Int. Conf. Centrifuge 91*, Boulder (eds H-Y Ko and FG McLean) AA Balkema, Rotterdam 3-9.

RAUDKIVI, AJ AND CALLANDER, RA (1976) *Analysis of groundwater flow.* Edward Arnold, London.

REESE, LC AND MATLOCK, H (1956) Non-dimensional solutions for laterally loaded piles with soil modulus assumed proportional to depth. *Proc. 8th Texas Conf. on Soil Mechanics and Foundations*, 41pp.

ROESLER, SK (1979) Anisotropic shear modulus due to stress anisotropy. *Proc. ASCE, Journal of the Geotechnical Engineering Division* **105** (GT7) 871-880.

ROSCOE, KH (1953) An apparatus for the application of simple shear to soil samples. *Proc. 3rd Int. Conf. on Soil Mechanics and Foundation Engineering* Organising committee of ICOSOMEF, Zurich **1** 186-191.

ROSCOE, KH AND POOROOSHASB, HB (1963) A fundamental principle of similarity in model tests for earth pressure problems. *Proc. 2nd Asian Conf. Soil mechanics and foundation engineering,* Tokyo **1** 134-140.

ROUAINIA, M AND MUIR WOOD, D (2000) A kinematic hardening constitutive model for natural clays with loss of structure. *Géotechnique* **50** (2) 153-164.

ROWE, PW (1952) Anchored sheet-pile walls. *Proc. ICE* Part 1, **1** 27-70.

ROWE, PW (1955) A theoretical and experimental analysis of sheet-pile walls. *Proc. ICE* Part 1, **4** 32-69.

SAADA, AS AND BAAH, AK (1967) Deformation and failure of a cross anisotropic clay under combined stresses. *Proc. 3rd Panamerican Conf. on Soil Mechanics and Foundation Engineering,* Caracas, Venezuela **1** 67-88.

SAADA, AS AND BIANCHINI, GF (1977) Discussion closure: Strength of one-dimensionally consolidated clays. *Proc. ASCE, Journal of the Geotechnical Engineering Division* **103** (GT6) 655-660.

SCHMERTMANN, JH (1970) Static cone to compute settlement over sand. *Proc. ASCE, Journal of the Soil Mechanics and Foundation Engineering Division,* **96** (SM3) 1011-1043.

SCHMERTMANN, JH, HARTMAN, JP AND BROWN, PR (1978) Improved strain influence factor diagrams. *Proc. ASCE, Journal of the Geotechnical Engineering Division* **104** (GT8) 1131-1135.

SCHOFIELD, AN (2000) Geotechnical centrifuge development can correct soil mechanics errors. *Proc. Int. Conf. Centrifuge 98,* Tokyo (eds T Kimura, O Kusakabe and J Takemura) AA Balkema, Rotterdam **2** 923-929.

SCHOFIELD, AN AND STEEDMAN, RS (1988) State-of-the-art report: Recent developments on dynamic model testing in geotechnical engineering. *Proc. 9th World Conf. on Earthquake Engineering,* Tokyo/Kyoto **8** 813-824.

SCHOFIELD, AN AND WROTH, CP (1968) *Critical state soil mechanics.* McGraw-Hill, London

SCHWEIGER, HF (2003) PLAXIS benchmark no. 2: Excavation 1: results. *Bulletin of the PLAXIS Users Association* Delft (13) 5-8.

SIMPSON, B (1992) Retaining structures: displacement and design. (32nd Rankine Lecture). *Géotechnique* **42** (4) 539-576.

SKEMPTON, AW AND BJERRUM, L (1957) A contribution to the settlement analysis of foundations on clay. *Géotechnique* **7** (4) 168-178.

SMITH, IM AND GRIFFITHS, DV (1988) *Programming the finite element method.* (2nd edition) John Wiley, Chichester.

SOKOLOVSKII, VV (1965) *Statics of granular media* (trans. JK Lusher, ed. AWT Daniel) Pergamon Press, Oxford.

SPRINGMAN, SM, LAUE, J, BOYLE, R, WHITE, J AND ZWEIDLER, A (2001) The ETH Zurich geotechnical drum centrifuge. *Int. J. of Physical Modelling in Geotechnics* **1** (1) 59-70.

SPRINGMAN, SM, NATER, P, CHIKATAMARLA, R AND LAUE, J (2002) Use of flexible tactile pressure sensors in geotechnical centrifuges. *Physical modelling in geotechnics: ICPMG'02*, Newfoundland (eds R Phillips, PJ Guo and R Popescu) AA Balkema Publishers, Lisse 113-118.

STEEDMAN, RS AND ZENG, X (1995) Dynamics (Chapter 5) *Geotechnical centrifuge technology* (ed RN Taylor) Blackie Academic and Professional, London 168-195.

STEWART, DP AND RANDOLPH, MF (1991) A new site investigation tool for the centrifuge. *Proc. Int. Conf. Centrifuge 91*, Boulder (eds H-Y Ko and FG McLean) AA Balkema, Rotterdam 531-538.

STONE, KJL AND MUIR WOOD, D (1988) Some observations of faulting in soft clays. *Proc. Int. Conf. on Geotechnical centrifuge modelling (Centrifuge '88)*, Paris (ed J-F Corté) AA Balkema, Rotterdam 547-552.

STURE, S, ALAWI, MM AND KO, H-Y (1988) *True triaxial and directional shear cell experiments on dry sand.* Final report GL-88-1 to Department of the Army, US Army Corps of Engineers, Vicksburg.

ŠUKLJE, L (1957) The analysis of the consolidation process by the isotache method. *Proc. 4th Int. Conf. on Soil Mechanics and Foundation Engineering*, Butterworths Scientific Publications, London **1** 200-206.

TAKE, WA AND BOLTON, MD (2002) An atmospheric chamber for the investigation of the effect of seasonal moisture changes on clay slopes. *Physical modelling in geotechnics: ICPMG'02*, Newfoundland (eds R Phillips, PJ Guo and R Popescu) AA Balkema Publishers, Lisse 765-770.

TAMATE, S AND TAKAHASHI, A (2000) Manual of basic centrifuge tests: 6, Slope stability test. *Proc. Int. Conf. Centrifuge 98*, Tokyo (eds T Kimura, O Kusakabe and J Takemura) AA Balkema, Rotterdam **2** 1077-1083.

TAYLOR, DW (1948) *Fundamentals of soil mechanics.* John Wiley, New York.

TAYLOR, GI AND QUINNEY, H (1931) The plastic distortion of metals. *Phil. Trans. Roy. Soc.* **A230** 323-362.

TAYLOR, RN (ED) (1995) *Geotechnical centrifuge technology.* Blackie Academic and Professional, London.

TAYLOR, RN, GRANT, RJ, ROBSON, S AND KUWANO, J (1998) An image analysis system for determining plane and 3-D displacements in soil models. *Proc. Int. Conf. Centrifuge 98*, Tokyo (eds T Kimura, O Kusakabe and J Takemura) AA Balkema, Rotterdam **1** 73-78.

TERZAGHI, K (1955) Evaluation of coefficients of subgrade reaction. *Géotechnique* **5** (4) 297-326.

THORNTON, C (2000) Microscopic approach contributions to constitutive modelling. *Constitutive modelling of granular materials* (ed D Kolymbas) Springer-Verlag, Berlin 193-208.

TIMOSHENKO, SP AND GOODIER, JN (1970) *Theory of elasticity.* (3rd edition) McGraw-Hill Kogakusha, Tokyo.

VARDOULAKIS, I (1978) Equilibrium bifurcation of granular earth bodies. *Advances in analysis of geotechnical instabilities* Waterloo, Ontario: University of Waterloo Press, SM study 13, paper 3, 65-119.

VESIĆ, AS AND CLOUGH, GW (1968) Behaviour of granular materials under high stresses. *Proc. ASCE, Journal of the Soil Mechanics and Foundations Division* **94** (SM3) 661-688.

WHITE, DJ AND BOLTON, MD (2002) Soil deformation around a displacement pile in sand. *Physical modelling in geotechnics: ICPMG'02,* Newfoundland (eds R Phillips, PJ Guo and R Popescu) AA Balkema Publishers, Lisse 649-654.

WINTERKORN, HF AND FANG, H-Y (1975) *Foundation engineering handbook.* Van Nostrand Reinhold, New York.

WOOD, DM (1984) Choice of models for geotechnical predictions. Chap. 32 in *Mechanics of engineering materials* (eds CS Desai and RH Gallagher) John Wiley, Chichester 633-654.

WOOD, DM (1985) Some fall-cone tests. *Géotechnique* **35** (1) 64-68.

WOOD, DM AND BUDHU, M (1980) The behaviour of Leighton Buzzard sand in cyclic simple shear tests. *Soils under cyclic and transient loading* (eds GN Pande and OC Zienkiewicz) AA Balkema, Rotterdam **1** 9-21.

YET, NS, LEUNG, CF AND LEE, FH (1994) Behaviour of axially loaded piles in sand. *Proc. Int. Conf. Centrifuge 94,* Singapore (eds CF Leung, FH Lee and TS Tan) AA Balkema, Rotterdam 461-466.

ZANGANEH, N AND POPESCU, R (2003) Displacement analysis of submarine slope using enhanced Newmark method. *Submarine mass movements and their consequences* (eds J Locat, J Mienert and L Boisvert) Kluwer Academic Publishers, Dordrecht 193-202.

ZIENKIEWICZ, OC AND TAYLOR, RL (2000) *The finite element method.* (5th edition) Butterworth-Heinemann, Oxford.

ZYTYNSKI, M, RANDOLPH, MF, NOVA, R AND WROTH, CP (1978) On modelling the unloading-reloading behaviour of soils. *Int. J. for Num. and Analyt. Methods in Geomechanics* **2** 87-94.

Index

9780419237303